TELEPEN
90 0045064 0

D1757714

Charles Seale-Hayne Library
**University of Plymouth**
**(01752) 588 588**
LibraryandITenquiries@plymouth.ac.uk

WITHDRAWN
FROM
UNIVERSITY OF PLYMOUTH
LIBRARY SERVICES

This book presents a historical investigation of great storms that have affected the North Sea and neighbouring northern seas, the British Isles and the fringe of northwest Europe. All the wind storms with serious effects that could be identified within the last 500–600 years are recorded and a few earlier cases discussed. In every possible case observations of weather and other circumstances reported during the storm have been used to produce a modern meteorological analysis. For the earliest cases (in 1570, 1588 1694–7) this has been done sketchily and in nearly every case from 1703 reasonably full meteorological analysis has been achieved. Such analysis facilitates wind strength estimates and aids diagnosis of origins. As a scientific study, this work takes advantage of the unequalled abundance in this region of reports and records going back so far in history. The book is designed not only to facilitate meteorological understanding, but also to look at trends and secular variations as well as impacts on human affairs – particularly damage to coasts, buildings, forests and other aspects of the landscape.

**Historic Storms
of the North Sea,
British Isles and
Northwest Europe**

# Historic Storms of the North Sea, British Isles and Northwest Europe

HUBERT LAMB

*Founder of the Climatic Research Unit,*
*University of East Anglia*

in collaboration with

KNUD FRYDENDAHL
*Danish Meteorological Institute*

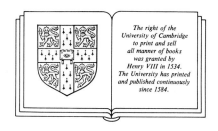

*The right of the*
*University of Cambridge*
*to print and sell*
*all manner of books*
*was granted by*
*Henry VIII in 1534.*
*The University has printed*
*and published continuously*
*since 1584.*

CAMBRIDGE UNIVERSITY PRESS

*Cambridge*
*New York   Port Chester*
*Melbourne   Sydney*

Published by the Press Syndicate of the University of Cambridge
The Pitt Building, Trumpington Street, Cambridge CB2 1RP
40 West 20th Street, New York, NY 10011–4211, USA
10 Stamford Road, Oakleigh, Melbourne 3166, Australia

© Cambridge University Press 1991

First published 1991

Printed in Great Britain by
BAS Printers Limited, Over Wallop, Hampshire

*British Library cataloguing in publication data*

Lamb, H.H. (Hubert Horace)
Historic storms of the North Sea, British Isles and Northwest Europe.
1. Storms
I. Title
551.55

*Library of Congress cataloging in publication data available*

ISBN 0 521 37522 3    hardback

POLYTECHNIC SOUTH WEST
LIBRARY SERVICES

| Item No. | 900045064-0 |
| Class No. | 551.55 LAM |
| Contl No. | 0521375223 |

BAS

# Contents

*Plates 1–13 are between pp. 138 and 139*

seventeenth century, held in the well-indexed data bank of the Danish Meteorological Institute was of outstanding importance, because the ships of the pre-1815 unified Danish–Norwegian navy and their merchant fleet were well distributed over the northern seas with which this study was necessarily mainly concerned. Examples of these ships' operations, outside their home waters around the Danish and Norwegian coasts, the Skagerrak and south Baltic, were ships on passage from Arendal in Norway to Limerick, from Iceland outward bound west of Ireland to the Mediterranean en route to the East Indies, and others plying between Denmark and the West Indies through the Channel or through the North Sea to the Faeroe Islands and Iceland. Observations relevant to storms between the late 1680s and about 1720 also came from ships in the North Sea and neighbouring waters thanks to the navies being out, even in winter, in those disturbed times. Examples were the Dutch

and English ships ferrying Prince William of Orange to England, Danish–Norwegian naval vessels taking supplies to the English operations in Ireland and also out on their own operations in the war against Sweden. The trade in grain from the Baltic to the east Scottish ports and the grain movements along the coast of Scotland to alleviate the long famine in the uplands of Scotland due to successive harvest failures in the 1690s (Smout 1963 and Dundee archives) produced more information.

Thus, in addition to the observation records already beginning at places on land – the work of interested individuals (some of them leading scientists, doctors, clergymen and gentlemen of leisure) – under the encouragement of the Royal Society in England or in a few cases even before 1700 a task of official observatories, there were reports to be found of observations at sea.

seventeenth century, held in the well-indexed data bank of the Danish Meteorological Institute was of outstanding importance, because the ships of the pre-1815 unified Danish–Norwegian navy and their merchant fleet were well distributed over the northern seas with which this study was necessarily mainly concerned. Examples of these ships' operations, outside their home waters around the Danish and Norwegian coasts, the Skagerrak and south Baltic, were ships on passage from Arendal in Norway to Limerick, from Iceland outward bound west of Ireland to the Mediterranean en route to the East Indies, and others plying between Denmark and the West Indies through the Channel or through the North Sea to the Faeroe Islands and Iceland. Observations relevant to storms between the late 1680s and about 1720 also came from ships in the North Sea and neighbouring waters thanks to the navies being out, even in winter, in those disturbed times. Examples were the Dutch and English ships ferrying Prince William of Orange to England, Danish–Norwegian naval vessels taking supplies to the English operations in Ireland and also out on their own operations in the war against Sweden. The trade in grain from the Baltic to the east Scottish ports and the grain movements along the coast of Scotland to alleviate the long famine in the uplands of Scotland due to successive harvest failures in the 1690s (Smout 1963 and Dundee archives) produced more information.

Thus, in addition to the observation records already beginning at places on land – the work of interested individuals (some of them leading scientists, doctors, clergymen and gentlemen of leisure) – under the encouragement of the Royal Society in England or in a few cases even before 1700 a task of official observatories, there were reports to be found of observations at sea.

# Preface

Investigation of the great storms of the past over the North Sea was desirable for a variety of reasons:

(i) The region has a record of storminess, involving innumerable shipping disasters and alterations of the coastlines through severe erosion of cliffs, incursions of the sea over low-lying land, and also through the building of sand-dune barriers and other deposits of blown sand.

(ii) From no other sea is there such a wealth of information as for the North Sea, crossed as it is and has long been by probably the busiest shipping traffic in the world and flanked by densely populated lands with a tradition of keeping records. Reports of storms at sea and in the lands round about go back several centuries.

(iii) This means that it should be worth the attempt to submit the storms of this period to synoptic meteorological analysis, gauging better the strengths, durations, extent, and other features of the storms, including their energy sources and the controls upon their incidence.

(iv) Some of the historical records seem to hint that the storminess in this region may have been greater at certain epochs in the past than in this century, although there have been signs of somewhat increased storminess again since about 1950. More detailed analysis should help to establish these points.

(v) Furthermore the North Sea is now the scene of greater and more economically important activities than ever. Besides the naval and merchant shipping and the fisheries, there are now oil and gas rigs and, in the northernmost North Sea, very deep drilling. Also, with rising sea level almost throughout the twentieth century the threat of storm damage to the low-lying coasts has been increasing.

The study was extended to the neighbouring sea and land areas for the sake of what might be learnt from the comparisons of storm frequency, severity and the directions of the main strong winds.

A few storms in the Norwegian Sea beyond the limit of our main area of attention have been included because (a) the N'ly cases illustrate developments liable to lead straight on into storms in the North Sea, and it must be of some interest to explore their violence in the region of their origin, and (b) to illustrate more clearly the effects of convergence of the wind along a mountainous coast. The famous storms over the Atlantic west of Portugal in October 1805 have been included, although they too were beyond the main limit of our survey, as a reminder illustrating the effects of occasional cut off pools of northern cold air in inducing storminess in lower latitudes. And the storm in 1986 over the central Atlantic, beyond our western limit, is included in order to register the deepest cyclone known to have occurred so near to northwest Europe, although the storm centre near the Hebrides in January 1839 seems to have attained nearly the same depth.

Extension of the sampling back to about the year 1500 was deemed worthwhile:

(a) to increase the samples and indicate the representativeness of different periods and storm patterns

(b) to investigate the storminess in another climatic regime, the colder climate of the so-called Little Ice Age

(c) to throw whatever further light might be possible on a number of reported great storm disasters in that period.

Estimation of possible future extremes can hardly be adequately based without such a survey of past centuries' experience.

Sampling of storms from among those reported in the last hundred years was increased so as to include as wide a variety of types and origins of storms in the region of interest as might be found, to improve understanding of earlier storms which seem to show some analogy.

The whole subject field is relevant to the practical concerns of the present day mentioned in (v) above as well as to coastal protection measures – from dyke barriers on the open coasts to the Thames Barrage and to the designing of wind and wave energy installations and other concerns of governments and of the insurance industry. It is also of interest to forestry and environmental concerns.

The investigation was begun by a research student (Mr C. Loader) in 1975, supported by a grant from Shell Exploration Ltd concerned with the planning and design of oil and gas operations in the North Sea. The problems of collecting the observation data needed and the multiplicity of ancient units and of European languages – including archaic forms – in which the data were reported, however, showed after just a few storms between 1791 and 1829 had been analysed (and the 1976 case for a modern comparison) that the task was far too big to be carried out in that way. Further effort was also needed to identify satisfactorily some older forms of place-names. Moreover, it was becoming known that many more important storms could, and should, be tackled and that a surprisingly great wealth of data could be acquired. A few years later the present author took over the project, generously supported by an Emeritus Fellowship from the Leverhulme Trust.

The sources of data used are indicated in the list of acknowledgements which follows. The contribution made by the voluminous data, notably ships' observations going back to the

# List of plates

# Contents

*Plates 1–13 are between pp. 138 and 139*

# Acknowledgements

The author expresses his thanks to Shell Exploration Ltd for the grant which started the project and especially to the Leverhulme Trust which made it possible to finish the project.

Particular thanks are due to my collaborator and friend Knud Frydendahl of the Danish Meteorological Institute, Copenhagen, for the supply of much of the material described above, as well as analysed weather maps from more recent times, and for valued discussion of basic aspects of the project.

The contribution of surveys, widened in scope by historical documentary research and radiocarbon dating, performed by Dr Poul Hauerbach of Skagen and Randers on the growth of the sandy point known as Skagen (or The Scaw) at the northern end of Jutland, Denmark – especially between the 1690s and 1790s – also turned out to be important to our tentative conclusions on the longer-term variations of storminess and are the subject of a section of this report. I have been very grateful for the opportunity to examine this work on site and for my discussions with Hauerbach and permission to use his material.

Acknowledgements are due to many individuals and institutions for photocopied observation data and other accounts, in many cases supplied free of charge, and my gratitude goes to many for the friendly and generous support which carried the investigation forward throughout: this includes the meteorological services in England, Ireland, the Netherlands, Denmark, Norway, Sweden, Germany, and Switzerland, the Meteorological Office, Edinburgh, and the National Maritime Museum, Greenwich, the Public Record Office, London and the Scottish Record Office, Edinburgh; also the more local archives in Dundee, Newcastle, Hull, Bristol and Truro, as well as in Kristiansand and Stavanger, Norway; the Observatory of Paris; the Royal Society Library, London (notably through the help of Dr Alan Clark); and the Directors of Meteorology in Oslo (Dr Kaare Langlo at the time), Prague, Warsaw and Vienna. Greatly valued personal assistance was given by Dr J. Vassie of the Tidal Institute, Birkenhead, calculating the tides at Aberdeen about the times of the storms in 1413 and the 1690s; by Professor Christian Pfister of the Historical Institute of the University, Berne, Switzerland, in the supply of reports including meteorological instrument readings from an old network of observing stations in central Europe coordinated in Breslau (now Wrocław) in the early eighteenth century; and by my friends and colleagues, R.A.S. Ratcliffe and M.J. Wood of the Meteorological Office, Bracknell, and D.A. Wheeler of Sunderland Polytechnic. In Barcelona and in Copenhagen staff were temporarily employed transcribing early observation reports thanks to Dr M. Puigcerver of the Academy of Arts and Sciences, Barcelona, and K. Frydendahl respectively. In Iceland Sjöfn Kristjansdottir of Reykjavik performed a similar service. In Trondheim, Norway, H. Nissen of the University Library supplied many eighteenth century meteorological reports photocopied from the local newspaper. I am much indebted to Dr Hans Rohde of the Bundesanstalt für Wasserbau, Hamburg, for copies of his works listing and discussing storm floods from 1661 to recent years and for facilitating the acquisition of related daily meteorological observations from an old Hamburg newspaper from 1786 onwards. I thank E. Hovmøller of the Swedish meteorological service for data from the same period from a network of places in Sweden and Finland, and P. Tallantire while working in Sweden for information from the diary of Maria Reenstierna near Stockholm. Acknowledgement is due to several Dutch archives, notably the Algemeen Rijksarchief, 's-Gravenhage, for very valuable data from the logs of ships at sea during the great storm in 1703.

I acknowledge the accounts of storms and storm damage about Spurn Head, near Hull, kindly supplied by G. De Boer and P. Spink, and analyses of a number of past historic weather events from Professor J. Neumann of Copenhagen (formerly of Helsinki and of the Hebrew University, Israel) and Erik Wishman of the Arkeologísk Museum, Stavanger.

Further observations of importance were supplied by M.G. Pearson, formerly of the Meteorological Office, Edinburgh, the farm diary from Kemnay on Donside, Aberdeenshire in the late 1700s, and by Dr Graeme Whittington of the University of St Andrews the farm diary of John Lamont in the 1650s and 1660s, reporting weather in Fife. In the case of the 1838 storm, reports were kindly supplied by Mr Calderwood, Curator of the Royal National Lifeboat Institution's Grace Darling Museum at Bamburgh, Northumberland. Very useful comment on my analysis of the Culbin Sands disaster in 1694 and its probable meteorological basis was received from D.P. Willis of Fortrose Academy.

A reference work on early American hurricanes by Ludlum (1963), kindly supplied by Dr K.C. Spengler of the American Meteorological Society, was consulted in connection with the origins of some storms.

Finally, I acknowledge gratefully the facilities I was able to use at the University of East Anglia, Norwich, the discussions and suggestions contributed by Professor T.M.L. Wigley, the present Director of the Climatic Research Unit, and the guidance supplied by Cambridge University Press in connection with this publication. And my warm thanks are due to Patty Banham and Julie Burgess who produced this text on the word processor with great patience and in frequently very busy situations. I also thank Phillip Judge and David Mew, who drew the maps and diagrams, and Malcolm Howard, formerly of the University of East Anglia Audio-Visual Centre, for all the photography involved.

# Part 1

## Introduction and analysis

Sediment
Transfer



Coast
Errosion

# 1

# Introduction

## Background and basis of the study

The regions here examined are close to, and to varying extents experience, the most frequented storm tracks of the northern hemisphere. The storminess of the North Sea in particular is known from the succession of storm flood disasters on its coasts catalogued by various compilers from Arends (1833) to the three-volume work of Gottschalk (1971, 1975, 1977), the latter particularly distinguished by the author's thorough critical examination of the original manuscript sources. The partly known history goes back at least to the Cymbrian flood of the coasts about the German Bight around 120 BC, which set off a migration of the Celtic tribes previously settled there. And both Aristotle and the early Greek navigator, Pytheas, who sailed round Britain in about 330 BC, had already reported the acquaintance of the Celtic tribes then living along the same part of the North Sea coast of the continent of Europe with its storms.

It is probable that, because of the seriousness of the disasters, lists of historic sea floods are the nearest approach we have to a homogeneous list of a series of great storms of the last 300 to 500 years. Before that the sea defences were so much less effective that the situation was hardly comparable at all, but ever since that time the effectiveness of dykes and sea walls and the dredging of channels must have affected the flow of tidal currents and storm surges.

There are other striking effects of great storms upon the landscape, particularly through blown sand, the formation and shifting of dunes sometimes forming a continuous coastal barrier, the scouring of sand or dry soil and spreading of drift-sand into nearly flat expanses; also the drift of sand and gravel by water currents at or along a coast and in other shallow-water areas – a process which is liable to produce offshore sandbanks and bars across harbour mouths that are open to scouring by any storm winds occurring at times of very low tide. But the effects along these lines of any individual severe storm are generally more local than the great sea floods.

Erosion of coasts takes place not only by wave battering and scouring in storms, but by slow continuing wastage through water current action. It is also affected at varying times and rates by processes that may have nothing to do with storms: by heavy rain and run-off, by frost and thaw, and by landslides. And the rates of these are clearly influenced by the topography and geological structure as well as by the variations of storminess in space and time. Exceptional cases of coastal recession due to marked erosion of cliffs or of low-lying promontories – as near river mouths – appear in the records from time to time, and these have been used to identify some of the severe storms in this com-

pilation. But these too are characteristically localized events and do not provide readily comparable evidence of the severity of a storm. In such cases this usually has to be established, if it can be at all, from meteorological evidence or reported wave heights etc.

The storms examined in this compilation were picked from the records of great sea floods and other coastal disasters in collections already known to the scientific literature (e.g. Gottschalk, 1971, 1975, 1977; Gram-Jensen, 1985; Petersen and Rohde, 1977) or reported soon after their occurrence in the standard journals (including the *Meteorological Magazine* and the *Monthly Weather Reports* of the Meteorological Office and similar sources in neighbouring countries). Others were discussed through commemorative articles in the *Meteorological Magazine*, in *Weather*, or in the equivalent journals in Denmark (*Vejret*) and Norway (*Vaeret*), or in the organs of shipping interests, recalling historic storms long after the event (as at hundredth and hundred and fiftieth anniversaries) and gathering together details from old newspapers and diaries. Others again were found in various local histories and archives, while a few were picked up from the early compilations of great weather events by Hennig (1904), Lowe (1870), Mossman (1898), Short (1749). Such reports need corroboration by independent accounts or other circumstantial evidence or – better by far – by collected simultaneous weather observations and weather map analysis. This has been the main policy of this investigation. Other sources used have been Brazell's *London Weather* (1968) and the manuscript detailing daily weather observations in London from 1723 to 1811 put together by the late Professor Gordon Manley and deposited with the Meteorological Office in the 1970s. Lists of extreme winds observed in the British Isles given by Bilham (1938) and Chandler and Gregory (1976) have also been used. Yet other information was gleaned from early newspapers, from port records and local archives from the countries around the North Sea, as well as some farm diaries kept at places near the coasts and the notes of scientific observers such as Gilbert White at Selborne in Hampshire and Thomas Barker in the east Midlands of England, whose weather journals between 1733 and 1795 are described and partly reproduced by Kington (1988). Storm reports from such miscellaneous sources need the verification and amplification that can best be provided by synoptic mapping of simultaneous observations, as has been possible in the great majority of cases in this study.

Circumspection is needed in accepting the reports even of very able men among the early observers when they were writing from memory of events witnessed many years before. Thus Defoe (1704), whose reporting on the great storm in southern England in 1703 is of great value, refers at length to what he remembered as a somewhat parallel case in February '1661' (1662 according to our modern reckoning of the beginning of the year); in fact, the records show that the 1662 storm passed across England on a quite similar west-to-east track, causing much destruction, but over a decidedly narrower belt than in 1703, and we have no word of effects on the continent. In another place, Defoe (1724) wrote of a storm encountered 'about the year 1692 (I think it was that year) . . . by 200 sail of light colliers bound northward out of Yarmouth Roads for Newcastle . . . when some turned back . . . some but very few rid it out . . . ; the rest being above 140 sail were all driven on shore, and dash'd to pieces . . .'. Luckily we have sufficient reports of the storm of 22 September 1695 (q.v.) for it to be reasonably clear that that was the storm which Defoe meant to refer to.

Non-meteorological evidence above all needs meteorological corroboration. In the eighteenth century, the great church at King's Lynn and the cathedral in Hereford were both extensively damaged when their southwest towers fell across the nave of the building, as if felled by a WSW'ly storm. The storm at King's Lynn on 19 September 1741 (New Style)* is abundantly supported by reports from Lynn and places in neighbouring counties. But the day when Hereford cathedral fell, 17 April 1786, was fine, with a light NE breeze! The cathedral records speculated that bad soil and weak foundations might have been the cause, but the place also has a history of earth tremors.

In a number of cases before about 1725 (see Lamb, 1977, pp. 48–50) the precise date of some severe storms seems never to have been recorded, and in others the calendar change introduced uncertainty into the date. When this is

*The modern (Gregorian or 'New Style') calendar was adopted first in the Catholic countries of Europe – in France, Italy, Portugal and Spain in 1582; and in the German Catholic states, Austria, Flanders and the (then Spanish) Netherlands from 1 January 1583. Poland adopted it in 1586, Hungary in 1587. The German Protestant states followed suit in 1700 as did the free Netherlands, Denmark and Norway. Sweden changed gradually by omitting 11 leap days between 1700 and 1740. Britain and her American colonies, and Ireland made the change in September 1752.

Bulgaria went over to the New Style calendar in 1915, Turkey and Soviet Russia in 1917, Jugoslavia and Rumania in 1919, Greece in 1923.

The corrections needed to convert the Old Style dates to New Style were:

> Between '29 February' 1400 and 28 February 1500 ADD 9 days.
> Between '29 February' 1500 and 28 February 1700 ADD 10 days.
> Between '29 February' 1700 and 28 February 1800 ADD 11 days.
> Between '29 February' 1800 and 28 February 1900 ADD 12 days.
> Between '29 February' 1900 and today ADD 13 days.

More information on this topic is given in Lamb (1977, p. 49).

The Russian observations (e.g. at St Petersburg) printed in the Ephemerides of the *Societas Meteorologica Palatina*, published for the years 1781 to 1792, appear to be dated on the Old Style calendar, unlike the other places where observations were made for, and assembled in, the Society's publication.

The only storm reported in the present survey about which there seemed to be any uncertainty over which calendar was in use in the brief historical account which is all that is here reprinted is the storm in September 1690. In the case of this storm, it is probable that the date given is an Old Style calendar date (since historians who are not dealing with a scientific problem customarily do not convert the dates).

not known, it is of course impossible to assemble and analyse simultaneous observations. These difficulties and the extent to which it may have been possible to resolve them are discussed in the text of our storm reports. The one remaining case of more than doubtful authenticity seems to be the 'violent gale' in London on 1 December (Old Style) 1737 reported by Lowe (1870) and quoted by Brazell (1968).

Greater difficulty surrounded the firm establishment of the date and details of the S'ly storm of blown sand which obliterated the centre of the medieval town of Forvie on the east coast of Scotland north of Aberdeen, reputedly on the 10 August (Old Style) 1413. However, the calculation by Dr J. Vassie of the Tidal Institute, Birkenhead that that date coincided with an unusually extreme low tide suggests that it may be the true date of the storm. The difficulties, however, indicate that the fifteenth century should be considered beyond the limit of our ability to establish certainty. In the case of at least the severest storm disasters after AD 1500 interpretation may be adjudged safer, although the reasonably firm evidence of conjunction with extreme exposure of sand to the wind in the 1413 case should be noted.

## Observations and instruments

The earliest storms mapped came within a few decades of the invention of the barometer and thermometer, and for over a century thereafter none of the available ships' observations at our disposal included instrument observations. In some cases, however, notably in the storm in 1717, the ships' reports of wind and weather, commonly recorded for each 4-hour watch of the day and night, were so numerous that only a small proportion of them could be shown on our maps, and in the 1703 case as well as in 1717 it will be seen that the wind reports received from ships at sea were of great value to the analysis. What could be done with such reports to determine the synoptic weather pattern will be understood from a careful reading of the report on the great storm in 1694 which is interpreted as the probable cause of the Culbin Sands disaster in northeast Scotland. A preliminary diagnosis of the meteorological situation was first sketched on the basis of the good series of weather observations reported in London from 29 October to 4 November 1694 (New Style) – partly listed in our text on that storm – taken together with the close descriptive account of the progress of the storm at the disaster site in northeast Scotland. The London observations indicated a long-lasting NW'ly wind, preceded by a light W'ly on the 29th–30th, the NW'ly outbreak becoming increasingly stormy and sleety until 2 November. This sequence suggests the arrival of fresh Arctic air from the north, which would presumably have been felt more severely in northeast Scotland. There the experiences of the people whose work in the fields (and whose homes) were overwhelmed by the storm at Culbin suggest that the rising wind may have been initially W'ly, and that it became strong at an early stage. The arguments for the direction then becoming N or NNW are given in the text of our analysis of the storm. Other, circumstantial evidence hinted that the storm affected the eastern side of the North Sea much less if at all. The wind pattern, revealed by the ships' reports received *after* the situation had been outlined, so far verified the diagnosis as to provide a test of the practicable extent of situation analysis with the evidence available.

Of the early instruments in use, the barometer readings were the least troublesome to interpret. Even so, the only values used in our analyses before the 1730s were those in the hands of leading scientists of the day: the Revd William Derham at Upminster, near London, Richard Towneley in the north of England, and the staff of the Observatory of Paris. The instruments were exposed indoors in an unheated environment and seem to have performed well. Many other early instruments gave more or less trouble with the fluid (usually 'quicksilver'/mercury) sticking to the sides of the glass. The units of measurement (length of the mercury column) have been converted to millibars and corrected from the approximately known height of the station to sea level and standard gravity (at 45° N). For the study of early instruments the reader is referred to Knowles Middleton (1969). The historical weather map reconstruction work at the Climatic Research Unit, Norwich by Mr J.A. Kington and the present author has led to the progressive compiling of a gazetteer with particulars of early observing stations, their positions and heights above sea level, instruments and units used, corrections needed, times of observation and some information about the observers.

By the late eighteenth century, with 35 to 40 observation points on our maps and up to 30 of them reporting barometric pressure, it was possible to establish the height above sea level, and correction to sea level appropriate in the cases where that information was missing, by finding the average correction needed to fit the maps drawn from the stations for which such information was known. In several of the months affected by great storms in the 1790s and in 1825, the situations every day for 2 to 5 weeks were analysed and the pressure values at places all over the map were studied for goodness of fit. The daily sequence of the discrepancies at each place from which the observations came and the means and standard deviations of the departures gave a numerical test of the reliability of the individual stations, showing the best performing ones and those with barometers sticking when pressure changed rapidly.

Exposure of the thermometers used was a difficult problem for the early observers and the best practices only really emerged in the late nineteenth century. Before that, the instruments were commonly positioned in unheated north rooms or outside on an open north wall, sheltered in some way from rain and direct sun. But these difficulties caused no serious problem for air mass identification and analysis of the stormy (largely cool-season) situations with which this study is concerned. Conversions had to be made from a great variety of old instrument scales to the centigrade scale used on all our maps.

The problems here discussed, and how they were dealt with, are the subject of special appendices following the texts of our accounts of the storms in 1717 and 1791, where it will be seen how the techniques described above were used also to test the reliability of our analysis and the limits of the area that could be satisfactorily covered by isobars and

gradient wind measurements from the network of observation measurements available. Another test of the analysis of an early storm is demonstrated in our account of the great storm in 1839 which was independently analysed in the Irish Meteorological Service in a recent paper.

The indications of wind strength on our maps are given in Beaufort wind force. The Beaufort scale, its equivalence with wind speeds and its correspondence with other, earlier wind scales,* are set out in the appendix note following our account of the great storm in 1703. The wind strengths plotted on our maps relating to storms earlier than about 1850 should not be taken as accurate to nearer than about two points on the Beaufort scale, and indeed nearly all the wind observations before 1900 were not instrument measurements but observers' estimates guided by the standard descriptions of the Beaufort scale. The gradual spread of anemometers after about 1880 is alluded to in a few storm accounts. It is only in the last few decades that actual measurements of surface winds at sea (apart from a few offshore lighthouse observations) enter the study. Over by far the greater part of the period of our survey the most dependable measures of wind are the gradient winds derived from the barometric pressure analysis. A sample estimate of error liability in gradient wind values measured off the maps appears in the appendix note following the account of the 1791 storms. (A description of a statistical method by which some indication of the probable gradient wind strengths in the Spanish Armada storms in 1588 was obtained will be found in an appendix following our account of the storms in that year.)

Tables for the conversion of ancient units have been given by the present author in Lamb (1986) and in Lamb and Johnson (1966), where many more details of the development of the world network of meteorological observations will be found. (For history of the development of the meteorological observation network see also Lamb, 1977.)

Weather diaries kept by people whose dedication is made clear by the regularity, terseness and completeness of their records were also used. The agreement between three independently kept diaries within 40 kms of each other in Norfolk over nearly a hundred days analysed in the 1790s was a tribute to the reliability of the best of such observation records, the only differences being such as fitted the weather situations, e.g. on days when winds were obviously stronger or showers more frequent near the coast.

Wheeler (1988), writing of his use of ships' records from the beginning of the nineteenth century, reports similarly that 'where ships are gathered within the same area their records of the weather conditions are consistent'. That was also true of by far the majority of the logs of the Scandinavian ships around the Skagerrak and neighbouring coasts here used in analysing the storm in December 1717, but it is a finding that depends on dedicated observers. Experience in another study showed, for instance, that the daily weather observations of the Salem, Massachusetts doctor, E.A.

---

*Frydendahl (1986) has written a useful short account of the historical development of a system for estimations of wind force by observers on ships at sea and of the Danish collection of early observations.

Holyoke, maintained over many decades between the 1750s and the early 1800s were more consistent than those kept by the staff of an official observatory in the area.

Another possible source of inhomogeneity in the early storm situations to be analysed in the present study was in the times of observation. This was resolved by usually mapping the situation as reported about 2 o'clock in the afternoon, this being the observers' commonest choice, and tolerance of an hour or so in either direction. This meant that the difference of natural time over the map area with reports analysed between about 20° W and 30° E could reasonably be ignored.

The plotting model and symbols used for entering the weather and instrument readings observed on the maps are explained in an appendix following our account of the earliest fully analysed storm, in 1703.

## Analysis

In approaching the analysis of the daily weather map sequences covering the storms in this compilation, the author was able to draw on much relevant earlier experience of working with sparse data, analysing weather patterns over the North Atlantic Ocean in wartime, then over the remotest regions of the Antarctic Ocean (Lamb 1956, especially pp. 22–3), and monthly situations back to the beginnings of a usable, instrument-observations network around 1700 – methods described in Lamb (1977), Lamb and Johnson (1966) – in all these cases subjected to subsequent tests. The tests were designed to show the magnitudes of the random errors to which the sparsely covered parts of the maps were liable and to identify where the effects of bias and misconceptions came in beyond the limits of any observation coverage. This determined the limits of the area which could be meaningfully analysed at all.

There is no doubt that, in the case of synoptic maps of the situation at a particular time on a particular day, modern understanding of frontal patterns primarily due to Bergeron (1928, 1930), Refsdal (1930), van Miegham (1936), and Pettersen (1936) (verified and continually demonstrated in recent years by satellite imagery (Lamb, 1988)) has materially improved the possibilities of outlining barometric pressure patterns over sparsely covered parts of a synoptic map and has reduced the error margins.

The magnitude of the present survey of historic storms, and of the task of gathering the data and reconstructing the meteorological situations involved, was too great to allow time for very extensive testing. Sporadic testing of various kinds was done in the earlier stages of the work and has been indicated on p. 5. The plan of work in this survey was to explore and demonstrate first how much interpretation was possible, and what its geographical limits were, by particularly thorough synoptic analysis of a rather small number of plainly very severe storms for which data coverage was particularly good: in 1703, 1717, 1791, 1792, 1795 and 1825. In most of these cases conditions at 40 to 60 places were entered on the original working charts. Tests applied to the resulting maps in the 1790s are among those reported with our storm accounts. Later, the difficult early storm in 1694 was analysed step by step (including diagnosis of the precise date of the storm) in a process that supplied some further tests of the method. For each of these storms, sequences generally of 10 days to a month (in 1694, just 7 days; in 1791, 36 days) were submitted to rigorous synoptic meteorological analysis. Developments were traced forwards and back and the maps drawn and corrected through much trial and error until, by successive approximations, the greatest possible continuity from day to day throughout the sequence was attained.

This work is illustrated here by texts covering just the most relevant runs of a few days to show the development of each of these storms. The great majority of other storms studied were analysed over shorter runs of just three to six consecutive days and are illustrated mostly by just one or two days analysed maps.

In the period since the first State meteorological services were founded,* and began publishing daily weather maps (which were soon increased to several times daily), these maps have been used and adapted as necessary to fit a satisfactory frontal analysis. The original maps used have been variously those of the British, Danish, and German weather services, including the remarkable series of daily weather maps stretching across the Atlantic from North America to eastern Europe produced between 1873 and the early part of the twentieth century by cooperation between the Danish Meteorological Institute and the Deutsche Seewarte, Hamburg.

In all, over 166 storms are studied here in the 144 reports. These include:

11 in the sixteenth century
25 in the seventeenth century
32 in the eighteenth century
35 in the nineteenth century
63 in the twentieth century

Among the twentieth century cases some less severe storms have been included to illustrate a greater variety of types of situation.

This draws attention to the need for some system or systems of grading the storms in all stages of the history.

*The first was the department of meteorology in the British Board of Trade, London, in 1855 under Admiral Fitzroy.

# 2

## Grading of storms

While the storms here catalogued were chosen for study because they had acquired historical note in one way or another,[1] it is obvious upon inspection that storms of a considerable range of severity are included. In what follows it has been thought useful to rank the storms separately in terms of:

(i) The *greatest wind speeds* indicated at the surface:
  (a) as measured in gusts
  (b) the greatest mean speeds over a period of 10 minutes or, here usually, over an hour
  In practice, measurements of the pressure gradient, and hence the gradient wind (the wind indicated by the pressure gradient) are often the most convenient way of obtaining representative values for comparisons between many different storms.
(ii) The *greatest area covered at any stage by winds causing widespread damage.* We have used gradient winds of more than 50 knots as our criterion for a damaging wind, since this indicates likelihood of some gusts of 50, to sometimes 60, knots or more at the surface. By definition storm force 9 on the Beaufort scale (38–44 knots) damages chimney pots and branches of trees, force 10 (45–52 knots) uproots trees and causes severe damage to buildings. (For the full Beaufort scale and other windscales see the appendix following the account of the great storm of 1703.)
  It seems, however, scarcely possible to assess the *greatest area* covered by damaging winds *at any stage* in the life of a storm in the historical past. It would also be of interest, and perhaps more feasible in some historical cases, to compare the total areas affected by damaging winds in the life of different storms, but this was not feasible in the present study.
(iii) The *total duration of the occurrence of damaging winds* during the life of different storms. For many purposes it must also be useful to compare the duration of winds of damaging strength at particular places of interest.

The severity of storms should also be considered in relation to the expectation (frequency) of high wind speeds in the area concerned. The great storm in 1987, which wrought widespread havoc in the south of England, reminiscent of that in 1703, and produced gusts of 119 knots on the north coast of France, 100 knots on the south coast of England at Shoreham, Sussex, and over 90 knots in the Thames estuary and far inland at Wittering near Stamford, is seen as a far greater extreme when viewed against the averages in that region. Figures in the Climatological Atlas of the British Isles (Meteorological Office, 1952) for the average number of days a year with gale[2] over a sample 20-year period 1918–37 are about two in the London area and most of central England including Wittering, five in the Thames estuary, and ten to 15 on the Sussex coast, compared with over 30 in the Shetland Isles and coasts of northwest Scotland as well as the most exposed coasts of southwest England, west Wales and west and northwest Ireland. Higher figures, up to 40 or more are believed to occur in two areas on the coast of Norway near 62° and 65° N (Børresen, 1987). Our Storm Severity Index figures indicate, however, that storms in the severest class occasionally bring very severe gales to places far outside the zone where such phenomena are most frequent. This is in line also with the fact that the differences between the strongest winds ever so far reported in northern Scotland or about the Norwegian Sea and on the coasts of the southern North Sea, the Channel, and elsewhere around the south of the British Isles is not so very great: about 120 to 160 knots in those northern areas – the values above 130 knots probably all associated with hill-top or other sites affected by convergent air flow – compared with 100 to 120 knots at the southern coasts mentioned.

### Storm Severity Indices

From another point of view there seems to be a requirement for *an overall Storm Severity Index*. This should presumably be of the following design:

$$V_{max}^3 \times A_{max} \times D$$

where $V_{max}$ is the greatest surface wind speed
$A_{max}$ is the greatest area affected by damaging winds
D is the overall duration of occurrence of damaging winds (or, alternatively, the duration in some place or area of interest).

We use the cube of the wind speed in conformity with Landsberg's 1941 definition of Wind Power.[3]

---

[1] The criterion of historic severity has been relaxed a little in the case of some storms, mainly since 1930, the inclusion of which seemed likely (a) to be valuable to demonstrate some further variety in the types of development of severe storms, and (b) to help establish the lower points of our scale of severity.

[2] *Gale* defined as mean wind over a 10-minute period exceeding 33 knots, i.e. Beaufort force 8 or over. The figures for numbers of days with gusts over 50 knots at the British stations here mentioned are almost identical (Meteorological Office, 1968).

[3] The dynamic pressure of the wind is proportional to the square of the wind speed, but the wind power (as, for instance, exercised in windmills or in the destruction wrought by storms) is a matter of the work done by the wind and thus involves the dynamic pressure multiplied by the run of the wind: hence the cube of the wind speed.

To obtain our desired numerical severity ratings it will be sufficient to express each item in the units in which they are commonly measured (knots for wind speed, hours for duration of winds of damaging strength, and metric units for area – in practice, it is convenient to use units of $10^5$ km$^2$ for the areas affected by damaging winds, since this produces a readily manageable range of numbers). There is nothing to gain in this case by conversion to a uniform system of units.*

Clearly, assessments of the areas and durations of damaging winds can only be rather rough in the case of storms from the historical past. They are in nearly all cases best made from measurements of the gradient winds at various stages in the life of the storm or from comparisons with more recent storms which seem to conform to a similar model.

It must be hardly surprising in the case of storms of the past, even of the fairly recent past, that however careful the meteorological analysis there is an awkward margin of uncertainty resulting from inevitably somewhat free-hand estimates of the areas covered by winds of damaging strength and of their duration. The calculations nevertheless do provide the basis for recognizing several grades of severity. They will also serve to indicate what measurements, and what methods of mapping them, should be made to assess the severity of storms occurring in the future.

In order to narrow the margins of uncertainty, the author spent some weeks examining critically the comparability of storms listed in this survey, repeating the assessments and calculations of index values up to five times before being satisfied that the best attainable series of estimates had been reached.

### Other assessments of severity

Altogether different ratings of storms which seem likely to be useful in some connexions include rankings in terms of:

(iv) Total damage to the landscape, particularly to coasts, e.g. by sea floods, erosion by wave action, or blown sand and sediment transported by the currents and by waves.

(v) Total damage to property (buildings, agricultural land, orchards and forests).

(vi) Numbers of human and animal lives lost.

(vii) Insurance claims arising and costs (over however many years) of restoration and measures for future protection.

It will be seen from the following lists that the storms found to be the most severe ones on these different criteria produce different lists. Also the different considerations in mind critically affect the perceptions of different observers and the reports they leave to posterity.

Were we to confine our attention to the storms affecting different smaller regions, again the ones appearing as most severe would be different. Here, for this first exploratory survey of the storms over a long period of time over the land and seas around northwest Europe, severity index calculations have been applied solely to the whole area between about 45° and 65° to 70° N and between 20° W and 15°

*To correct square kilometres to square nautical miles one could divide the areas here quoted by 3.5 approximately and the resulting Storm Severity Index figures would all be reduced in the same proportion.

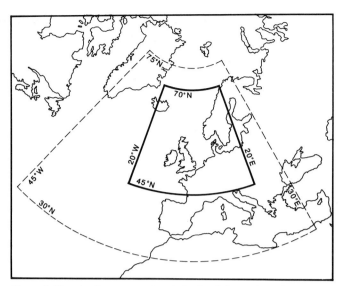

*Figure 1.* Map to show approximate limits of the main region of interest used in Storm Intensity Index assessments and of the extended area surveyed when the situation permitted.

to 20° E (Figure 1). Nevertheless, it should be useful to recalculate the severity of each storm in terms of the wind speeds, duration, and the areas affected by winds of damaging strength, within each smaller region of interest, such as Ireland, Scotland, England, North Sea, Denmark and so on.

### The resulting lists

A. *The storms surveyed in order of Severity Index ratings*

In this list some early storms with insufficient data for analysis have been given supposed ratings on the basis of analogies with later storms apparently similar in type or the scale of their effects. Such suggested ratings have been placed in brackets. Brackets are also used in this list to indicate uncertainties about the precise date of some early storms. All the dates here listed are on the modern (Gregorian or 'New Style') calendar.

The time distribution of these storms is shown in Figure 2.

| Date | Storm Severity Index | Remarks |
|---|---|---|
| 15.12.1986 | About 20 000 | Not within our central region of interest |
| 10–12.12.1792 | 10 000–20 000 (taken as 12 000) | |
| 4.2.1825 | About 12 000 | |
| (31.10–2.11.1694) | (About 10 000) | Much indirect deduction – some basic assumptions open to question |
| 7–8.12.1703 | 9000 | |
| 22.10.1634 | (About 8000) | |
| 6–7.1.1839 | About 8000 | |
| 16.10.1987 | 8000 | |
| 14–16.10.1886 | 7000 | |
| 11–12.11.1570 | (About 6000) | Inadequate data – some necessary assumptions about length of fetch of |

## STORM SEVERITY INDEX ASSESSMENTS

Open circles conservative estimates of storms before 1730
where some non-meteorological data had to be used

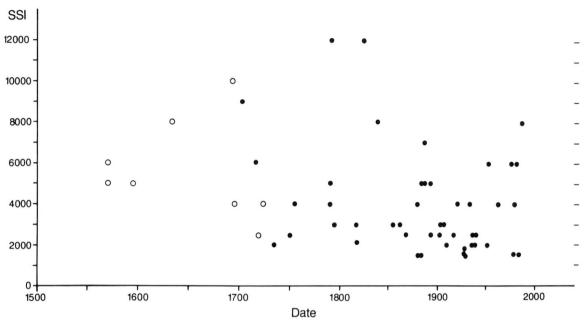

*Figure 2.* Time distribution of storms among those studied in this survey with Severity Index assessments over 1500.

| Date | Storm Severity Index | Remarks | Date | Storm Severity Index | Remarks |
|---|---|---|---|---|---|
| | | the storm over the Norwegian Sea | 9.4.1933 | About 4000 | Mainly north of 65° N |
| | | | 16–17.2.1962 | 4000 | |
| 24–25.12.1717 | About 6000 | | 4–5.12.1979 | 4000 | Largely coast of Norway north of 60° N |
| 31.1.–1.2.1953 | 6000 | | | | |
| 2–3.1.1976 | 6000 | | 6–9.5.1795 | About 3000 | |
| 23–25.11.1981 | 6000 | | 12–16.1.1818 | About 3000 | Overall severity rating of prolonged storm with several separate phases |
| (About 1570) | (About 5000) | Inadequate data: first of the great inland sand drift episodes in the Norfolk and Suffolk Breckland | | | |
| | | | 1.1.1855 | 3000 | |
| | | | 26–27.12.1862 | 3000 | |
| | | | 26–27.2.1903 | 3000 | |
| 21–22.3.1791 | About 5000 | NW to NNE'ly part of this storm – over the northern North Sea | 12–13.3.1906 | 3000 | |
| | | | (About 1720) | (About 2500) | Imprecise data and no surrounding coverage |
| 25–27.1.1884 | 5000 | | | | |
| 8–9.12.1886 | 5000 | | 11.9.1751 | 2500 | |
| 17–19.11.1893 | 5000 | | 24.1.1868 | 2500 | |
| (Jan. to Feb. 1595) | (Probably 5000 approx.) | | 11–12.2.1894 | 2500 | |
| | | | 25–26.12.1902 | 2500 | |
| 1–2.10.1697 | (About 4000) | Assumptions about coherence of the observation data open to question | 16.12.1916 | 2500 | |
| | | | 28.1.1927 | 2500 | |
| | | | 17–19.1.1937 | 2500 | |
| | | | 23–24.11.1938 | 2500 | |
| (In or about 1725) | (About 4000) | Inadequate data | 15–16.1.1818 | About 2200 | |
| 7.10.1756 | 4000 | | 19.1.1735 | 2000 | |
| 21–22.3.1791 | About 4000 | The W'ly to NW'ly part of this storm – over the southern North Sea | 3.12.1909 | 2000 | |
| | | | 26–27.10.1936 | 2000 | |
| | | | 15–17.12.1938 | 2000 | |
| 28.12.1879 | 4000 | | 30.12.1951 | 2000 | |
| 26–27.1.1920 | 4000 | | | | |

| Date | Strongest wind indicated (knots) | Region |
|---|---|---|
| | northwest coasts of Ireland | |
| 16–17.2.1962 | Gust to 154 – by far the highest ever measured in the British Isles – in a NW'ly gale on a hill in Shetland. Gradient winds up to 85–95 | Britain and North Sea |
| 6.3.1967 | Gust to 124 on the top of Cairngorm, eastern Scotland | Atlantic fringe NW Ireland, all Scotland and later Norway near 62° N |
| 14–15.1.1968 | Gusts to 108 on Lowther Hill (55°23′ N 3°45′ W, 736 m) and 102 on Tiree in the Inner Hebrides | Eastern Atlantic, northern parts of British Isles, central North Sea and Denmark |
| 7.2.1969 | Gust to 118 near Kirkwall, Orkney in a N'ly gale. Gradient winds about 90 | British Isles, western North Sea and part of Norwegian Sea |
| 4.1.1976 | Gust to 88 on the island Sylt off Schleswig Holstein, the second highest ever measured there. Gusts to 91 at two airfields in eastern England. Gradient winds to about 100 | British Isles and North Sea, especially southern parts, and neighbouring parts of the continent and southern Scandinavia |
| 12.1.1978 | Gradient wind from about N 100 to probably 125, supported by wind measurements | Western, central and southern North Sea, with eastern England and other neighbouring land areas |
| 4–5.2.1979 | Gust to 111 on coast of Norway near 62° N in SW'ly gale, which later turned W'ly. Gradient winds 90 to 100 and possibly later over 100 | Atlantic fringe of the British Isles and Norway to 67° N and later the sea area between Iceland and Norway |
| 24.11.1981 | Gradient winds in W'ly to NW'ly storm reached 120 to 130. Gusts over 100 measured in central North Sea | Whole North Sea from Faeroe Islands to the continent and including Denmark |
| 16.10.1987 | Gust to 119 at Quimper on the west coast of Brittany and 117 on the west-facing coast of Normandy. Gradient winds in SSW to WSW'ly gale up to 100–120 | A belt within the fringe of Europe from northwest France and southeast England to Norway |
| 13.2.1989 | Gust to 126 at Fraserburgh lighthouse, northeast Scotland – the highest ever measured at a low-lying site in the | Northern Scotland and neighbouring sea areas |

British Isles – in a W to NW'ly gale

## C. Storms which affected the greatest areas with destructive winds among those studied

Areas are in units of $10^5$ km². Square brackets indicate cases with uncertainties explained in the descriptive texts.

| Date | Estimated extent of destructive winds in climax phase | Estimated total extent of destructive winds during the storm |
|---|---|---|
| 11–12.11.1570 | ? | ? |
| 22.10.1634 | ? | ? |
| [29.10–2.11.1694] | ? | ? |
| [1–2.10.1697] | ? | ? |
| 7–8.12.1703 | 10–13 | 15–20 |
| 24–25.12.1717 | 10–15 | 20–25 |
| 19.1.1735 | 7–10 | About 15 |
| 10–11.12.1792 | 7–10 | 15–20 |
| 19–23.12.1792 | ? | 15 or more |
| 6–9.5.1795 | 7–10 | Probably about 15 |
| 13–16.1.1818 | 7–8 | 10–15 |
| 11.3.1822 | 7–9 | ? |
| 3–4.2.1825 | 8–10 | ? |
| 6–7.1.1839 | About 10 | ? |
| 1–2.1.1855 | ? | About 12–15 |
| 28.12.1879 | Probably 5–7 | About 7–10 |
| 14–15.10.1881 | 5–7 | About 10 |
| 6.3.1883 | 6–8 | About 10 |
| 25–27.1.1884 | 8–10 | Probably 12–15 |
| 14–16.10.1886 | 5–8 | Probably 10–12 |
| 8–9.12.1886 | Probably 5–7 | Probably about 10 |
| 17–19.11.1893 | About 7 | Probably about 10 but perhaps 25 including Atlantic areas |
| 26–27.1.1920 | Up to 5 | 12–15 or more |
| 9.4.1933 | 7–10 | ? |
| 17–19.1.1937 | 7–9 | ? |
| 31.1–1.2.1953 | Perhaps up to 9 | 10–12 |
| 21–23.12.1954 | Up to 10 | Probably 20 or more |
| 16–17.2.1962 | 12–15 | ? |
| 2–3.1.1976 | About 10 | 12–15 |
| 4–5.12.1979 | 5–7 | ? |
| | | Over 10 |
| 23–25.11.1981 | About 7 | 20 |
| 15.12.1986 | ? | ? |
| | Perhaps as much as 60 to 70 mainly over the Atlantic | |
| 16.10.1987 | About 15 | Probably 15–25 including Atlantic areas |

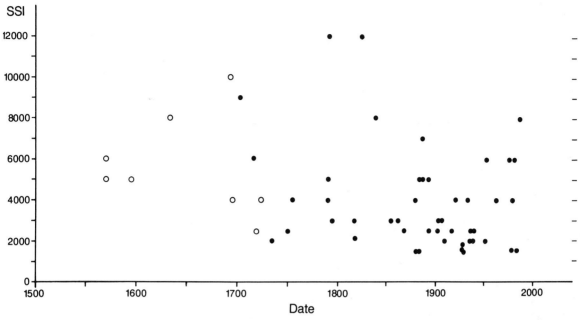

*Figure 2.* Time distribution of storms among those studied in this survey with Severity Index assessments over 1500.

| Date | Storm Severity Index | Remarks |
|------|------|---------|
| | | the storm over the Norwegian Sea |
| 24–25.12.1717 | About 6000 | |
| 31.1.–1.2.1953 | 6000 | |
| 2–3.1.1976 | 6000 | |
| 23–25.11.1981 | 6000 | |
| (About 1570) | (About 5000) | Inadequate data: first of the great inland sand drift episodes in the Norfolk and Suffolk Breckland |
| 21–22.3.1791 | About 5000 | NW to NNE'ly part of this storm – over the northern North Sea |
| 25–27.1.1884 | 5000 | |
| 8–9.12.1886 | 5000 | |
| 17–19.11.1893 | 5000 | |
| (Jan. to Feb. 1595) | (Probably 5000 approx.) | |
| 1–2.10.1697 | (About 4000) | Assumptions about coherence of the observation data open to question |
| (In or about 1725) | (About 4000) | Inadequate data |
| 7.10.1756 | 4000 | |
| 21–22.3.1791 | About 4000 | The W'ly to NW'ly part of this storm – over the southern North Sea |
| 28.12.1879 | 4000 | |
| 26–27.1.1920 | 4000 | |

| Date | Storm Severity Index | Remarks |
|------|------|---------|
| 9.4.1933 | About 4000 | Mainly north of 65° N |
| 16–17.2.1962 | 4000 | |
| 4–5.12.1979 | 4000 | Largely coast of Norway north of 60° N |
| 6–9.5.1795 | About 3000 | |
| 12–16.1.1818 | About 3000 | Overall severity rating of prolonged storm with several separate phases |
| 1.1.1855 | 3000 | |
| 26–27.12.1862 | 3000 | |
| 26–27.2.1903 | 3000 | |
| 12–13.3.1906 | 3000 | |
| (About 1720) | (About 2500) | Imprecise data and no surrounding coverage |
| 11.9.1751 | 2500 | |
| 24.1.1868 | 2500 | |
| 11–12.2.1894 | 2500 | |
| 25–26.12.1902 | 2500 | |
| 16.12.1916 | 2500 | |
| 28.1.1927 | 2500 | |
| 17–19.1.1937 | 2500 | |
| 23–24.11.1938 | 2500 | |
| 15–16.1.1818 | About 2200 | |
| 19.1.1735 | 2000 | |
| 3.12.1909 | 2000 | |
| 26–27.10.1936 | 2000 | |
| 15–17.12.1938 | 2000 | |
| 30.12.1951 | 2000 | |

| Date | Storm Severity Index | Remarks | Date | Storm Severity Index | Remarks |
|---|---|---|---|---|---|
| 23–25.11.1928 | 1800 | | 7.2.1969 | 300 | |
| 6.1.1928 | About 1600 | | 13–14.2.1979 | 300 | |
| 14–15.10.1881 | 1500 | | 28.12.1849 | 200–300 | |
| 6.3.1883 | 1500 | | 9–10.3.1891 | About 250 | |
| 5–7.12.1929 | 1500 | | 5.12.1937 | 250 | |
| 11–12.1.1978 | 1500 | | 9–10.2.1949 | 250 | |
| 1.2.1983 | About 1500 | | 23–25.10.1949 | 250 | |
| 21.9.1588 | (About 1200) | Some indirect deduction of wind strengths | 23.2.1967 | 250 | |
| | | | 4.9.1967 | 250 | |
| 27.2.1736 | 1200 | | 27–29.10.1859 | 200 | |
| 5.12.1792 | 1200 | | 24.3.1895 | 200 | |
| 11.3.1822 | 1200 | | 26.10.1949 | 200 | |
| 27–29.11.1836 | 1200 | | 17.12.1952 | 200 | |
| 17–19.10.1936 | 1200 | | 13–14.8.1979 | 200 | |
| 14–15.9.1786 | About 1000 | | 9–10.2.1988 | 200 | |
| 21–22.12.1792 | 1000 | | 3–4.3.1988 | 200 | |
| 23.12.1954 | 1000 | | 10.1.1849 | 150–200 | Data hardly adequate |
| 18.1.1983 | 1000 | | 1.3.1791 | 150 | |
| 25.1.1739 | 800–900 | | 3–4.8.1829 | About 150 | |
| 12.11.1740 | (About 800) | Data hardly adequate | 15–16.6.1869 | About 150 | |
| 19.9.1741 | (About 800) | Data hardly adequate | 27.3.1878 | 150 | |
| 12–13.1.1818 | About 800 | | 1–2.6.1938 | 150 | |
| 21.2.1861 | 800 | | 14–15.1.1968 | 150 | |
| 28–29.10.1927 | 800 | | 19.11.1973 | 150 | |
| 13.2.1989 | 800 | | 6.12.1773 | 150 | |
| 22.9.1695 | (About 700) | | 27.3.1980 | 150 | |
| 12–13.11.1972 | 700 | | 7.9.1838 | About 100 | |
| 2.4.1973 | 700 | | 15–16.1.1818 | Under 100 | |
| 14–18.8.1588 | (About 600) | Some indirect deduction of wind strengths | 22–23.12.1937 | About 50 | |
| | | | 23–24.2.1939 | Under 50 | |
| 14.8.1737 | 600 | | 6.3.1967 | Under 50 | |
| 16–17.9.1961 | About 600 | | | | |
| 22.10.1702 | (About 500) | | | | |
| 18–19.9.1740 | 500 | | | | |
| 28–29.11.1897 | 500 | | | | |
| 16–17.11.1928 | 500 | | | | |
| 10–13.2.1938 | 500 | | | | |
| 9–10.3.1751 | About 400 | | | | |
| 2–3.1.1784 | About 400 | | | | |
| 14.10.1829 | 400 | | | | |
| 20–21.10.1846 | About 400 | | | | |
| 5.12.1938 | 400 | | | | |
| 21–22.12.1954 | 400 | | | | |
| 29.7.1956 | 400 | | | | |
| 23–25.8.1957 | 400 | | | | |
| 20.4.1773 | (About 300) | Data hardly adequate | | | |
| 25–26.12.1783 | About 300 | | | | |
| 7–8.12.1792 | 300 | | | | |
| 25.11.1829 | 300 | | | | |
| 18–19.11.1835 | 300 | | | | |
| 23.11.1836 | 300 | | | | |
| 10–11.11.1931 | 300 | | | | |
| 15.1.1938 | 300 | | | | |

## Classes of severity: suggested definitions

While remembering that one must remain sceptical about the placing of storms assessed near the arbitrarily chosen boundary of any class, it seems reasonable to divide the above list of storms into broadly the following grades or classes of severity:

| | Severity Index | Number in class |
|---|---|---|
| Class I | 5000 or over | 14 storms fully analysed (15 including the 1986 example over the Atlantic, 20 including indirectly assessed early cases) |
| Class II | Between about 4000 and 1800 | 28 storms analysed (31 including early cases indicated) |
| Class III | Between about 1600 and 700 | 24 storms analysed (26 including early cases indicated) |
| Class IV | Between about 600 and 300 | 24 storms analysed (26 including early cases indicated) |

| Index | | |
|---|---|---|
| Class V | Between about 250 and 150 | 24 storms |
| Class VI | 100 or less | |

B. *Greatest wind speeds noted in the storms studied since the earliest cases with either measurable pressure gradients or measured wind observations*

These cases (34 in all) are placed in chronological order rather than in order of wind speeds, since the indications in many cases gave about the same maximum wind speeds, within the accuracy of which the data and methods of analysis allow.

| Date | Strongest wind indicated (knots) | Region |
|---|---|---|
| 7–8.12.1703 | Gradient winds from SW to W c. 150 | England and southern North Sea to Denmark |
| 25.12.1717 | Gradient winds between SW and NW up to 130 | Eastern North Sea and Denmark |
| 19.1.1735 | Gradient winds SW'ly perhaps 110 | Southeast England and fringe of continent |
| 25.1.1739 | Gradient winds from WSW up to 110 suspected. (No map) | Limited zone across British Isles and North Sea 53° to 57° N |
| 11.9.1751 | Gradient winds NW'ly over 100, possibly up to 150 | North Sea |
| 7.10.1756 | Gradient winds from SW to NW over 100, possibly up to 150 | North Sea |
| 21–22.3.1791 | Gradient winds from WSW probably up to 150 | Central North Sea to German Bight |
| 22.3.1791 | Gradient winds from N to NNE about 120 to 130 | Northern North Sea and Britain's east coast |
| 10–11.12.1792 | Gradient winds from WNW to NNW probably 140 to 150 | Widely over England and the continental fringe from Flanders to the German Bight |
| 10.12.1792 | Gradient winds W'ly 150 ± 30 | Faeroes–Shetland region of northeast Atlantic |
| 15.1.1818 | Gradient winds W'ly up to 120 | Over and around Denmark |
| 1.2.1825 | Gradient winds NW'ly about 140 | Over Scotland and neighbouring sea areas to north and east |
| 3–4.2.1825 | Gradient winds mainly NNW to NNE up to 120 to 140 | Over northern, western, central, and southeastern parts of the North Sea |
| 6–7.1.1839 | Gradient winds between | Over Britain north |

| Date | Strongest wind indicated (knots) | Region |
|---|---|---|
| | SW and NW, SE to S at first in southern Scandinavia: 90 to 100, perhaps stronger later in Denmark and Baltic | of 54° N, the North Sea and Denmark |
| 1.1.1855 | Gradient winds NW'ly 100 to 120 | Over Britain and the North Sea |
| 26–27.12.1862 | Gradient winds W to NW'ly possibly up to 150 | Over northern Britain, North Sea and Denmark |
| 24.1.1868 | Gradient winds S to SW'ly perhaps reaching 140 | Over western and northern parts of the British Isles and northern North Sea |
| 25–26.1.1884 | Gradient winds from about W probably 120 to 140 | Across England and Wales and the Channel and later (SW'ly) over the continental fringe and the North Sea |
| 9–10.3.1891 | Gradient winds from E, in a very localized strip near the front, perhaps about 100 | Southernmost England and the Channel |
| 27.1.1920 | Gust in a S'ly gale at Spanish Point near Quilty, Co. Clare (Ireland) reached 97. Gradient wind about 80 | Atlantic fringe of the British Isles |
| 6.12.1929 | Gust in a SSW gale in the Scilly Isles reached 96. Gradient wind about 75 | All British Isles affected |
| 9.4.1933 | Gust in a N'ly gale at Jan Mayen (71° N 8° W) 163 – highest value ever recorded. Gradient wind apparently about 80. Extreme gust probably not widely representative, probably attributable to topographical effects of the great mountain Beerenberg on the island – lateral convergence or lee waves | Greenland Sea–Norwegian Sea |
| 30.12.1951 | Gradient wind W'ly perhaps 130 to 150. Gust to 94 at Millport on Bute in the inner isles of western Scotland | Scotland, northernmost Ireland and northern North Sea |
| 31.1.–1.2.1953 | Gradient winds from about N 100–130 (regarded as 'phenomenal' measurements at the time). Gust to 109 at Costa Hill, Orkney | North Sea and its coasts and most of the British Isles |
| 16–17.9.1961 | Gusts to 98 at Malin Head and over 90 some way inland from the west and | Atlantic fringe of Ireland and Scotland |

11

| Date | Strongest wind indicated (knots) | Region |
|---|---|---|
| | | northwest coasts of Ireland |
| 16–17.2.1962 | Gust to 154 – by far the highest ever measured in the British Isles – in a NW'ly gale on a hill in Shetland. Gradient winds up to 85–95 | Britain and North Sea |
| 6.3.1967 | Gust to 124 on the top of Cairngorm, eastern Scotland | Atlantic fringe NW Ireland, all Scotland and later Norway near 62° N |
| 14–15.1.1968 | Gusts to 108 on Lowther Hill (55°23′ N 3°45′ W, 736 m) and 102 on Tiree in the Inner Hebrides | Eastern Atlantic, northern parts of British Isles, central North Sea and Denmark |
| 7.2.1969 | Gust to 118 near Kirkwall, Orkney in a N'ly gale. Gradient winds about 90 | British Isles, western North Sea and part of Norwegian Sea |
| 4.1.1976 | Gust to 88 on the island Sylt off Schleswig Holstein, the second highest ever measured there. Gusts to 91 at two airfields in eastern England. Gradient winds to about 100 | British Isles and North Sea, especially southern parts, and neighbouring parts of the continent and southern Scandinavia |
| 12.1.1978 | Gradient wind from about N 100 to probably 125, supported by wind measurements | Western, central and southern North Sea, with eastern England and other neighbouring land areas |
| 4–5.2.1979 | Gust to 111 on coast of Norway near 62° N in SW'ly gale, which later turned W'ly. Gradient winds 90 to 100 and possibly later over 100 | Atlantic fringe of the British Isles and Norway to 67° N and later the sea area between Iceland and Norway |
| 24.11.1981 | Gradient winds in W'ly to NW'ly storm reached 120 to 130. Gusts over 100 measured in central North Sea | Whole North Sea from Faeroe Islands to the continent and including Denmark |
| 16.10.1987 | Gust to 119 at Quimper on the west coast of Brittany and 117 on the west-facing coast of Normandy. Gradient winds in SSW to WSW'ly gale up to 100–120 | A belt within the fringe of Europe from northwest France and southeast England to Norway |
| 13.2.1989 | Gust to 126 at Fraserburgh lighthouse, northeast Scotland – the highest ever measured at a low-lying site in the British Isles – in a W to NW'ly gale | Northern Scotland and neighbouring sea areas |

## C. Storms which affected the greatest areas with destructive winds among those studied

Areas are in units of $10^5$ km². Square brackets indicate cases with uncertainties explained in the descriptive texts.

| Date | Estimated extent of destructive winds in climax phase | Estimated total extent of destructive winds during the storm |
|---|---|---|
| 11–12.11.1570 | ? | ? |
| 22.10.1634 | ? | ? |
| [29.10–2.11.1694] | ? | ? |
| [1–2.10.1697] | ? | ? |
| 7–8.12.1703 | 10–13 | 15–20 |
| 24–25.12.1717 | 10–15 | 20–25 |
| 19.1.1735 | 7–10 | About 15 |
| 10–11.12.1792 | 7–10 | 15–20 |
| 19–23.12.1792 | ? | 15 or more |
| 6–9.5.1795 | 7–10 | Probably about 15 |
| 13–16.1.1818 | 7–8 | 10–15 |
| 11.3.1822 | 7–9 | ? |
| 3–4.2.1825 | 8–10 | ? |
| 6–7.1.1839 | About 10 | ? |
| 1–2.1.1855 | ? | About 12–15 |
| 28.12.1879 | Probably 5–7 | About 7–10 |
| 14–15.10.1881 | 5–7 | About 10 |
| 6.3.1883 | 6–8 | About 10 |
| 25–27.1.1884 | 8–10 | Probably 12–15 |
| 14–16.10.1886 | 5–8 | Probably 10–12 |
| 8–9.12.1886 | Probably 5–7 | Probably about 10 |
| 17–19.11.1893 | About 7 | Probably about 10 but perhaps 25 including Atlantic areas |
| 26–27.1.1920 | Up to 5 | 12–15 or more |
| 9.4.1933 | 7–10 | ? |
| 17–19.1.1937 | 7–9 | ? |
| 31.1–1.2.1953 | Perhaps up to 9 | 10–12 |
| 21–23.12.1954 | Up to 10 | Probably 20 or more |
| 16–17.2.1962 | 12–15 | ? |
| 2–3.1.1976 | About 10 | 12–15 |
| 4–5.12.1979 | 5–7 | ? |
| | | Over 10 |
| 23–25.11.1981 | About 7 | 20 |
| 15.12.1986 | Perhaps as much as 60 to 70 mainly over the Atlantic | ? |
| 16.10.1987 | About 15 | Probably 15–25 including Atlantic areas |

D. *Storms with longest durations of damaging winds among those studied*
Square brackets indicate cases with uncertainties explained in the descriptive texts.

| Date | Wind direction | Duration | Remarks |
|---|---|---|---|
| 15 and 23.1.1553 | Mainly NW'ly | 2 gales 3–5 days | North Sea and its coasts |
| 11–12.11.1570 | SW becoming NW | Probably over 12 hours | Whole North Sea |
| 2.11.1592 | Mainly S'ly | 3–4 days with one-day interruption | Skagerrak and presumably North Sea |
| 22.10.1634 | About WNW | Perhaps 24 hours | North Sea, especially southeastern parts and nearby lands |
| 28.2.1662 | SW–W becoming N | 15–20 hours | Southern and central England and North Sea |
| [Autumn 1676] | NW'ly | Presumably 12–18 hours or more | Northeast Scotland (Nairn) – serious sand drift |
| [29.10–2.11.1694] | W'ly, then NW to N | 36–48 hours with little intermission | Northeast Scotland (proposed date of Culbin Sands sand drift) |
| [1–2.10.1697] | Mainly NW'ly | About 24 hours | North Sea (sea flood and proposed link with great sand drift in Outer Hebrides) |
| 24–25.12.1717 | Mainly N | 24–30 hours | North Sea (sea flood) |
| 19.1.1735 | S to SW | 18–24 hours perhaps locally 30 hours | Southeast England and fringe of Europe |
| 27.2.1736 | N'ly | 18–24 hours | North Sea and eastern Britain |
| 14.8.1737 | E'ly and NW'ly | 24–30 hours | Eastern Britain and North Sea |
| 18–19.9.1740 | S and SW | 12–24 hours | North Sea with English and Norwegian coasts |
| 12.11.1740 | Mainly N–NE but also a narrower belt S–SW | Perhaps locally 24 hours | North Sea and England |
| 25–26.12.1783 | E'ly | 36–48 hours | Northern Scotland and North Sea |
| 2–3.1.1784 | SE'ly | 36–48 hours | Northern Scotland and North Sea |
| 14–15.9.1786 | WSW becoming NW | 30–36 hours | Southern parts of British Isles, North Sea, and later Denmark and Norway coasts |
| 1.3.1791 | NW–N | 12–24 hours | Southernmost North Sea, Kent and Channel |
| 21–22.3.1791 | WSW becoming NW | About 24 hours | Southern North Sea |
| 21–22.3.1791 | NW–NNE | Up to 50 hours | Northern North Sea and Scotland |
| 10–12.12.1792 | W–NW | 50–60 hours | North Sea |
| 19–23.12.1792 | W–NW | Up to 3 periods of about 18 hours | North Sea |
| 6–9.5.1795 | N'ly | 70–80 hours | Sweden and at times northern Norway and Norwegian Sea |
| 23–28.10.1805 | SW–W | [2 periods of 48–60 hours of Beaufort 8 or more] | Atlantic Ocean west of Gibraltar |
| 15–16.1.1818 | SW–NW | 18–24 hours | Denmark and south Baltic |
| 2–5.2.1825 | WSW–WNW | 50–60 hours | North Sea |
| 3–4.8.1829 | N–NE | 12–24 hours | Northeast Scotland |
| 14.10.1829 | N–E | 12–18 hours | Much of North Sea with Scotland and Denmark |
| 25.11.1829 | E | 18–24 hours | North Sea and its coasts |
| 7.9.1838 | N–NE | Up to 24 hours | Northern North Sea and British coasts |
| 6–8.1.1839 | SW–WNW | About 24 hours | A zone across Scotland and northern England |
| 20–21.10.1846 | SW–W | About 24 hours | Coasts of Ireland, southern Britain and the Channel |
| 10.1.1849 | E | Perhaps 20 hours | Northern Scotland |
| 1.1.1855 | NW'ly | 12–15 hours | North Sea and much of Europe |
| 26–27.12.1862 | WSW–NW | Perhaps 10–15 hours | North Sea and its coasts |
| 24.1.1868 | S'ly becoming WSW | About 18 hours | Mainly Scotland and northern North Sea |

| Date | Wind direction | Duration | Remarks |
|---|---|---|---|
| 14–15.10.1881 | All round the compass | 12–20 hours | North Sea and surrounding lands with most of the British Isles |
| 25–27.1.1884 | SW–W | 15–20 hours | Atlantic Ocean 45–60° N, Britain, North Sea, and parts of continent |
| 14–16.10.1886 | SW–W later NW–N | Probably 30–36 hours | British Isles, Biscay and surroundings, then on to Denmark and central Europe |
| 8–9.12.1886 | SW–NW | About 30 hours | The eastern Atlantic, Biscay, the British Isles, and parts of the continent |
| 9–10.3.1891 | NE | Perhaps 18–24 hours | Southern England and Wales and the Channel |
| 17–19.11.1893 | NW–N | 30 hours or more, locally 40–42 | British Isles and surroundings, the eastern Atlantic, Biscay, the North Sea, Denmark and south Baltic |
| 11–12.2.1894 | SW–NW | 24 hours, with intermissions | North Sea and surrounding lands, later also south Baltic |
| 28–29.11.1897 | Mainly NW–N | 20–24 hours | British Isles, the North Sea and Denmark |
| 12–13.3.1906 | N | Up to 30 hours | N'ly over Britain and the North Sea, S–SW east of the North Sea |
| 16.12.1916 | WSW–WNW | 15–20 hours | British Isles, the Channel, southern North Sea and Baltic |
| 26–27.1.1920 | Mainly S–SSE | Up to 24 hours | Eastern Atlantic, British Isles, parts of the North Sea, Denmark and the south Baltic |
| 5–7.12.1929 | SSW–W | Up to 20 hours | British Isles and surrounding seas |
| 9.4.1933 | N'ly | 24–30 hours | Greenland Sea and Norwegian Sea |
| 17–19.10.1936 | W–NW | 12–15 hours | British Isles, North Sea and its coasts, later also Denmark and south Baltic |
| 26–27.10.1936 | Mainly W–NW but S'ly on continental fringe | 15–20 hours | British Isles and North Sea with its continental coast |
| 17–19.1.1937 | SE–S | 30–40 hours | Coast of southwest Norway, North Sea and much of the British Isles |
| 22–23.12.1937 | S'ly | 24–30 hours | West coast of south Norway |
| 15.1.1938 | SE–S | Up to 15 hours | West coast of south Norway |
| 15–17.12.1938 | Mainly SE | Perhaps 30–40 hours | Coasts of south Norway, North Sea, and British Isles |
| 31.1–1.2.1953 | N'ly | 12–18 hours | North Sea and its coasts |
| 21–23.12.1954 | W–NW | 15–18 hours | Mainly North Sea and its coasts |
| 29.7.1956 | S, W and NE | Up to 12–15 hours | Southern England, Wales and the Channel |
| 23–25.8.1957 | W'ly | Up to 30–36 hours | British Isles and surroundings, southern North Sea and Denmark |
| 16–17.2.1962 | Mainly NW | 12–15 hours | North Sea and surrounding lands |
| 2–4.1.1976 | SW–W becoming NW–N | Up to 24 hours in the eastern parts of the region affected | British Isles, North Sea – especially the eastern parts, Denmark and parts of the continent |
| 13–14.2.1979 | E'ly | 15–24 hours | The Channel and southern England |
| 4–5.12.1979 | SW–W'ly | Up to 24 hours | Norway coast 63–67° N, Norwegian Sea and northeast Atlantic |
| 23–25.11.1981 | NW–W'ly | Up to 20–30 hours | North Sea from Faeroe Islands to Denmark and the south Baltic |
| 18.1.1983 | W–NW | Up to 12–20 hours | Northern, central and eastern North Sea, Denmark, and north Germany |
| 1.2.1983 | Between W and N | Up to 15–20 hours | North Sea and Britain and parts of the continental coast |
| 15.12.1986 | All round the compass | About 15–20 hours | Central, northern and eastern North Atlantic |

| Date | Wind direction | Duration | Remarks |
|------|----------------|----------|---------|
| 3–4.3.1988 | N'ly | Perhaps 12–15 hours | Northern North Sea and its coasts and eastern Britain |
| 13.2.1989 | W–NW | Perhaps 10–15 hours | Northern Scotland and neighbouring waters |

## E. *The most destructive storms*

We have listed the storms that rated highest on the Severity Index, those that produced evidence of the strongest winds, and those that seem to have brought damaging winds over the widest areas and longest durations.

Many interests would value a listing of the storms that caused the greatest destruction, and a study of their incidence, if it were possible to define this satisfactorily and identify the storms concerned. This might be attempted in terms of money cost or insurance losses, and these are mentioned in the accounts of a few storms for which figures have been published. But great difficulty arises in any attempt to estimate these for storms of long ago. Not only must the almost continuous decline in the value of money over the years be allowed for; but in times when nobody possessed electrical appliances, central heating, motor transport or washing machines, the importance to the possessor of whatever things he did own doubtless ranked as high as, or in some cases perhaps much higher than, the same articles in today's world. And, in some of the worst storms on unprotected coasts in the middle ages and earlier, land was lost for good – or, at any rate, for many centuries – as when the Jadebusen and Zuyder Zee (the relics of this are now known as the Ijsselmeer) were formed. And the people of some areas lost all they had. This was clearly the experience of the coast-dwellers when land was inundated by the sea, usually (we may suppose) in storms, during times of rising sea level, as when the North Sea was re-created after the last ice age. Such losses in the main exceeded the deprivations caused by the storms of recent centuries, regardless of the costliness of modern technological aids and comforts that had not been invented then. To a peasant in former centuries living in a remote place in a shack, with a hole in the roof instead of a chimney to let the smoke out, the loss of his house would be as severe as the loss to us of a modern labour-saving thermally insulated home. Similarly, losses of many ships, as in some past storms in this compilation (e.g. in 1695), whether naval or merchantmen, should be considered not just in terms of the cost of them but in relation to implications of the reduction of the whole fleet. Such considerations also arise in cases where the sea-banks or dykes, and lighthouses etc., that protect the coast are damaged or destroyed. Nevertheless, it may be instructive to attempt assessment of the money costs of damage in the case of a few of the severest storms in the earlier centuries covered in this study. No attempt will be made to rank them in order of the apparent magnitude of the losses but what is known in some sample cases will be listed below.

It must be understood that the causes which make some storms much more destructive than others are by no means all meteorological: in most cases other things besides the weather are involved. This is obviously the case in the highest sea floods with storm surges on exposed coasts: clearly the state of the tide and whether the storm comes with neap tides or spring tides is important, also the precise timing of whether the surge nearly coincides with the peak of the tide or not. The outcome will be greatly affected too if the protective banks and dykes happen to be in a state of disrepair after an earlier storm or some other calamity.

What can be done in the way of estimating money values of the losses in major storms of the past, and the problems and limitations that are encountered, may be illustrated briefly by consideration of the Culbin Sands disaster in 1694. The overwhelming by sand of 16 farms, including the manor house, and 2000–3000 hectares (5000 to over 7000 acres) of the very best fertile grainlands in that part of Scotland, the so-called 'bread-basket of Moray' – where forest, moor and mountain, with sandy flats near the coast are far more general – must have been a very serious blow to the local economy. Best croplands in the best districts of Aberdeenshire and elsewhere today command prices of about £2000 an acre and perhaps rather more in some cases. Thus the land alone that was lost would probably be valued at 10 to £15 million in today's money (or in the values of that time perhaps one hundredth of today's figures). The houses for such farms would today in total command £800 000 to £1 million.* Additionally, the workers' cottages, the barns and equipment, seem impossible to value now, since no details of them are known. Also the total loss of the 1694 barley harvest must be reckoned the equivalent of another £1 million today, and was doubtless more serious, and entailed more costs and anguish, in a year when – as every year between 1693 and 1698 in northern Scotland – harvests were failing generally in the country and grain was being imported from the eastern Baltic states. Thus, the immediate losses in the area about the Culbin lands were probably the equivalent of £20 to £25 million today. Further to this, the (natural) coastal defence barrier of the dunes was so damaged, and the River Findhorn forced to alter course, that the wrecking of the little town of Findhorn at the river mouth by another storm eight years later (in 1702, q.v.) can be looked upon as a consequence of the 1694 disaster. No cost accounting can be put down for this, since it is not known whether any protective works had been undertaken in the 8-year interim, but the town and harbour of Findhorn had to be re-built (presumably over some quite long term of years). Finally, unknown and presumably considerable sums were spent at various times over the 230 years between 1694 and the 1920s on various attempts to prevent further movement of the sands and, only at the end of that time successfully, to reclaim the former grainlands for forest. Thus, despite our lack of detailed knowledge, and taking no account

*I am indebted to Dr D. Glass, MD of Dinnet, near Ballater in Deeside, and recently of Peterhead, for the up-to-date information about current land and property values in this part of northeast Scotland.

of the probability of damage by the 1694 storm to shipping and harbours and buildings elsewhere within range of the North Sea and Channel coasts, it is reasonable to rank this storm with others that produced total losses of £100 million and more in 1990 money.

## F. *Coastal changes due to sea floods*

Those sea floods which have been most destructive of the coastlands can logically be seen as just the later chapters in the long record of recovery of the North Sea associated with the world-wide rise of sea levels as the glaciers of the ice age melted and the Earth's crust in the region of northwest and northeast Europe continued to adjust to changes in the load of ice and water both within the region and beyond. There have also been continual further, but much smaller, changes of sea level as glaciers and the total volume of water in the world's oceans wax and wane in response to smaller changes of climate.

There is fairly general, but not universal, agreement that after the great melting of ice-sheets and glaciers the highest post-glacial sea level stand (probably not more than 1–2 metres above present) was reached about 4000 years ago, around 2000 BC (Lamb, 1977, p. 117; Mörner, 1980; Tooley, 1974, 1978). Others think that the rise has continued, but at a greatly diminished rate, broadly up to the present day. It is generally agreed that the fluctuations that have undoubtedly occurred within the last four millennia have been relatively small, amounting to no more than a few decimetres (Pränge, 1971). Sea level seems to have been lower than at present, perhaps by as much as 1 m, around 500 BC, but rose during Roman times, particularly between AD 200 and AD 400 (evidence and sources are summarized in Lamb, 1988, p. 91). Sea level was then stationary for some time, but may once more have been a little below present around AD 600 to AD 800. There was another high-level phase around the year AD 1000 possibly up to 40–50 cm above the present level and again around AD 1300 to AD 1400, and at least in the North Sea possibly for most of the time between those dates (cf. Funnell 1979). At some stage during the development of the colder climate of the so-called Little Ice Age over the centuries that followed sea level fell and was registered by a tide-gauge at Amsterdam in 1682 as 17 cm lower than in 1930. The level is believed to have risen about 15 cm since 1890. These figures seem to apply quite well all around the southern North Sea, but in the region of the German Bight and Schleswig-Holstein the changes have been consistently greater. Petersen and Rohde (1977, p. 30) report that the sea levels on both coasts of Schleswig-Holstein have risen as much as 25 to 30 cm per century since about AD 1550, by now having regained there as elsewhere about the position they held around AD 1000.

Coasts, particularly low-lying coasts, have inevitably been affected by the sequence of changes of world sea level. This is somewhat complicated around the North Sea basin by unequal bending of the Earth's crust. Despite such local variations, correspondence with the general history is plain enough. There is evidence (Jelgersma *et al.*, 1970) that the coastline of the Netherlands between 4000 and 5000 years ago lay about 50 km, locally even about 75 km, farther east

than now. Similarly, on the German coast between about 2000 BC and 1000 BC the tides from the North Sea penetrated farther up the River Elbe, to beyond where Hamburg stands today (Petersen and Rohde, 1977, p. 34). After that stage, sand barriers – the so-called 'Older Dunes' – were built up 10 m high only a little east of the present Dutch coast and the tides failed to reach Hamburg. Next, during a centuries-long time of evidently less windy climate and little sand movement that followed from around AD 200 to AD 1000, those dunes became forested. Presumably the process was aided by periods of once more higher water-table when the sea level was relatively high.

The variations of the coasts since about the year AD 1000 betray the effects of the further variations of sea level and of apparent changes of storminess (Lamb, 1980, 1988).

Manuscript records tell us of at least the main events on the coasts of northwest Europe during the last thousand years but need to be interpreted with caution. There is argument, for instance, as to whether the losses of land west of Schleswig-Holstein in the great storms of the thirteenth and fourteenth centuries were in fact real and long-lasting (Gram-Jensen, 1985, pp. 9–20) though in the case of the greatest of those storm floods, in 1362, it seems to be verified that about 50 parishes in those areas were swallowed by the North Sea. The situation becomes clouded by the effects of various human efforts to protect and improve the lands for settlement. Early attempts at dyke-building seem to have had little lasting success before about AD 1500, and drainage channels provided to adapt the precariously protected marshlands for agriculture probably made them all the more vulnerable when serious sea floods came. Some endyked areas were also weakened by peat cutting. Historians also believe it possible to show that the greatest reported losses of people by drowning were several times as great as the probable total population of the coastal marshlands. But such calculations may be as much wrapped in uncertainty as the original reports. From about the year AD 1500 onwards the reports seem verifiable.

There is, in fact, a remarkable map of the broad zone of marshlands in about AD 1240 west of the present coast of Schleswig-Holstein by cartographer Johannes Mejer (who lived in the area at Husum) put together in 1652 on the basis of an exhaustive exploration of the region of the lost lands by boat consulting the memories of 'old reliable men', from folk tradition, and from a medieval list of parishes. Gram-Jensen more or less rejects Mejer's map, but it is largely accepted by Petersen and Rohde (1977, p. 35) who print a reproduction.

What seem certain as the major events in the history of the coasts of the North Sea are:

| | |
|---|---|
| *Between about 120 and 114 BC* | A great storm flood of the coasts of modern Jutland and northwest Germany, the 'Cymbrian flood', which caused a migration of the former (Celtic) population out of the area. |
| *About AD 800* | The first recorded one of a series of storm floods (this, like many others, also affected the Dutch coast) which reduced the size of |

16

the island of Heligoland by more than half by the year 1300. (Continued erosion of the cliffs has since reduced the island to 1–4 km across compared with about 60 km in AD 800.)

*February 1164*

Flood described by Petersen and Rohde (1977) as the first great, damaging flood after the building of the dykes on the German coast. This flood began the formation of the now considerable Jadebusen bay in Oldenburg. Many thousands drowned.

*From 1200 (and especially from 1212) to 1219*

An exceptional series of great North Sea floods, mainly on the Dutch and German coasts. The earliest eye-witness report is from that in January 1219, which produced great losses of land and people in the marshlands between Hamburg and Jutland. Some of these storm floods are believed also to have affected the English coast.

*1287*

Many storm floods on the East Anglian, Kent and Sussex, and continental coasts. The Dollart Bay, in the region of the River Ems mouth at the Dutch–German border, and the Norfolk Broads, began forming.

*January 1362*

The 'Grote Mandrenke' (drowning). The high point of the sea flood disasters and losses of land in the thirteenth and fourteenth centuries. Many parishes disappeared off Schleswig. The chronicles tell of 100 000 deaths, but Gram-Jensen (1985) estimates them at between 11 000 and (hardly as much as) 30 000.

This storm was also felt across southern England as an outstanding SW'ly gale which destroyed bell towers in London, Bury St Edmunds and Norwich.

*1436*

The greatest flood of the fifteenth century. The big island of Nordstrand off the coast of Schleswig divided in two.

*1509*

This flood brought the Dollart Bay to its greatest extent.

*1570*

One of the greatest floods among the accounts in this study.

*October 1634*

A great storm flood which caused losses of land comparable with those in 1362. With this storm the islands of Nordstrand and Pellworm, off Schleswig-Holstein, were finally separated. The peak water levels were among the highest ever recorded on marks near the sea, and in view of the rather lower sea levels then generally prevailing may have marked a storm surge never since equalled.

Records of other notably high North Sea floods will be found in our accounts of storms in October 1663, 1697, 1717, 1751,

1791, 1792, 1825, 1881, 1894, 1916, January and November 1928, 1953, 1962, 1976, and 1978.

Rohde (1964) has found records of a remarkable history of the sea floods at Tönning on the west coast of Schleswig-Holstein near 54°20′ N, including former flood marks engraved on a building – the house of a shippers' guild – that was demolished and rebuilt in 1808, recording flood levels back to the year 1625 with sufficient detail to fix their heights relative to the 1717 and 1756 floods. Since it is also recorded that the peak flood level in February 1825 was 1 foot 5 inches higher than that in 1717, Rohde (1964, p. 128) is able to give a diagram showing the levels of seventeenth and eighteenth century floods at Tönning compared to the later floods (reproduced here as Figure 3 by kind permission). The great flood on 10–11th December 1792, which is not included, reached 2 cm higher than the 1756 one at Tönning. In the same article Rohde reprints the descriptive accounts of many earlier floods at Tönning from AD 1020 on, taken from two sources: Heimreich (1666) and a manuscript dated 1724 found in the archives of St Laurentius church and attributed to a Tönning organist, Johann Hasse, who witnessed the 1717 flood there. Sea floods there seem to have been most numerous in the fourteenth century (15 cases) in the eighteenth century (17 cases), and perhaps also in the seventeenth century (13 or more).

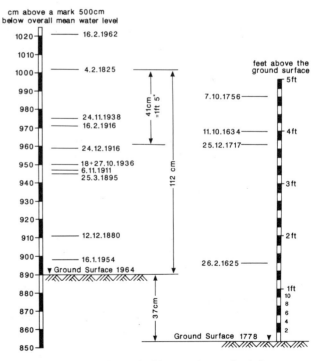

*Figure 3.* Peak levels reached by North Sea floods between 1625 and 1962 as recorded on the walls of buildings at Tönning on the west coast of Schleswig-Holstein (54°19′ N 8°55′ E). (Adapted from Rohde (1964).) Note that the great flood on 10–11 December 1792, here omitted, reached 2 cm higher than the 1756 one. One must also allow for the fact that Tönning is in the region of anomalously great sea level changes specified on p. 16.

G. *Changes of landscape at the coast and elsewhere caused by shifting sands*

Not only sea floods, storms and rough seas which bit by bit erode cliffs and sand barriers, cause changes to coastlines. Blown sand, no doubt supported in many cases by water-current drifts shifting sediments in river estuaries and firths as well as on open coasts, can also bring about important changes – as already noticed (p. 16) in the formation of the line of dunes (the 'Older Dunes') on the Dutch coast, which helped to stabilize a great recession of the North Sea in that area some three to four thousand or so years ago. Tooley (1985) reports that dunes began to form on the west coast of northern England (Morecambe Bay) about the same time. Incidents of this kind have been relatively frequent in times of low, or falling, sea level. And in at least one case, in 1413 at Forvie on the northeast coast of Scotland, it seems that unusually extreme tides may have been important. This could operate through rough seas around high water transporting sand and shingle up, as well as along, the shore and the stormy winds at extreme low water scouring the sand. Some reported cases of remarkable scouring of beaches are mentioned in connexion with storms in this compilation in 1837–8 and in 1862.

One can trace alternate periods of coast recession/marine transgression and of sand-dune building and dunes invading the landscape in the past. Brooks (1949) drew attention to a period notable for sand drift, altering the coasts and occasionally advancing some kilometres inland, in South Wales and, it is believed, parts of Brittany, Cornwall, Scilly and the Channel Islands between about 400 and 200 BC.\* Sand layers in a raised bog in eastern Jutland, Denmark, also indicate a period marked by blowing dust and sand, about the same time, between about 600 and 100 BC (Bahnson, 1972). In some places, in the Channel Islands and East Anglia and again in the same bog in Jutland, there are hints of a similar tendency around AD 500 to 800.

Incidents amounting to another major period remarkable for blowing sand changing the aspect of the coasts of northwest Europe in many places are recorded frequently from around AD 1300 to the eighteenth century or later. A presumably related event, the formation of the 'Younger Dunes' along the present coastline of the Netherlands, which have since been strengthened by coastal engineering works, began as early as the twelfth century (Jelgersma and van Regteren Altena, 1969). In those times Amsterdam harbour had continual troubles through the silting up of the entrance by sand blown and drifted from a nearby sandbank, and even much later – in the eighteenth and nineteenth centuries, when the problem was aggravated by the bigger size and deeper draught of ships – 'scheepskamelen', literally 'ship camels', a kind of floating dock were developed and long in use to transfer ships over the entrance sands into the harbour (I am indebted to Dr A. van Engelen, Koninklijk Nederlands Meteorologisch Instituut, De Bilt for most of this information : personal communication, 20 January, 1990). In the main

climax in and around the seventeenth century of this recent sand-blow period, drifting sands and dune formation occurred even far inland in East Anglia and the Netherlands (see our report on the storms between 1570 and 1668). A number of the coastal troubles with blown sand are also mentioned in the accounts of storms in this compilation.

Examples of severe storm incidents of the category here discussed from the centuries spanned by our list of particularly great sea floods are :

| | |
|---|---|
| *About 1316* | Storms causing sand-dune movements closed a medieval port at Kenfig, near Port Talbot on the coast of South Wales. Further sand movements caused by storms between 1344 and 1480 finally buried the former Roman coast road which had remained in use in the area, and in 1553 an Act of Parliament was brought in 'touching the sea sands of Glamorgan'. |
| *Some time after 1385* | Formation of a great belt of sand-dunes – presumably by S'ly or SW'ly storm winds – enclosed and protected the flat land now known as Morfa Harlech at the north end of the west-facing coast of Cardigan Bay, closing for good the medieval port of Harlech which had been in use until then. |
| *19.8.1413* | The medieval township of Forvie on the coast of Aberdeenshire in northeast Scotland buried by sand now constituting a 30-metres-high dune, reportedly by a great S'ly storm. This, together with some less precisely recorded storms in 1401 to 1404, however, seems rather to have been the beginning of a process which ended a thousand-year-long period of stability of the local landscape at Forvie (Landsberg 1955). |
| | The Forvie sand-dune, which advanced 50 to 250 m to the north in this storm, advanced about 200 m farther before the end of the fifteenth century; but still greater advances occurred later, in the sixteenth and seventeenth centuries. Calculations kindly undertaken by Dr J. Vassie at the Institute of Oceanographic Sciences, Bidston Observatory, Birkenhead, revealed that the tides at Aberdeen on the date given for this 1413 storm were very near to the extreme astronomical tide. With a S'ly storm superposed on the extreme tidal range it would be the low tide that was probably most extreme. |
| *1427 and 1479 and after* | Earliest references found in the Danish archives by K. Frydendahl to blowing sand incidents, which became so rich thereafter that search had to be restricted to the northernmost county in Jutland (Hjørring Amt (county)). |
| *About 1570 and after* | Dune-forming massive sand movements driven by SW'ly or WSW'ly winds inland in |

\*I am indebted to Dr D.S. Ranwell of the Nature Conservancy Council and the University of East Anglia, Norwich for some of this information.

| | |
|---|---|
| *1573* | A W'ly or SW'ly storm drove a line of high sand-dunes 3 km inland at Kenfig from the coast of South Wales near Port Talbot. |
| *In and around the seventeenth century* | The greatest advance northward of the huge sand-dune burying Forvie on the Aberdeenshire coast took place in or about this period so that by 1688 the northern limit was 450 m north of the position about 1500. There has been much less movement since 1700; at some points a 100 to 150 m advance in the eighteenth century was partly lost afterwards and elsewhere fixed by vegetation. |
| *Seventeenth century* | In the Faeroe Islands, at the northern end of the west coast of the biggest island, Streymoy, a great storm one February drove so much sand in through the narrow, mountain-girt entrance (only 20 to 30 m wide) to the formerly fine sheltered natural harbour at Saksun that the harbour was choked and destroyed. This is surely a case where water transport carried much more of the shifted material than the wind. |
| *1694* | After storms which had brought preliminary intimations of disaster in April 1663 and in the autumn of 1676, the great storm of autumn 1694 in our compilation finally obliterated a large area of the Culbin estate, the most fertile area of farmland in northern Scotland. |
| *1697* | Great sand-blow disaster at Udal on the island of North Uist in the Outer Hebrides obliterated a site that had been inhabited for 4000 years. |
| *About 1720* | Storm demolished a coastal hill of sand and deposited it, blocking for good a formerly useful fishing harbour, near Rattray Head on the northeast coast of Scotland. |
| *1725* | Final event in a series of drifting sand difficulties starting in the late sixteenth century caused abandonment of the village of Tibirke in northern Sjaelland, Denmark. |
| *1745* | The Lensmann (sheriff) of Stavanger Amt reported between 10 and 20 farms near the sea in Jaeren suffered serious damage from drifting sand. |
| | (Some damage from 'flying sand' was again reported by Lensmann M.A. Grude at his farm in the same district as late as the 1870s in an article describing successful experiments in Denmark at stilling driftsands by planting marram grass (Grude, 1914).) |
| *1840s approx. and earlier* | There are similar records (and legends) of blown sand along the west and north-facing |

coast of Cornwall. At St Enodoch, 8 km northwest of Wadebridge, the church built in 1430 was buried by sand during 'a fortnight of terrible storms' in or about the 1840s. It was then dug out by the parishioners who regained access through the roof of the steeple.

At another point, Perranzabuloe near Perranporth, the first church built in the middle ages, according to legend, was overwhelmed by sand. A second church was built and it too was later threatened by further sand advances.

Particular interest in connexion with the variations of storminess and the lasting effects of individual storms attaches to the growth of the point of Skagen (or The Scaw in older English literature) at the extreme north of Jutland, Denmark. Through a series of local disasters due to storms which caused sea floods and much drifting sand from about 1591 onwards, the old town of Skagen was abandoned and was still 'awash' with sand partly burying many of the buildings, including the church, when painted by Vilhelm Melbye in 1848. Studies pursued over a number of years by the surveyor P. Hauerbach who lives in the present town of Skagen, with the aid of radiocarbon dates and documentary records, provide a record of the growth of the land in the area over a very long history. The point is formed from sedimentary materials drifted northwards and northeast from the west coast of Jutland and deposited where the current swirls round the point (Skagen Odde). The growth rate is greatest in periods of enhanced winds and current drifts. Sometimes storms disrupt the newer part of the promontory and cut off the point, which later grows again. But, in general, great storms throw up a new shingle and sand platform, in the form of a step which may be up to some metres wide and a decimetre or more in height, and is later consolidated by sand blown along by the continual winds and later storms. Finally, grasses and other vegetation gain a hold. Examples of such steps have been identified, for instance, with the 1703 and 1987 storms. About 20 km back along this remarkable spit, at Råbjerg Mile, active dunes with a surface of open loose sand have built up to a height of 41 metres. According to this writer's measurements from Hauerbach's maps, the spit, continually growing northeastwards, has caused the point Skagen Odde to advance between 20 and 25 km in the last 7000 years, an average rate of between 3 and 4 metres a year. About the same rate applies to the last 2000 years, but between positions dated to AD 1355 and 1695 the advance averaged 8 m/year and just between 1695 and 1795 the growth seems to have been quickest, averaging 23 m/year. The variations (see Figure 4) since 1695, when the still standing second Skagen lighthouse was built, may be listed as follows:

| | |
|---|---|
| 1695–1795 | 23 m/year |
| 1795–1850 | 5.5 m/year |
| 1850–87 | 15 m/year |
| 1887–1945 | about 7 m/year |

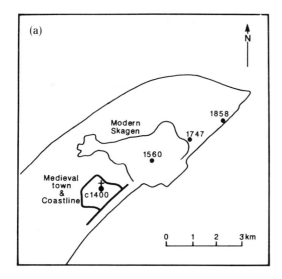

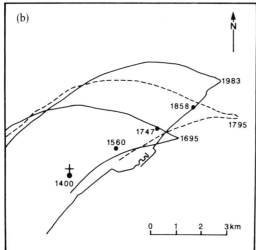

*Figure 4.* Maps of the north point of Jutland, showing: (a) The old town of Skagen clustered round its church built about AD 1400 (when its tower, no doubt, served as a useful navigation mark for boats at sea) and the positions of three lighthouse towers (still standing) built to mark the point Skagen Odde at the successive later dates shown. A more primitive type of light erected in 1695 to mark the end of the spit at that time has been reconstructed inland from its original position, (b) Positions of the spit in 1695, 1795 and 1983.

1945–83      some net retreat: rapid advance 1954 to 1964, then a retreat presumably as a result of severance of the tip of the spit.

The coincidence of the main period of rapid growth with a time that included five of the fifteen greatest storms in our compilation affecting Europe, and specifically Denmark and the North Sea, suggests that this registers notably enhanced windiness and sea movement as characteristic of the Little Ice Age period, from the storms of 1570 or thereabouts to 1825. Note also that four more of the 22 highest ranking storms came within that period (see our list of the grading of storms).

It is curious that, additionally, the great storm in 1694 came just before the period of most rapid growth of the Skagen spit and the storm of the disastrous sea floods in 1953 similarly came just before the second period of most rapid

growth began. One wonders whether these two storms shifted sediment in the North Sea which was indirectly connected with the advance of the Skagen spit in the years immediately following. The 1962 storm is also likely to have been effective in the area and to have played its part in the growth from 1954 to 1964. The lack of growth in recent years must surely be accounted for by a severance, and later storm-destruction, of the spit.

The third period of rather rapid growth includes the severe storms of 1879 and the 1880s.

H. *Storms notable for destruction of trees and forests*
Cases distinguished by reports of particularly great damage to trees and losses in forests are found in our accounts of the storms in 1703, 1737, 1739, 1756, 1773, 1786, 1795, 1839, 1884, October and December 1886, 1893, 1894, 1895, 1902, 1903, 1953, 1972, 1973, 1976, and 1987.

I. *Storms with the greatest reported losses of human life*

*January 1362*      Probably the greatest North Sea flood disaster in historical times with more than half the population of the marshland districts along the North Sea coast of Jutland and Slesvig (modern Schleswig) drowned – the variety of estimates is discussed on pp. 16–17 – and some parishes lost to the sea for good. It was 50 years before the dykes could be repaired (Gram-Jensen 1985).

*1570*      Another very great North Sea flood with no accepted figure for the undoubtedly very big death toll, suggested as high as 400 000 in some sources but probably very much lower.

*1588*      The greatest losses of life in the tale of the Spanish Armada were due to the storms encountered, particularly the very great storm of 21 September 1588 in which certainly not less than 17 of the 130 ships originally in the Armada were wrecked. It is estimated that in all the wrecks that occurred on the Atlantic coasts of Ireland and Scotland in August, September and October of that year some 4000 of the Armada's men were drowned; there were perhaps 2000 who got ashore besides.

*1634*      A storm flood of the North Sea on the continental coast rivalling that in 1362. The losses of life are reasonably assessed at 6000.

*22.9.1695*      Perhaps more than 1000 perished in shipwrecks (see our account of this storm).

*7–8.12.1703*      Estimates of the numbers who died in the great storm over southern England and the Channel reasonably run up to 8000, about half of them navy men on ships blown out to sea and wrecked.

*25.12.1717*      About 11 000 drowned in storm flood of the North Sea on the German coast.

| | |
|---|---|
| *1953* | Great storm accompanied by tidal surge flood of the continental and English North Sea coasts and shipping losses elsewhere: about 2000 died. |
| *1962* | North Sea storm flood on the continental coast: about 300 drowned near Hamburg. |

J. *Losses of shipping and lives at sea*

Cases when particularly serious losses of ships and ships' crews were recorded will be found in our accounts of the storms in 1588, 1695, 1703, 1756, 1783, 1786, 1822, October and November 1829, 1839, 1881, 1883.

Disasters to ships and their crews at sea also caused great concern in the storms included in our survey in 1838, 1859, 1878, 1893, 1903, 1909, 1927, 1937, 1953. A new alarm in recent years has been disasters caused by storm damage to oil and gas rigs in the North Sea, the first of these costing 123 lives in 1980.

K. *Storms with the heaviest financial/insurance losses determined among those surveyed in this study*

Attempts to estimate these have been made in the probably particularly serious cases in 1694 (pp. 15–16), 1839, 1976, 1981 and 1987 (see our accounts of these storms).

Conspicuous structural damage to buildings was reported in the storms in 1662, 1669, 1703, 1735, 1737, 1739, November 1740, 1741, 1756, 1786, 1818, 1836, 1839, 1861, 1868, 1879, 1881, 1894, 1895, 1897, 1903, January and October 1927, October 1949, 1962, December 1979, 1981, 1987, February 1988 and February 1989.

# 3

# Results

In discussing the results of such a broad-ranging historical survey, which also spans a particularly long period of time, we must be clear about its complications and our reasons for examining this long period.

It is not to be expected that our list of historic storms can be complete. They have come to our knowledge through the particular sources of information available to us and, in the case of the storms that produce great sea floods, only through the accidental occurrence of a suitable conjunction between wind and tide.

It seems evident that our coverage is most nearly complete as regards storms in the severest class, though we must suppose that there may be gaps in our knowledge even of these.

The signs of our failing knowledge of the less intense storms as we go back in time are seen in the numbers of storms within our broad region of interest assessed in each grade in each half century since the earliest one which could be (however approximately) assessed, in 1570 (Table 1).

If we include the numbers of storms which we have here studied from 1570 onwards that can be tentatively, but fairly confidently, assessed in each grade on the basis of indirect deductions from the damage caused etc., the figures (which may still be incomplete in the earlier years at any time before 1800 and perhaps later) become those seen in Table 2. These figures confront us with an indication of heightened frequency of storms in the severest class in the earliest years surveyed – apparently before about 1650. It is also apparent in the plot (Figure 5) of the numbers of storms with various Severity Index ratings in our catalogue: the frequency distribution shown by the open circles which mark the tentative assessments of the early storms, which were difficult to assess because of the necessary consideration of non-meteorological data, is clearly centred about higher values of the index than the whole survey is.

This could be an important conclusion. The difficulty of proving it is in line with similar difficulties encountered in establishing the nature of the apparently colder climates of the so-called Little Ice Age, which ruled about that time, mainly just before the era of meteorological instrument measurements. The cold, disturbed climate seems to have had several culminating phases between the early fourteenth and late nineteenth centuries but perhaps most of all between about 1560 and 1720. We detect the occurrence most easily by the well-attested greater size of glaciers all over the world (Grove, 1988), the greater extent and longer duration of snow-cover, and the greater spread of the Arctic sea-ice. It is also demonstrable in the longest series of daily temperature observations (e.g. Lamb and Johnson, 1959, 1961; Manley, 1974), though the lowest average temperature levels shown by the longest series of observations were found in the late seventeenth and early eighteenth centuries – before the beginning of most long records – when they had to be deduced from the behaviour of very early instruments with now-obsolete scales and in non-standard and somewhat unsatisfactory exposures. Similarly with the storms, there are indications of especially great intensity, and perhaps of a greater frequency of Class I storms rather generally in the times before about 1840, which rest partly, or in the earliest years here surveyed largely, on non-meteorological evidence. For this reason, it is fortunate that the storms with notably high Severity Index values in 1791, 1792 and 1825 as well as in 1839 and, albeit with very few barometers, in 1703 can be supported by thorough synoptic meteorological analysis with measured pressure gradients. More indirect deduction from the thorough analysis of the storm studied in 1717, using measurements of the rapid progress of the meteorological systems and a statistical basis provided by the works of Palmén (1928) and Chromow (1942), again indi-

Table 1. *Numbers of storms in different Severity Grades assessed within the region 45° to 65–70° N and 20° W to 15–20° E*

| Dates | Severity Grade | | | | | |
| | I | II | III | IV | V | VI |
|---|---|---|---|---|---|---|
| 1570–99 | (1) | 0 | (1) | (1) | 0 | 0 |
| 1600–49 | (1) | 0 | 0 | 0 | 0 | 0 |
| 1650–99 | (1) | (1) | (1) | 0 | 0 | 0 |
| 1700–49 | 2 | 1 | 2 | 1 | 0 | 0 |
| 1750–99 | 2 | 4 | 2 | 3 | 1 | 0 |
| 1800–49 | 2 | 2 | 3 | 5 | 3 | 2 |
| 1850–99 | 4 | 5 | 3 | 1 | 5 | 0 |
| 1900–49 | 0 | 12 | 4 | 5 | 5 | 2 |
| 1950–89 | 4 | 3 | 7 | 6 | 10 | 1 |

Table 2. *Numbers of storms in different Severity Grades tentatively indicated within the region 45° to 65–70° N and 20° W to 15–20° E*

| Dates | Severity Grade | | | | | |
| | I | II | III | IV | V | VI |
|---|---|---|---|---|---|---|
| 1570–99 | 6 | 1 | 1 | 1 | 0 | 0 |
| 1600–49 | 3 | 0 | 0 | 0 | 0 | 0 |
| 1650–99 | 1 | 1 | 1 | 0 | 0 | 0 |
| 1700–49 | 2 | 3 | 4 | 3 | 0 | 0 |
| 1750–99 | 2 | 4 | 2 | 4 | 1 | 0 |
| 1800–49 | 2 | 2 | 3 | 5 | 3 | 2 |
| 1850–99 | 4 | 5 | 3 | 1 | 5 | 0 |
| 1900–49 | 0 | 12 | 4 | 5 | 5 | 2 |
| 1950–89 | 4 | 3 | 7 | 6 | 10 | 1 |

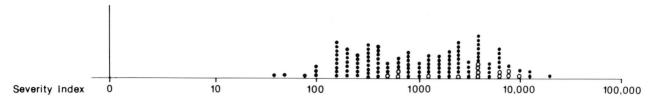

*Figure 5.* Numbers of storms assessed at various Storm Severity Index ratings. Each dot represents one storm. The open circles mark the assessments of storms before AD 1725.

cated very high wind speeds. The same method and reasoning was used in the case of the famous Spanish Armada storms in 1588 (q.v.) and it indicated wind speed levels, and a frequency of high wind speeds, probably unmatched in August and September in modern times.

Petersen and Rohde (1977, p. 52) remark that the frequency of storms on the German North Sea coast in the 1792–3 winter was unmatched thereafter until 1972–3 though the storms of 1972–3 were not quite so severe.

Non-meteorological indications of the severity of storms in the earlier centuries of our survey come chiefly from reported occasions of (i) blown sand and shifting dunes changing the landscape, even far inland, as well as (ii) great erosion of the coasts, and (iii) in some areas permanent flooding of coastal lowlands after seas breached the previously existing barriers. Cases will be found in our accounts of the storms which shifted huge quantities of sand inland in the heart of East Anglia in the Lakenheath–Santon Downham area between about 1570 and 1668 and in the Veluwe region of the eastern Netherlands about the same time, in the history of the Culbin sands near the northeast coast of Scotland particularly in the disastrous storm in 1694, and in our account of the storm in 1697 with another sand-blow disaster in the Hebrides (island of North Uist), also the storms in northeast Scotland (Loch of Strathbeg) about 1720, and in Denmark between 1591 and 1725 at Skagen and at Tibirke in Sjaelland. We have referred in a previous section (on pp. 18–19) to several incidents of the same kind and similar magnitude in the fourteenth and fifteenth centuries. On (pp. 17–18) we have also listed some of the great changes near the continental shores of the North Sea through historic storm surge floods. Of most specific interest, however, are Hauerbach's dated surveys (on pp. 19–20) of the successively farther advanced limit of the Skagen spit, which is built of sand and gravel transported by the wind and water currents at the north point of Denmark. Its rate of advance shows a strong maximum in or about the eighteenth century.

There are also indications from the logs of ships of the nations around the North Sea, and from harbour records, that the 1690s – the climax decade of the cold climate in Britain, Iceland, Scandinavia and central Europe – may have been particularly stormy as well, with more frequent severe storms than we have found details of and with a notably high frequency of N'ly and NW'ly winds over the North Sea. There are perhaps valuable indirect testimonies to this in a great number of reports (Paterson, 1880) of the general disrepair of the harbours and coastal towns all down the east coast of Scotland, from Tain and Dingwall in the north to North Berwick on the Firth of Forth, between 1694 and the early

years of the next century. The Records of the Convention of Royal Burghs, quoted by Paterson (1880) contain petitions to the parliament in Edinburgh about these matters, noting for example the ruinous condition of the tolbooths in Banff, Forres and Tain (where it was 'lyck to fall doune') and other burghs. The harbours at Arbroath and Burntisland, and later Dysart, received subventions for repair, and there were pleas in 1695 from five coastal towns in Fife for tax relief.* Anstruther Wester in Fife, like most of the others, was described as in a 'mean and low estate . . . in need of many reparations', and similar descriptions were applied in 1695 to Tain, Dornoch and Dingwall. We read in the same source first in 1718 of the beginnings of progress in some of these places, 'but that much is yet wanting'.

There have been a number of signs that the frequency and severity of storms in our region, and specifically over the North Sea, have increased again since about 1950 (see Lamb, 1988, chapter 11). To enumerate a few of these:

(i) The pressure gradients measured in the great North Sea storm flood situation in 1953 (q.v.), giving gradient winds in the range 100 to 130 knots, were regarded as 'phenomenal' at the time by meteorologists writing about the disaster. Such values have been measured again in this region several times since.

(ii) The great storm over southern England in 1987 was considered the severest in that region since 1703.

(iii) The German navy reported a 50% to 100% increase in the frequency of wave-heights greater than 4–5 m in the North Sea from about 1953 onwards and commissioned studies of the phenomenon published in recent years (Weiss and Lamb, 1970; Lamb and Weiss, 1979). It was noted that the changes seemed to be associated with a decreased frequency of direct W'ly wind situations and increases of other directions, particularly winds from NW and N.

(iv) The British Meteorological Office is reported (King, 1979) to have found it necessary to make successive

*No doubt some will be inclined to attribute this picture of disrepair and neglect along the coast to the impoverished state of the Scots economy on the eve of the Union with England. But these were not the districts in which the long years of successive harvest failures and famine in the 1690s struck really hard. That disaster was concentrated most in the upland parishes all over Scotland, and Andrew Fletcher of Saltoun in Midlothian addressed the Scottish parliament in 1698 attacking the well-to-do in the more prosperous lowlands about the Firth of Forth and elsewhere for their unconcern with the plight of people in the hinterland. The east coast ports must at times have benefited from the shipping trade bringing in emergency supplies of grain from the eastern Baltic (Graham, 1899; Lamb, 1977; Sinclair, 1791–9).

revisions shortening the official 'return periods' of very high wind speeds over the British Isles.

(v) The Swedish Meteorological Service has reported an 86% increase in the average number of days a year with storm force winds (Beaufort 9 or over) at points along Sweden's southwest coast in 1966–80 as compared with the preceding years, 1951–65.

(vi) The variations of an index of the frequency of gale situations over the North Sea and over the British Isles (Jenkinson, 1977; Lamb, 1988, pp. 93–4), based on gradient wind measurements, show (Figure 6) a declining frequency of gale situations from the 1880s to the period 1910 to 1939 and then a recovery in the 1960s and 1970s to a level about the same as in the 1880s.* Other investigations (e.g. Lamb and Johnson, 1959, 1961) have suggested that in the period of highest frequency of W'ly winds over the British Isles, between 1900 and 1939, the cyclone centres tended to be farther (north) away from the British Isles than in either the 1880s and 1890s or the decades since 1955 to 1960.

(vii) These findings are at least partly paralleled by the analysis of 109 years of wind observations on the island Fanø, at the west coast of Denmark from 1872 to 1980 (Peterson and Larsen, 1984). The overall mean wind speed was lowest (about 11 knots) in the 1920s and 1930s and for some years about 1910, and was about 15 knots in the 1880s and 1890s and on average from 1940 to 1980, but was highest (of the order of 17 knots) in the late 1940s and early 1950s and was again rather low for a few years about 1970. However, this is a difficult, and in some ways unsatisfactory, record involving several changes of observation site and estimates of wind force unsupported by instruments. It also has one unique and unexplained feature: an almost continuous decline in the reported frequency of west winds from about the 1880s to the last decades of the record.

There may be an indication in our observations of a tendency for storms in our region of interest to be more severe and more frequent in or about the second halves of each century, perhaps most of all in the eighties and nineties in each case. The sixteenth century seems probably to have been most affected in this way, coinciding with the time of a well-established temperature decline in Europe (Flohn, 1949; Pfister, 1984). The figures we have for the eighties of the seventeenth and eighteenth centuries do not show it, although we know of the very severe stormy record of the 1790s and we have mentioned incomplete indications of a similar climax of storminess in the 1690s.

There are details here that remain uncertain, but it may be safe to conclude that there was a high level of storm activity in the early centuries of our survey, and perhaps as late as 1825 to 1840, and again about the 1880s and 1890s, followed by a decline to about the first 40 years of the present century. The increase of storminess since 1940 to 1950 is

*The index values in the 1980s do not seem to have been calculated yet. The areas of reference are both between 50° and 60° N, respectively 0–10° E ('North Sea') and 0–10° W ('British Isles').

GALE INDEX SURVEY BASED ON
PRESSURE DISTRIBUTION

(a) Over the North Sea

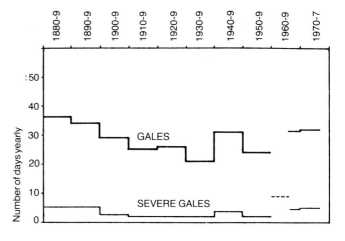

Very severe gales averaged
1 per year in the 1890s
1 in 2 years in the 1880s and 1970s
1 per decade between 1910 and 1939

(b) Over the British Isles/East Atlantic

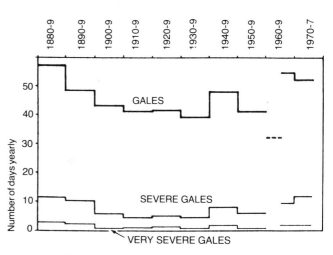

*Figure 6.* Changing frequency of gale and strong gale situations over the North Sea and over the British Isles decade by decade from the 1880s to 1970–7, according to an index (based on pressure gradient measurements) due to Jenkinson (1977).

also well attested but may not yet have reached the level of the period 1570 to 1840. Thus, the maximum gradient winds measured (see List B in Chapter 2) on the maps of the storms in 1703, 1717, 1751, 1756, 1791, 1792 and 1825 and perhaps 1839 seem to be at speeds since matched approximately only in 1862, 1868, 1884, perhaps December 1886, 1951, 1953, 1981 and 1987 among the later storms in our compilation.

We have found hints in our compilation of some repeating shorter-term fluctuations in the incidence of the severest class of storms on period lengths of about a hundred years and less: thus the second half of each century showed more

Grade I storms than the preceding half-century, and it seems that the nineties decade (or perhaps the eighties and nineties) may have been the stormiest part of each century in the region chiefly studied. But clearly our compilation is not complete enough to give a really firm ground to this.

*Seasonal variation* in the number of storms in our compilation within the area of our main concern is shown by the monthly figures in Table 3. This seasonal pattern, with its concentration of severe storms between late September or October and February to March, seems a constant feature throughout the period surveyed. Equally, at all stages of this history, it is only June and July that are nearly immune. There have been storms even in those months, but they have been of more local extent and not of the greatest severity.

### Wind directions aloft and the severity of storms
The figures noted were as in Table 4.

As might be expected, the distribution of strongest surface winds measured was clearly related to that of the gradient wind directions but a little backed (i.e. shifted in an anticlockwise sense), though this shift amounted to very little in the case of gusts, especially over the sea. The figures in Tables 5 and 6 seemed to be the highest in our compilation.

We see that the strongest gusts have been registered in winds from about NW or N. The same seems to apply to the strongest surface winds averaged over periods of ten minutes or an hour. There is, however, an important exception to these statements in the case of tornadoes and whirlwinds, where the most violent winds are concentrated within a band only some metres or less in width. These are features in which the greatest wind speeds ever measured near the Earth's surface occur – up to 200–400 knots according to Blüthgen's (1966) survey of the literature – albeit in extremely narrow confined bands within the system. The peak values are generally found on the side of the whirling storm that is in line with the strongest upper winds associated with the shear that is generating the system: in the case of Europe and North America this means the strongest wind in the tornado is generally from SW or S.

### Gustiness
Since it is the buffeting caused by the sudden stress of the strongest gusts of wind that, apart from the extraordinary twisting in tornadoes and whirlwinds, causes the most damage, it is the peak wind speeds that must especially concern us. Gustiness arises from the turbulent eddies formed by, and carried along with, the wind. These may be pictured

as eddy variations about the mean wind speed and expressed as a gustiness ratio or index. Bilham (1938, p. 54) defines this *gustiness index* as the ratio of the range of the oscillations to the mean velocity of the wind. The ratio may be as high as 2 in rough surface locations and is commonly greater over towns than either over flat lands and even low-lying coasts or the open sea, though it is liable to increase as the wind strengthens and wave-heights increase. Values exceeding 2 at rugged coasts and at sea seem to be indicated in some of the storms in our compilation. The *gust factor* defined by Chandler and Gregory (1976, p. 61) as the ratio of the maximum gust speed to the mean wind speed is simpler to calculate. These authors quote 1.35 over the open sea, 1.6 over flat country, 1.8 in well-wooded country, and 2.1 over the centres of big towns, as typical values of their *gust factor*. Values of this factor calculated from data on storms studied in this compilation are in keeping with these figures, but commonly 1.5 to 1.75 over and near rugged coasts where convergence effects may come in, and more locally perhaps 1.9 to 2.0 in some extreme cases especially near coasts.

### The directions of the winds in the storms
Study of the wind directions in the storms for which weather maps could be analysed over the British Isles and North Sea, with the immediately surrounding land and sea areas, over the survey period from 1703 to now, showed that, as expected, most of the strongest gradient winds were from westerly points. But there seemed to be differences in different parts of this wide region, with some areas where strong winds from other directions were prominent.

Our results in this aspect must be considered provisional, since the storms were generally identified for study and graded on the basis of their prominence in the records of the British Isles and North Sea, just in the central parts of the region surveyed.

Among the most obviously different cases were the occasional, even rather rare, E'ly storms in the Channel. (E'ly winds there are not prominent in most of the wind-roses published.) Instances will be found in our compilation in January 1740, March 1891 and February 1979. They seem to be a characteristic liability in severe winters with a great, blocking anticyclone to the north of the area affected. Other examples are known to have occurred in January 1658, in the great winter of 1683–4, in January 1709, 24–25 December 1794, and mid-February 1795, in the winter of 1829–30, in January 1880, as well as in 1940, 1947, on 19 Janu-

Table 3. *Numbers of storms since 1570 in the compilation by months of the year, within the region 45° to 65–70° N and 20° W to 15–20° E*

| Severity Index | J | F | M | A | M | J | J | A | S | O | N | D |
|---|---|---|---|---|---|---|---|---|---|---|---|---|
| Severity Index 1500 | 12 | 7 | 3 | 1 | 1 | 0 | 0 | 0 | 1 | 8 | 5 | 12 |
| All grades over 100 | 23 | 21 | 14 | 4 | 1 | 1 | 1 | 5 | 11 | 19 | 18 | 30 |

Table 4. *Strongest gradient winds noted in storms from various directions*

| Speeds up to | Numbers of cases identified in the compilation | | | | | | | |
|---|---|---|---|---|---|---|---|---|
| | SW | W | NW | N | NE | E | SE | S |
| 150 knots or over | 1–2 | 3 | 1–2 | | | | | |
| 130 or 140 knots | 1 | 4 | 2 | 1 | | | | |
| Over 100, up to 120 knots | 1 | 4 | 2 | 1 | | | | |
| About 100 knots | 5 | 8 | 3 | 3 | 2 | 1 | 1 | 3 |

Table 5. *Strongest mean surface winds measured by instruments (wind speed in Knots)*

| SW | W | NW | N | NE | E | SE | S |
|---|---|---|---|---|---|---|---|
| About 70 Norway coast 5.12.1979 | 80 Orkney 20.1.1884 | 90 Tiree 83 Paisley | Over 100 Jan Mayen (71° N) 9.4.1933 | | | | 55 Norway Coast near 62° N 15.2.1962 |
| 69.5 Cornwall 1.6.1938 | 56 Copenhagen 26.12.1902 | 82 Bell Rock all on 27.10.1936 | 78 Aberdeen 18.11.1893 | | | | |
| 69 Hamburg 12.2.1894 | | 85 at oil rigs in Ekofisk field, North Sea West of Norway 83 in north Jutland 80 west of Stavanger and 200 km west of Denmark – all these on 24.11.81 | | | | | |
| 64 Coast of NW Ireland 28.1.1927 | | | | | | | |
| 63 Thames Estuary 1.6.1938 | | | | | | | |
| 61 Falmouth 5.12.1929 | | | | | | | |
| 59 Thorney Is. 16.10.1987 | | | | | | | |
| 59 Southport 27.12.1903 | | | | | | | |

ary 1963 and 11–12 January 1978. In a number of these cases the wind was NE'ly and the severest effects were on the French side of the Channel. But the wind strengths were seldom comparable with the greatest storms in our survey, except rather locally where convergence effects at a rugged coast were at work. These storms are chiefly known for the bitter weather, for rapid onset of severe frost, problems with drifting snow, freezing rain, ice encrustations on trees, bushes, overhead wires, and the upper works and rigging of ships. In the extreme case, in February 1684, there was a great deal of thick sea-ice on both sides of the Channel and its westward drift along the English coast was rapid (Lamb, 1988, p. 112). Analogous situations are by no means so uncommon at the south coast of Norway and even on the Danish side of the Skagerrak. The cases here analysed in 1937 and 1938 are typical examples. Again in that area, the occurrences are related to blocking anticyclones to the north and are particularly liable to appear in severe winters. The rugged and mountainous coasts of Norway produce stronger convergence effects in the winds before the coast and some of these cases become undeniably major wind storms.

**The storm wind directions in different parts of the region surveyed**

Let us now consider the more general picture of the wind direction in great storms as seen in the meteorologically analysed storms in this survey, area by area, from 1703 to date. We confine our attention to those storms that attained Grade I or Grade II severity, since our coverage of the rather less severe storms is much less nearly complete.

*Storms passing through, or across, the Atlantic fringe of the British Isles*
Of the storms studied, 43 were in this category. Just 24 of these attained Grade I or II severity: among these the main directions of the storm winds were:

SW, 31%; W, 23%; NW, 20%; S, 16%; N, 6%

The pattern was different among those storms that moved *along* the Atlantic fringe from southwest to northeast without crossing the British Isles. There were just 6 cases of Grade I or II storms, of which:

S, 45%; SW, 37%; W, 8%; SE, 8%

Table 6. *Strongest gusts of the surface wind measured (wind speed in Knots)*

| SW | W | NW | N | NE | E | SE | S |
|---|---|---|---|---|---|---|---|
| 119 Quimper, Brittany 117 Normandy coast 100 Sussex coast – all these on 16.10.87 | 94 Millport, Bute 30.12.1951 94 St Ann's Head, S. Wales 23.11.1928 | 154 hill in Shetland 16.2.1962 126 harbour at Fraserburgh 13.2.1989 | 163 at Jan Mayen 9.4.1933 118 at Kirkwall, Orkney 7.2.1969 | | | | 97 at Spanish Point, Quilty, Co. Clare 27.1.1920 |
| 111 at Haramøy 107 at Svinøy both on Norway coast near 62° N 5.12.1979 95 at Dunfanaghy Head, NW Ireland 28.1.1927 | 93 Belmullet, Co. Galway 9.2.1988 | 109 hill in Orkney 1.2.1953 103 hill in Dumfries 98 at Kinloss both on 16.2.1962 96 Cranwell, Lincs 17.12.1952 | | | | | |

*Storms over the Bay of Biscay*
34 of the storms studied were in this group, 15 of which attained Grade I or II rating. The storm winds in these were from:

SW, 53%; W, 22%; NW, 18%; S, 7%

The pattern might be different if *all* storms over Biscay, including those that never entered our central regions of interest, had been surveyed.

*Storms passing through, or over, the Channel*
68 of the storms studied were in this group, 27 of them reaching Grade I or II rating. The storm winds in these were from:

SW, 35%; W, 30%; NW, 17%; S, 7.5%; N, 3%

*Storms affecting a zone along the fringe of the continent from Biscay to near Denmark*
84 of the storms studied were in this category; 37 of them reached Grade I or II. The storm winds in these were from:

SW, 32%; W, 23%; NW, 25%; N, 10%; S, 8%

24 other storms moved *along* this zone from southwest to northeast rather than crossing it. In the nine cases where these reached Grade I or II severity, the storm winds were from:

SW, 65%; W, 23%; S, 8%; NW, 5%

*Storms over England and Wales between 50° and 53° N*
83 of the storms studied were in this category, 35 of them reaching Grade I or II severity. The storm winds in these were from:

SW, 29%; W, 30%; NW, 21%; S, 12%

*Storms crossing Britain and Ireland between 53° and 56° N*
75 of the storms studied were in this category, 37 of them rated as Grade I or II severity. The storm winds in these were from:

W, 30%; NW, 26%; SW, 20%; S, 10%; N, 9%

*Storms over northern Scotland north of 56° to 57° N*
65 of the storms studied were in this category, 34 of them reaching Grade I or II severity. The storm winds in these were from:

NW, 36%; N, 19%; W, 11·5%; SE, 9%; SW, 8%; S, 7%

*Storms crossing the southern North Sea south of 55° N*
78 of the storms studied were in this category, 42 of them attaining Grade I or II severity. The storm winds in these were from:

SW, 28%; W, 28%; NW, 21.5%; S, 9.5%; SE, 7%

*Storms crossing the northern North Sea north of 56° to 57° N*
71 of the storms studied were in this category, 36 of them attaining Grade I or II severity. The storm winds in these were from:

NW, 27.5%; N, 25%; W, 13.5%; SE, 11%; S, 10%; SW, 9%

*Storms over the Norwegian Sea near and west of the Greenwich meridian*
54 of the storms studied came into this category, 33 of them of Grade I or II severity. The storm winds in these were from:

N, 48%; NW, 12%; NE, 9.5%; S, 8.5%; SW, 6%

*Storms of the Norway coast zone between 58° and 62° N*
61 of the storms studied were at some stage of their history in this group, 37 of these ranked as of Grade I or II severity. Evidently because of effects of the mountainous coast, the characteristic distribution of the storm winds in these was quite different from those in storms over the open North Sea. The storm winds in this group were from:

S, 24%; N, 21%; SW, 17.5%; W, 16%; NW, 9%; SE 8%

*Storms over Denmark*
72 of the storms studied came into this group, 37 of them reaching Grade I or II severity. The storm winds in these were from:

SW, 29.5%; W, 28%; NW, 15%; S, 11%; SE, 7.5%

Several of these results require comment.

(i) The different distributions of storm wind directions characterizing storms moving *along* the two parallel southwest-to-northeast orientated zones of the Atlantic fringe and the fringe of the continent from the storms which crossed those zones are well in line with what should be expected. The strong representation of winds from about SW in the case of both zones corresponds to the steering of the storm itself.

(ii) The stronger representation of winds from W and NW, N and SE, and the somewhat reduced frequency of winds from SW, in storms over the two more northern sections of the British Isles as compared with the southern part of England and Wales, and especially as compared with southeast and central England, seem to be in line with published statistics and wind-roses at the best exposed observing stations for winds of all strengths (Meteorological Office, 1952, 1968). This presumably corresponds to the greater openness of the more northern regions to storm winds from the northern Atlantic and the Norwegian Sea.

(iii) The distribution we find for northern Scotland is so different from that among storms of the Atlantic fringe of the British Isles that there must be a sharp change in the wind direction frequencies in the main storms of northern Scotland on the one hand and those of the Atlantic zone reaching towards the Hebrides.

(iv) The marked difference we have noticed between the Norway coast zone south of 62° N and the areas to the west, with far greater representation of SW and S winds in the coast zone, also applies to the Norwegian coast zone north of 62° N (cf. Børresen, 1987).

(v) Our figures for storms over Denmark are in some conflict with the Danish statistics (Frydendahl, 1971) of the directions of the strongest winds there, based on the years 1931–60. These show a higher frequency of NW and a lower frequency of SW cases than our figures. The discrepancy may well be due to our collection of storms from sources primarily concerned with Britain and the North Sea and assessed for severity by a system centred on the Greenwich meridian, while Denmark lies so much farther east.

*Wind direction frequencies (%) in Danish storms 1931–60*
*(after Frydendahl, 1971)*

|  | SW | W | NW | N | NE | E | SE | S |
|---|---|---|---|---|---|---|---|---|
| Beaufort force 9 or over | 10 | 41 | 32 | 3 | 2 | 8 | 4 | 0 |
| 8 or over | 15 | 29 | 24 | 4 | 5 | 10 | 9 | 4 |

A separate study was therefore made of the severe storms over Denmark in our survey from 1861 to date, during which period storms notified from Danish sources were more regularly recruited to our collection, although

*Wind direction frequencies (%) in Danish storms since 1860*

|  | SW | W | NW | N | NE | E | SE | S |
|---|---|---|---|---|---|---|---|---|
| Grade I only (7 storms) | 35 |  | 43 | 7 | 7 |  |  | 7 |
| Grades I and II (24 storms) | 27 | 23 | 21 | 2 | 2 |  | 15 | 10 |

the storms were still graded by reference to the big region centred on the Greenwich meridian. This selection of storms did indeed show a greater proportion of NW'ly storm winds over Denmark, particularly among the severest (Grade I) storms.

There is a suggestion therefore both in our figures and in the much more extensive Danish statistics that the strongest storms may be more liable to blow from the NW than the generality of gales. Frydendahl has independently found (personal communication, 23 February, 1990) that excess of NW'ly over SW'ly winds – up to three times – has been a feature of the strong storms of Beaufort force 9 or over in most of Denmark, at least in the figures for most periods since about 1860, with the notable exception of the southeastern and easternmost part of the country, particularly Bornholm in the Baltic where SW and W winds predominate. With less strong winds W and SW are more frequent in all parts of Denmark.

**Apparent changes with time in the direction frequencies for storm winds**
It is curious that the figures obtained in this study (p. 27) from the 37 Grade I or II storms from 1703 to the present over Denmark show a distribution of frequency of the storm winds with more SW'ly and fewer NW'ly than the Danish

Table 7. *Storms of Grade I or II: wind direction frequencies (%)*

|  |  | SW | W | NW | N | NE | E | SE | S |
|---|---|---|---|---|---|---|---|---|---|
| England | 1703–1860 (10 storms) | 10 | 35 | 35 | 5 | | | | 15 |
| & Wales | 1860–1989 (25 storms) | 36 | 27 | 15 | 5 | 1 | | 4 | 10 |
| 50° to 53° N | | | | | | | | | |
| British | 1703–1860 (9 storms) | 6 | 22 | 56 | 11 | | | | 6 |
| Isles | 1860–1989 (28 storms) | 23 | 33 | 18 | 10 | 2 | 2 | 4 | 8 |
| 53° to 56° N | | | | | | | | | |
| Northern | 1703–1860 (6 storms) | 17 | 8 | 50 | 25 | | | | |
| Scotland | 1860–1989 (28 storms) | 6 | 1 | 33 | 18 | 4 | 6 | 11 | 9 |
| Southern | 1703–1860 (10 storms) | 20 | 25 | 30 | 15 | | | | 10 |
| North Sea | 1860–1989 (32 storms) | 30 | 28 | 19 | 2 | 2 | | 9 | 9 |
| Northern | 1703–1860 (6 storms) | 8 | 8 | 42 | 42 | | | | |
| North Sea | 1860–1989 (30 storms) | 8 | 14 | 24 | 21 | 2 | 5 | 13 | 12 |

late nineteenth and twentieth century figures, i.e. similar to the overall distribution for winds of all strengths in the period 1931–60. This seems to imply that the severe storms in our collection between 1703 and 1860 were more inclined to produce SW'ly winds over Denmark than those in the later period, although this could be an artefact of our selection of storms in that period. That changes of distribution *are* liable to occur when different periods of this length are considered is indicated by the figures yielded by our survey for storms over the British Isles and the North Sea in Table 7.

Our samples, particularly from the earlier period, are rather small and the figures in Table 7 should not be relied on closely, but it seems safe to conclude that the distribution in the long periods covered before and after 1860 were in some respects significantly different.* Moreover, the changes indicated in the relative frequencies of SW and NW storm winds seem to have been opposite in sense over the British Isles and Denmark, with more frequent N'ly components in the earlier than in the later period over Britain. The changes over the North Sea are shown as smaller, but more like the changes over Britain than over Denmark.

An attempted further breakdown of the storm wind direction frequencies in six subdivisions of the overall period surveyed, including this time Grade II/III borderline cases with Severity Index assessed at 1500 or over, in order to increase the size of the samples, and including the indirectly argued assessments of those severe storms back to 1570 which could not be fully analysed, suggested some further changes of tendency which are thought to be more or less in line with changes in the frequencies of winds of all strengths. All the regions within our survey seemed to show increased representation of S'ly components in the storms

*It is unfortunate that our samples of severe storms are too small, and the listing not guaranteeable as complete, so that year by year or decadal variances of frequency cannot be established and hence statistical significance tests cannot be applied.

between 1900 and 1949, increased representation of N'ly components in the storms between 1570 and 1725, and again between 1790 and 1839 or perhaps 1850, and since 1950. The tendencies seemed more mixed in different regions between 1726 and the 1780s and between about 1840 or 1850 and the end of the nineteenth century.

The best guide we have to such long-term variations is in the long history of the frequency of different directions of the winds in England over past centuries, considering winds of all strengths. This was first explored by Brooks and Hunt (1933), whose survey included the early Lincolnshire weather diary of the Revd W. Merle from 1340–3, and with amplification from later unpublished works by Manley can be based on daily observations in the London area continuously from 1667 to the present date. Brooks and Hunt already noticed that there were some long periods in the past record when the frequency of the usually prevalent SW'ly and W'ly winds was much reduced, and although they lacked London data for two of the most interesting periods (1686–1712 and 1748–86) they mentioned that observations in Edinburgh and Perth between 1770 and 1792 showed variable winds, often with a NW'ly resultant direction and that the frequency of SW to W'ly winds in London was low till about 1810. This history was studied further by Weiss and Lamb (1970, 1979), giving figures decade by decade from the 1720s to date which show some statistically significant variations. Thus, the raised frequency of surface winds from northerly directions (NW, N and NE) in London over the 120 years from 1730 to 1849 as compared with the immediately following 30 years (1850–79) appears significant at the 1% level, the great positive deviation of the frequency of NW'ly winds in particular in the 1770s being also of about that level of significance. A curious feature, noticed also by Brooks and Hunt, is that just in a period of about 15 to 20 years around 1786 to 1800 there was also an exceptional peak frequency of SE'ly winds (at 88 days/ year), which became for that one period the commonest gradient wind direction over London. (This perhaps marks the culmination of a period of weak development of the usual circumpolar westerlies over this sector of the northern hemisphere, producing a shortened wave length in the upper westerlies and so a westward displacement just then of the trough, which also could be responsible for increased S'ly winds in Denmark.) A similar phase of SE as the commonest single wind direction appears in Tycho Brahe's daily observations in Denmark just 200 years earlier, from 1582 to 1597 (La Cour, 1876). The deviations in frequency of winds over the British Isles, shown by Lamb's (1972) daily weather type classification since about 1960 show similar features to those indicated just 200 years earlier in the 1780s and in terms of winds in London in the period from 1770 onwards, with some aspects apparently attaining statistical significance (viz. the reduced frequency of the W'ly type, increased frequency of NW and cyclonic types).

We may conclude that the general tendency for W'ly and SW'ly winds has undergone significant variations and that these seem to have been more or less in line with the changes in the frequencies of storm winds from different directions noticed in this compilation of storm occurrences. The decade average frequency of days of general W'ly weather type (Lamb, 1972) was 65% greater in the 1920s (109 days/year) than in the least W'ly time in the 1780s (66 days). The decade average frequency of the NW'ly type doubled in the peak decades (1780s and 1970s) as compared with its lowest frequency (13 days/year) in 1910–19. Against the long-term background variations in the winds of all strengths discussed in these paragraphs, it seems there may also be a real tendency for unlike long-term variations in the frequency of northerly and southerly components over Denmark as compared with the British Isles and the North Sea, as our compilation of historic storms has suggested.

An outline picture of the history of the varying frequency/predominance of general westerly situations, as indicated by the SW'ly surface winds in England, may be seen in Figure 7. It is during the periods of relatively infrequent westerly situations that other wind directions, notably NW, N and SE, have come to prominence.

### Tracks and types of development of the great storms

A first count of the wind directions involved, primarily over the British Isles and North Sea, in developments of the 49 storms with Severity Index ratings of 1500 and more showed a noteworthy prominence of situations involving a marked swing of the strong winds from S or SW to NW or N (29% of all cases). Winds from W to NW and from S or SW swinging to just WSW or thereabouts came next in frequency, each accounting for 12%, while winds from S or SE accounted for 10%. These last were much more prominent on the eastern side of the North Sea, especially the Norwegian coast. Another class, with NW to N'ly gales over the western North Sea and strong winds or gales from between SSW and W over the coasts of Denmark and the German Bight accounted for 8%. Storms from NW or N over the whole area were another 8%. This frequency pattern was well supported when six more great storms from among those early cases more difficult to analyse were added into the count; those

cases with big swings of the storm winds from about S or SW to NW or N were then 31% of the whole sample.

When the experiment was tried of counting the recurrences of the wind direction patterns in storms of successively lower grades of severity, the most obvious result was that the patterns became successively more various as each lower grade was considered. The overall frequencies for 128 storms of grades I to V (i.e. all cases with a Severity Index rating of 100 or more) were:

| | |
|---|---|
| Big shift of the strong winds from S or SW to NW or N | 20% of all cases |
| Swings from W to NW or N | 13% |
| Winds from NW to N throughout | 9.5% |
| Swings from NW to N | 4% |
| Winds from about N (NW to NE) | 8% |
| Winds from about E | 6.5% |
| Winds from SE to S | 8% |
| Swings from S or SW to about WSW | 13% |
| Winds from WSW or W throughout | 8% |
| Swings from S or SW to W | 3% |
| Swings from SW to about WNW | 4% |

The most general suggestion derived from this is that developing meridionality,* more particularly with intrusion of Arctic cold air from the north, usually from the Norwegian Sea, plays a part in most cases. Only the last four of the eleven patterns of development in the list above are apparently free of this meridionality or developing meridionality element: they account for just 28% of the cases. And, of course, it is nearly stationary, meridional patterns of wind-flow which produce most of the longest-lasting storms: 59% of the 17 longest and 72% of a longer list of 70 long-lasting storms.

The origins of the storm systems could be categorized in at least the following ways:

*Tropical storms.* Although often suggested, this seems to be rarely the origin of the severe storms in our region (but see p. 62). It seems to be substantiated in only two cases in our list: the storms in 1846 and 1961. Neither of these penetrated far east into our region. There may, nevertheless, be more cases in which an infusion of warm, moist tropical air impelled northward at some stage into the zone of prevailing westerlies over the North Atlantic by a tropical disturbance played some part in the energy supply for storms developing in our region.

*Fast-travelling frontal wave depressions.* These, carried east across the ocean by a strong jet stream, sometimes produce successive storms of only rather moderate extent which arrive only a day or two apart. Examples in our compilation are the storms in January 1818, November 1836, February 1894, March 1895, October 1927, November 1972 and April 1973.

---

*Meridional patterns of the wind circulation, and 'meridionality', are terms used to describe patterns with prominent northerly and/or southerly windstreams, or prominent northerly or southerly components of the main wind flow. These patterns are, of course, notably effective in transporting warm and cold air to other latitudes.

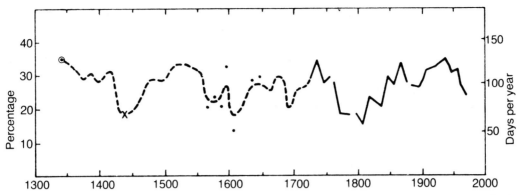

*Figure 7.* Frequency of SW'ly surface winds in England at various times since AD 1340: ten-year averages. The diagram is based on daily observations in London from 1669 to 1977 and indirect indications before that, including some weather diaries (e.g. Rev. William Merle in Lincolnshire 1340–3).

Similar cases which begin to occlude and slow their progress in the British Isles–northwest Europe region. The storm, or storms, in January 1735 seem to illustrate this.

*Storms which advance more slowly across our region with deepening or rejuvenation of the cyclone centre and turn somewhat to the left.* Examples are the storms in October 1756, 7–12 December 1792, January 1884, November 1897, January 1927 and August 1979. The storm centres in January 1920 and 1961 turned left farther west but were still of importance to our region.

*Occluding, or occluded, storms which nevertheless continue to travel steadily, and sometimes fast, across the Atlantic and Europe.* Examples in our compilation are the great storms in 1588, 1703, 1839, 1881, January and November 1928, and January 1976. This seems to be an important class where great strength of the jet stream continues to add energy to the storm system when other processes have completed their course.

*Meridionality developing: storms becoming nearly stationary near the north and northwest Europe region, with more or less N'ly winds predominating over the British Isles and North Sea.* This seems the commonest case among severe storms, as indicated by our exploration of the wind directions in the beginning of this section (p. 30). Examples seem to be provided by the storms in 1570, 1694, 1697, 1717, 1791, 1792, 1795, 1825, 1829, 1855, 1859, 1869, 1878, 1883, 1893, February 1938, 1952, 1953, 1954, February 1969, 1981 and 1983.

*Meridionality (or blocking anticyclones over northern Europe) developing, with S'ly and SE'ly or E'ly winds dominating at least some parts of the fringe of Europe (e.g. Norway and Denmark).* Examples were found in 1783–4, 1937–8 and other cases.

*Meridionality developing with a slow-moving, north–south extended low pressure system over our region, with N'ly and S'ly winds in the areas on either side of the system.* Examples are given by the storms in 1736, September 1751, December 1783, 21–3 December 1792, 1859, 1878, and January 1978.

*Storms which turned right while travelling over the middle or eastern Atlantic, or over our region at the fringe of the ocean, upon encountering the jetstream to the north of them, and* deepened with an indraught of cold air from the Norwegian Sea. Examples of this may be seen in the storms in December 1902, December 1973, January 1976, and October 1987.

*Coalescence of Lows from different origins.* These cases are liable to concentrate potential energy by bringing together airmasses of quite different origins, temperatures, and moisture content. Examples are the storms in 1717, several of the storms in our compilation in December 1886, 1897, August 1957, and January 1968.

*Deepening of Lows connected with a build-up of extremely contrasted airmass types within a short distance (airmasses of widely different origins).* This is noted in many cases in the catalogue: e.g. May 1795, December 1951, January 1976, December 1979, and October 1987.

*Explosive deepening of Lows* also occurs in some cases that are not easily explained without plentiful upper air temperature and humidity information. The explanation is probably always in terms of latent instability of the airmasses for motions in the vertical plane. The storm in December 1951 is believed to be an example.

*Severe wind speeds (and often vertical motion) sometimes developed in localized squalls at (usually) cold fronts.* Pronounced cases in our compilation produced some severe, but localized, disasters: e.g. in 1822, September 1838, October 1859, June 1869, March 1878, December 1879, and March 1980. The sometimes very damaging whirlwinds, tornadoes, and waterspouts at sea, come under this heading.

*Topographically induced confluences and convergences in the wind flow.* These enhance wind speeds in suitably directed windstreams at coasts, along the flanks of hills and mountain walls, particularly when a front is approaching, and over flat or rounded hill-tops and plateaux. Cases may be noted particularly along the west-facing coast of Norway in S'ly and SE'ly winds and along the east coast of Greenland in N'ly winds, as well as before the south coast of Iceland in strong E'ly winds and similarly with winds from between SE and NE at the south coast of Norway in, and near, the Skagerrak. There are examples in this compilation in storms in 1698, in the early part of the storms in 1751, and in 1786, 1879 and 1909, in 1936, and repeatedly in 1937 and 1938. Funnelling of the wind in valleys and straits acts similarly.

*Mountain waves.* These produce concentrations of stronger wind flow at intervals downwind from hills or mountains – a series of waves or undulating flow giving more or less regularly spaced convergences downstream – which give rise to occasionally, and locally, very damaging winds, as in the 1962 storm in Sheffield and, probably, in the 1868 storm in Edinburgh. (These occurrences are a special case of topographically induced local anomalies in the flow of the winds.)

*Topography, including quite minor hills and valleys.* When meteorological conditions are suitable (see e.g. Lamb, 1957) topography also sometimes plays a part in unleashing rotating wind systems (tornadoes, whirlwinds, waterspouts). The effects of the extraordinary shear and pressure reductions in these rotating storms make them among the most destructive of all atmospheric phenomena.

As with other meteorological phenomena, those storms in which more than one, or sometimes even several, of these conducive factors are at work are liable to develop particular severity.

# 4

## Concluding summary

This work places on record the known details of a wide-ranging collection of storms of the last 500 years which were regarded by contemporary reporters as historic in one way or another. These are supplemented by a number of twentieth century storms to serve as a basis for comparisons. The destructiveness of the storms is briefly catalogued where known, but there can be no systematic list of either the death tolls or the financial costs. Certainly all the biggest death tolls were produced by North Sea floods.

We notice differences in the prevailing storm wind directions in different parts of the regions studied. As is well known, there are also characteristic differences in the frequency of gales and greater storms in the different geographical areas and more local settings within the regions surveyed.

It becomes clear that there have also been variations of storminess with time, not only from year to year but also from decade to decade and over various other, longer and shorter, periods. Some of these variations are caused by changes in the frequency of storm cyclones travelling on different tracks or in this or that latitude. Other variations appear to be due to changes in the general intensity of development of the storm systems – at least those affecting the regions surveyed.

These more general differences of storm intensity seem to have occurred on different time-scales. The recent increase in frequency of great storms and very deep low pressure centres – since about 1950, but particularly noticed in the latest decades – has some parallel in the very intense storms and lowest pressures noted in some of the storms in our region in the eighteen-eighties and nineties. And there is an appearance in our collection of storms that there were other impressive climaxes of storminess in this region in the 1570 to 1620s period and in the 1690s to early 1700s and again in the 1790s. But the great storm reported in 1839 may also have produced an extreme of low pressure at the climax stage of development of the system similar to those in the 1880s and in recent years. Indeed, there is a general impression from the evidence surveyed in this compilation that from the beginning of our survey period there was a greater intensity, and a greater frequency of intense storm development, during the long time of mostly colder climate, the so-called Little Ice Age, of which the years between about 1780 and 1850 might be regarded as the last main stage.

It must remain debatable how far the evidence which we have been able to muster substantiates this impression of greater storminess, and greater violence of the storms, in the Little Ice Age than in later times especially between about 1900 and 1950. The barometric pressure maps here presented of some of the great storms in the 1790s and one in 1825 seem firmly enough based on barometric pressure measurements to guarantee the extreme pressure gradients and gradient winds deduced. But before that – and unfortunately including all through the severest phases of the cold Little Ice Age climate – too few barometric pressure measurements are available to establish the point in the same way. We are driven therefore to examine other types of evidence. And, for that reason, this study has paid considerable attention to several other approaches.

The indirect deduction of jet stream strengths in the storms 'accidentally surveyed' by the Spanish Armada in 1588 point to jet streams, and therefore probably also surface gradient winds, on as many as six occasions (i.e. not just in the case of the two storms in that year illustrated as historic storms in this compilation) in the months of July to September 1588 of speeds that could probably not be matched in those months in any year in the present century. (This point is developed more fully by Douglas, Lamb and Loader (1978).) The same argument, based on the speeds of travel of the storm cyclones, pointed to an extremely strong jet stream in the development of the disastrous North Sea storm at Christmas 1717.

More indirect evidence, in the magnitude of the sand-blow events in the Little Ice Age cold period, also points to winds – in this case, specifically surface winds – of strengths probably unparalleled in the present century. This surely applies to the great Culbin Sands storm in 1694 and to the sand-drift occurrences far inland in the Breckland of East Anglia reported several times between 1570 and 1668, perhaps also to the sand drift in that period far inland in the eastern part of the Netherlands, and probably to other occasions here recorded in Scotland (e.g. 1697, as well as during an earlier, cold climate phase in the fifteenth century) and in Denmark both in Jutland and in Sjaelland, north of Copenhagen. The overwhelming by sand of the township of Forvie at the east coast of Scotland during an earlier cold climate phase, in a summer month, in August 1413, seems to be a parallel event.

We also have the indication (cf. pp. 19–20) from the history of the growth of the spit at the northern tip of Jutland (Skagen Odde) of a peculiarly active period as regards winds and water currents between about 1695 and 1795, unmatched except during part of the recent period of increased storminess since about 1950.

It seems likely that the increased intensity of storms in the Little Ice Age period had to do with the source of potential energy in the, at that time, enhanced thermal gradient between the colder ocean surface in the seas about Iceland – particularly to the north, east and southeast of Iceland (Lamb, 1979) – and the ocean south of 50–55° N and the Bay of Biscay. In the latter regions departures from twentieth century conditions were certainly smaller.

It is not easy to point to a firmly established reason for the increase in storminess affecting northern and north-western Europe, and the off-lying seas, since 1950. Some would probably attribute it with little more ado to the greenhouse effect expected from the increase of carbon dioxide and other gases in the atmosphere caused by human activities. There has certainly been an increase since about 1950–60 of a few tenths of a degree (perhaps best estimated at 0.3 to 0.5 °C) in the overall average temperature of the surface of the tropical oceans between latitudes 2–5° N and 10–20° S, according to data published by Flohn and Kapala (1989) and Folland and Parker (1989 and C. Folland, personal communication, 1985). At the same time, for reasons not yet established, prevailing ocean surface temperatures in the North Atlantic generally north of about 50° N have become lower by 0.5 °C or more than they were in the 1950s (see Figure 8).* Thus, an enhanced thermal gradient as compared with conditions between about 1900 or 1920 and 1950 has once more come into existence. This enhancement appears smaller than that affecting the temperature gradient in the North Atlantic between the latitudes of Biscay and the Faeroe Islands–Iceland in the seventeenth century, probably no more than about one fifth of the magnitude of the anomaly in the Little Ice Age. On the other hand, the increased warmth – albeit small – over the vast area of the ocean in the equatorial zone between 5° N and 20° S has presumably increased the amount of water vapour evaporated into the atmosphere and therewith the amount of energy that can be released as latent heat in condensation processes in the atmosphere in all latitudes.

The suggestion of a roughly 100-year cyclic variation in storminess superposed on the major trend may have to do with a secular variation in solar activity that must surely be related to the 200-year and 400-year oscillations in atmospheric radiocarbon: these seem now firmly established and are registered in dating records that go back 7000 years or more (see Lamb 1977, pp. 65–6, and H.E. Suess in *Radiocarbon variations and absolute chronology: Proceedings of the Twelfth Nobel Symposium, Uppsala 1969*, edited by I.U. Olsson, Stockholm (Almqvist & Wiksell) and New York (Wiley), 1970; and more recent papers by Suess and others). Frydendahl (1990) has lately also identified an oscillation

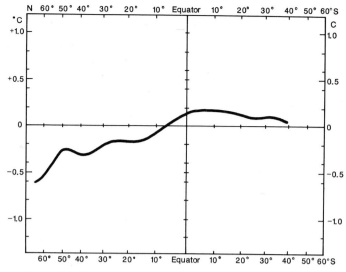

*Figure 8.* Differences of prevailing ocean surface temperatures at latitudes between 65° N and 40° S in 1968 to 1984 as compared with 1951–60. (Adapted from a diagram by C. Folland by kind permission.)

of close to 100 years length in ocean surface temperature levels world-wide from the earliest reasonably full coverage in the 1850s to the present and in their differences from air temperatures over land regions. The ocean temperature oscillation lags by about twenty years behind the over land changes because of the inertia of large masses of water.

As regards lessons of a non-meteorological, or partly not meteorological, nature coming out of this study:

(i) All the greatest death tolls due to storms have occurred in those that produced either great sea floods or many shipwrecks. Latterly, accidents to oil and gas rigs in storms have produced many deaths.

(ii) The numbers of deaths, and amounts of damage, caused by great storms at several points in history have clearly been notably reduced for 50 to a 100 years after a thorough campaign triggered by some previous disaster of building up and restoring the sea defence barriers. This even applied to cases, as on the German coast in the 1750s and 1790s and in eastern England (Norfolk) in 1978, when later storms brought higher flood tides than the one that caused the previous disaster.

The most obvious practical recommendation, therefore, from this and from the types of damage to buildings seen in some recent storms, is that lives – sometimes many lives – can be saved when all buildings and sea defence barriers and river bank protections are well designed, soundly constructed, and kept in good repair.

*Weber (1990) identifies corresponding changes in the temperatures in the overlying atmosphere, as indicated by 300 to 1000 mb thickness (representing all the troposphere below about 10 km height), when averages for 1977–86 are compared with 1951–60. A rise south of latitude 40° N, averaged around the world, contrasts with a fall north of 60° N. The overall increase in equator to pole temperature difference amounts to about 1 °C. The northern hemisphere changes were greatest near the equator and near 80° N.

# Part 2

# Catalogue and descriptions of great storms reported in the North Sea and neighbouring regions

## 26 September (Old Style)/6 October (New Style) 1509

*Area*

Holland, Zealand and Friesland.

*Observations*

A disastrous sea flood, with the tide rising far above the then existing coastal dykes (or earth-banks of possibly natural – or at least much earlier – origin). There was much loss of life among people and cattle. In the areas affected it was one of the most disastrous floods in history, certainly unmatched until that date since the flood in 1362 in the same areas. The river Ems near Emden changed course and many villages had to be abandoned. There is, however, no mention of this flood from the Schleswig-Holstein or Danish coast farther north: those areas presumably escaped or were much less seriously affected. They had experienced a much worse disaster with dividing of the islands of Nordstrand and Pellworm (off Schleswig) in 1436.

The 1509 flood was produced by a 'mighty storm', according to the record used by Woebcken (1924).

*Interpretation*

Judging by the areas affected, the storm wind was probably from nearly N rather than a NW'ly or W'ly storm. Possibly the cyclonic centre moved into the German Bight, producing E'ly winds over coasts mentioned which seem to have escaped the disaster.

*Data Source*

*Chronyk van Ostfrieslant* (Beninga), published by Harkenroth, Emden 1723; quoted by Carl Woebcken in *Deiche und Sturmfluten an der deutschen Nordseeküste*, Bremen-Wilhelmshaven (Friesen Verlag) 1924 and reissued by the Sändig Reprint Verlag, Schaan/Liechtenstein (1981).

## 4–5 November (Old Style)/14–15 November (New Style) 1530

*Area*

Storm surge flooding of coasts and estuaries in the southern Netherlands and southern parts of the English North Sea coast.

*Observations*

Flooding severe on the coasts of south Holland and, especially in the eastern parts of the estuaries in the southern Netherlands (Gottschalk, 1975), where 25 towns and many villages are alleged to have been lost. Sea flood also on the coasts of Essex and Kent (Thanet) in England.

The surge coincided with the spring tide.

The flooding in the Netherlands is well recorded in official reports, but the disaster seems not to have extended to northern parts of the country.

*Interpretation*

The distribution of flooding is consistent with a general surge of the North Sea, rendered severe in the southernmost parts by a strong N'ly storm south of about latitude 52° to 53° N.

Evidently the water level in the North Sea had been raised, presumably by Atlantic gales from about W over latitudes between Scotland and Iceland and the situation in the southernmost part of the North Sea was made disastrous by a very vigorous secondary storm centre there.

## 2 November (Old Style)/12 November (New Style) 1532

*Area*

Storm over at least the southern half of the North Sea. Tidal flooding of the continental coasts from Denmark to the southern Netherlands and Flanders.

*Observations*

This storm produced one of the great North Sea storm surges, although it came with a neap tide.

Gottschalk (1975) reports that there were many breaches of the dykes on the Dutch, German and Danish coasts, and the repairs which had been made after the 1530 floods were destroyed. Islands in the southern Netherlands area (e.g. North Beveland) which had been mostly flooded in 1530 were completely submerged this time. In one place it is recorded that the highest flood level in 1532 was 2 feet above the 1530 level. Nevertheless a few of the islands (Walcheren and Voorne) were less affected in 1532, presumably because of effective repairs after the previous flood.

As in 1530, flooding was particularly severe in the eastern (innermost) parts of the estuaries in the southern Netherlands.

*Interpretation*

Again, we must suppose a broadly N'ly storm, but of greater extent than in 1530. This time a bigger depression must have produced storm winds over probably most of the North Sea. Since the coasts of England seem to have been spared, the precise direction of the storm winds may have been about NNW over the southernmost North Sea and probably NW farther north, perhaps between W and NW on the coast of Denmark.

## 5 and 8–15 January (Old Style)/15 and 18–25 January (New Style) 1552 or more likely 1553*

*Area*

North Sea (all the British and continental coasts) and western Europe.

*Observations*

Great wind storms, accompanied by hail, snow and thunder, North Sea tidal surges and widespread sea flooding, especially 5th and 13th (Old Style). Also river floods in the Wesertal in western Germany (11th Old Style). (Gottschalk (1975) reports a further storm surge flood on 15 February (Old Style). She adds (p. 534) that this noteworthy succession of storm flood incidents ended a remarkable period of calm without much storminess in the southern North Sea that had lasted more or less since around 1520, apart from storms in 1530 and 1532.) On the German coast only the January storm floods are mentioned in 1552 or 1553, and these also affected the coasts of England and Scotland, according to Gottschalk.

*Interpretation*

The weather sequences suggest that the January 1553 events each began with mild southwesterly winds, ended by a sharp cold front bringing a strong gale from about NW in the southern, and from N in the northern, North Sea.

---

*At this date, and for long afterwards, there was no complete uniformity about which day should be taken as starting a new year. The commonest usages were to take either January or March as the first months, perhaps most frequently March. (This is why September, October, November and December bear the Latin names for seventh, eighth, ninth and tenth month and the Julian calendar began with the 'Ides of March'.) Gottschalk (1975, p. 572 mentions that the year 1553 is frequently mentioned in the literature as the date of the flood which she, perhaps inadvisedly, attributes to 1552.

---

## 1–2 November (Old Style)/11–12 November (New Style) 1570. (Long famous as the 'All Saints Flood')

*Area*

Storm over the whole North Sea and severe inundation of the continental coast.

*Observations*

Hennig (1904) notes: 'the greatest North Sea flood after (that of) 4.2.1825'. Amsterdam, Muyden, Rotterdam, Dordrecht, and many other cities flooded. The numbers drowned were from 100 000 to 400 000 according to different sources.

Gottschalk (1975, p. 700ff.) reports 'This was one of the great storm surges . . . The effects were felt along the whole North Sea coast of the European mainland from France to Denmark. England was most probably not affected by it, although . . . the English east coast had already been struck by a surge on 5 October (Old Style)/15 October (New Style) 1570 . . . During a great part of the year [1570] western Europe was troubled with floods'. 'The All Saints flood occurred 2 days after new moon, which means it was a spring tide. The gale was SW'ly, but, according to a contemporary Groningen witness, the wind veered to NW.' A contemporary Brussels witness states that the storm was already blowing up on 31 October (Old Style)/10 November (New Style). [This means that the prefrontal winds were strong as well as those after the front.]

'The earliest moment at which dyke breaches are mentioned is the afternoon of 1 November (Old Style) at between 4 and 5 o'clock . . . the dykes gave way . . . on the island of Walcheren. In Middelburg too there was extensive flooding during the afternoon.'

'At about 6pm, the situation was very critical on the coast of Brabant north of Antwerp. [This all sounds as if it means when the wind later veered NW.] Elsewhere in the southwest Netherlands the highest tidal levels and dyke breaches occurred mainly between 8 and 9pm, as near South Beveland and near Rotterdam.'

'On the coast of Groningen [in the north and northeast of the Netherlands], the dykes gave way at 10 o'clock in the evening . . . Along the coasts of the Zuyder Zee and in Friesland . . . [breaches of the dykes occurred] during the night of 1–2 November (Old Style)/11–12 November (New Style).' This fits the idea of a N–S orientated cold front making very slow progress eastwards – presumably in a V-shaped depression.

'Our [i.e. Gottschalk's] impression is that Flanders suffered less from the All Saints flood of 1570 than from the earlier surges of 1530 and 1532. The west coast of Brabant between Antwerp and the Hollands Diep was very seriously affected by dyke breaches and floods in 1570. In Zeeland not a single island was spared by the All Saints flood . . . .'

Petersen and Rohde (1977, p. 40) report for Hamburg and the NW German coast: 'Severest storm floods of the period from the fourteenth century to the end of the eighteenth century were the All Saints floods in 1532 and 1570. The flood of 1532 was worst in Schleswig-Holstein; that of 1570 produced most damage west of the Elbe.' This suggests that the storm wind was more N'ly in the 1570 case than in 1532.

Petersen and Rohde add (1977, p. 41) that the peak level of the storm flood in 1570 seems to have been between 4.40 and 4.50 m above normal sea level in the River Ems at Emden. In the Jadebusen, at Dengast a height of 4.41 m above (German) normal sea level was reported.

*Comparisons of the All Saints flood levels in 1570 with later floods* are easiest in the case of *Scheveningen*, Holland (Gottschalk, 1975, p. 708):

1570: 4.15 m above Amsterdam datum
1894: 3.53 m above Amsterdam datum
1953: 3.90 m above Amsterdam datum

The All Saints flood in 1570 is said by van Meteren, writing in 1599, to have been one foot higher than a flood in 1530 and 2 feet higher than another in 1552. Comparisons between the flood levels of the sixteenth and seventeenth centuries and those of modern times are not likely to be vitiated by later improvements in the dykes and sea walls, but the liability to floods must be affected by such changes as have occurred in the general sea level (probably no more than about half a metre).

The details quoted indicate that there was probably some westerly component in the strong winds at all times during the passage of this storm system in 1570, apart from possibly a few hours of S'ly wind in the last stages before it veered NW. This possibility may be suggested by the very slow eastward march of the system, although SW was the direction mainly reported. The lack of effect on the English coast indicates no easterly component at any time. There clearly *was* a North Sea surge.

*Interpretation (meteorology)*

A map of the progress eastwards of the 'wave' of flooding over the Dutch and northwest German coastlands (see inset) suggests the slow eastward march of a cold front dividing a SSW windstream ahead of the front from NW or NNW winds advancing behind it. It does not seem possible that such a regular progression of the flooding, although slow, could have been produced by the surely more haphazard incidence of dyke breaches controlled only by structural failure. Nor could this result be produced by the timing of the tides. The front's progress eastwards seems to have been at a rather steady rate close to 5 knots (2.5 m/sec) over the roughly 24-hour period covered by the map.

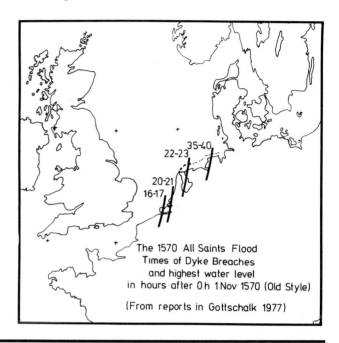

The 1570 All Saints Flood
Times of Dyke Breaches
and highest water level
in hours after 0 h 1 Nov 1570 (Old Style)

(From reports in Gottschalk 1977)

## Circa 1570 to 1668 or after

*Area*

Storms in the Breckland of Norfolk and Suffolk, England and in the Netherlands.

*Observations*

Testimony to the occurrence, and indeed some frequency, of stormy winds stronger than experienced in the twentieth century may be seen in records of the transport of great quantities of sand in inland districts in the centre of East Anglia, and an apparently parallel situation inland in the eastern Netherlands, in and about the seventeenth century. It seems to be implied by the reports that the winds exceeded any commonly experienced before about 1570. The areas concerned are known respectively as the Breckland, in Norfolk and Suffolk, and the Veluwe, near Apeldoorn, in the Netherlands.

John Evelyn mentioned this phenomenon as a sight to see in East Anglia, while on a visit to Euston, near Thetford, in his famous diary (10 September (Old Style) 1677): 'The Travelling Sands . . . that have so damaged the country, rouling from place to place . . . like the sands in the Deserts of Libya, quite overwhelmed some gentlemen's whole estates . . .'

An extenso account by one of those affected, Thomas Wright, Esquire, is given in the *Philosophical Transactions of the Royal Society, London*, vol. III, No. 37 (in the issue of 13 July 1668), pp. 722–5. The following are extracts from what the author described as 'the account you required of those prodigious Sands, which I have the

unhappiness to be almost buried in and by which a considerable part of my small fortune is quite swallowed up'.

'. . . these wonderful Sands . . . have not yet exceeded one Century since they first broke prison . . .' i.e. broke loose. With some difficulty he traced their origin to 'a Warren in Lakenheath', about 5 miles (8 km) southwest by west from his house, 'where some great sand-hills (whereof there is still a remainder) having [their surface] . . . broken by the impetuous South-west winds, blew upon some of the adjacent grounds; which being much of the same nature, and having nothing but a thin crust of barren earth . . . easily fitted to increase the Mass, and to bear it company . . .'

The first eruption of the sands, this report states, 'does not much exceed the memories of some persons still living'. From its source to Mr Wright's residence at Santon Downham 'all the ground it passed over was almost of as mutinous a nature as itself'. About 2 km from the source the moving sands built up against a farmhouse, which was partly buried but later released as the bulk of the sands moved on. The moving mass reached the edge of Santon Downham about 1630 ('between thirty and forty years since'), then halted for 10 or 12 years before entering the place. Then, moving uphill, it buried and destroyed various houses and overwhelmed the cornfields. A great part of this first advance over Santon Downham took place in just two months. Attempts to put up effective barriers in some places produced 'sandbanks near 20 yards high'. The river at Santon was nearly blocked for 3 miles (i.e. over a 5 km length) and the pastures and meadows were overwhelmed.

The writer attributed the disaster to the light, sandy nature of the soil and to the position of his land 'east-north-east of a part of the great level of the Fenns . . . fully exposed to the rage of those Impetuous blasts we yearly receive . . . which, I suppose, acquire more than ordinary vigour by the winds passing . . . so long a Tract without any check'. He then refers to (legal) actions . . . brought in the neighbouring county of Norfolk (to the north) for grounds blown out of the owners' possession. At the date of his writing, some success had been achieved by various techniques, using fir hedges as barriers to divert the sand and replacing them as soon as they were levelled with the sand, and also 'laying hundreds of loads of muck and good earth upon' the loose sands.

From these descriptions, it seems that something of the order of 50–100 million cubic metres of sand were involved in the main mass of the 'Sand-floud' or 'Wandering Sands' that overwhelmed Santon Downham at this time. And their total progress was not less than 10 km. The total mass indicated for the main wave of moving sand that invaded the immediate Santon Downham area may have been of the order of 100 000–250 000 tonnes.

The area is now in the heart of Thetford Forest, thanks to successful afforestation since the 1920s. (The village – 'sand-town' – probably owes part of its name to the events here reported. Before the sands came it was known as a town.)

A report by E.A. Koster (1978) indicates a seemingly parallel course of historical development about the same period in the Veluwe region (52° N, 5.5° to 6° E) of loose sands west of Apeldoorn in the eastern Netherlands, which is also an inland district far from the sea. In that area sand drift seems to have begun about 5000 years ago when Neolithic farmers cleared the previous vegetation. But drifting on a larger scale set in around the end of the Middle Ages and continued until the area was more recently stabilized by afforestation with pine. In the Veluwe the prevailing height of the dunes then formed, and now largely fixed by vegetation, is 2.5 to 3.5 m above the level of the surrounding land.

*Inference*

These reports suggest that there was a significant increase of wind stress in these inland districts near latitudes 52° N on either side of the North Sea around the sixteenth century (probably the late sixteenth century), which lasted a long time, but with some decrease of vigour, until afforestation was successfully used to stabilize the landscape in the early twentieth century. Formerly stable sand surfaces had broken loose (around 1570) without any change of land use, and in the Netherlands case where there had long been some blowing sand the situation had become much aggravated around the same epoch. Today, less than 5% of the drift-sands of the Veluwe are active. In the English Breckland the proportion is doubtless very much less than that. In both areas, the prevailing direction of the strongest drift seems to have been from between about WSW and SW as is the case with the strong winds of today. The Netherlands reports indicate that the winds in the twentieth century have been more from the SW, in the eighteenth and nineteenth centuries the wind data available for Amsterdam and (after 1850) for Utrecht indicate WSW as the prevailing direction of the strong winds. The seventeenth century drift recorded at Santon Downham in the Breckland seems to have been from between SW by W and WSW.

It is implied, by Thomas Wright's account of the history of the advance of the sand to Santon Downham in this area of eastern England, that not less than three very severe storms of wind from about WSW were involved – about 1570, 1630, sometime in the 1640s, and perhaps in the 1660s.

# 14–18 August 1588 (New Style)

*Area*

Northern North Sea off Scotland's east coast and among the Northern Isles.

*Observations*

Ships of the Spanish Armada reported (14 August): 'A very great gale at SW'; 'A great storm at WSW' and 'From 13th–18th we experienced squalls, rain and fogs with heavy sea and it was impossible to distinguish one ship from another'.

Elsewhere in the North Sea (probably southern North Sea) Sir Francis Drake reported (15 August): 'A great storm considering the time of year'.

*Meteorology*

The wind directions reported in the northern North Sea were from virtually every point of the compass, during those days NE, SE, SSE, S, SW, WSW, W, NW and N. The strongest winds mentioned were from WSW on the 14th and SSE on the 16th.

It seems clear that two depressions were successively involved and that the centres crossed the British Isles and North Sea region, broadly from west to east. The depressions may be identified with the centres marked Low g and Low h on the map analyses. The centres are shown on the maps moving from SW or SW by W to the NE, coming from the Atlantic across or very close to the British Isles and probably continuing across Scandinavia to near the White Sea. One was centred east of Berwick on the 13th and the next between Orkney and the Faeroes on the 16th, followed by NW and N winds.

*Maximum wind strengths*

Before the existence of barometers the pressure gradients cannot of course be accurately indicated on the maps.

From the distances covered in 24 hours by the depression centres Low g to Low i on the map sequence, a jet-stream of great to exceptional strength for the time of year (up to 60 m/sec or 115–120 knots) seems to be implied by the statistical association between depression speeds and jet stream strength (Douglas *et al.*, 1978, following

Palmén, 1928). Surface pressure gradient winds in the warmer air around 20 m/sec (39 knots) and at the rear side of the depressions up to 35 m/sec (70 knots), with the strongest surface gusts certainly reaching this order of speed, could be expected to develop. From the descriptions by both Spanish and English captains, it seems likely that the surface pressure gradients in the warm air over the North Sea, especially the southern part, were actually rather stronger than these values which represent the statistically most probable figures. Gradient winds approaching 50 knots in the warmer air stream and 80–85 knots in the NW'ly rear-side air stream seem indicated, with strongest gusts at the surface of about that strength.

*Data source*

Details of the weather reports available for the storms in 1588 from the ships of the Spanish Armada and other records, and on how the analysis was performed, are given in *The Spanish Armada Storms* by K.S. Douglas, H.H. Lamb and C. Loader, Climatic Research Unit Research Publication CRURP6. Norwich (University of East Anglia), 1978.

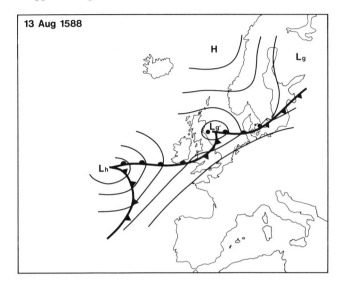

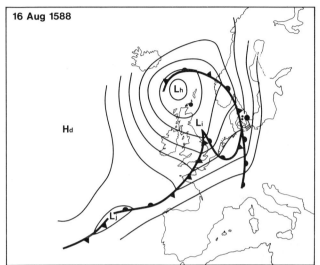

## 21 September 1588 (New Style)

*Area*

West coast of Ireland (and, presumably Hebrides, NW and N Scotland on 21–23 September).

*Observations*

Ships of the Spanish Armada in Blasket Sound and at Streedagh Strand, Co. Sligo report (21st): 'On the morning of the 21st it began to blow with most terrible fury. It was bright and with little rain.'

'There sprang up so great a storm on our beam (i.e. W'ly) with a sea up to the heavens so that the cables could not hold nor the sails serve us and we were driven ashore with all three ships upon a beach covered with fine sand, shut in on one side and the other by great rocks . . .'

Many of the Armada ships, including three in each of these places were wrecked upon the rocks or the open shores, a greater loss than had occurred in the battle with the English at Gravelines in the Channel in August.

The date of the storm, also in Co. Mayo reckoned as 'a great gale', seems confirmed by various written reports of the Clerk to the Council of Connaught.

On 21st Tycho Brahe reported the wind in Denmark as a S'ly storm and 'at night sometimes slashing rain, wind SW'.

*Meteorology*

This severe storm can be confidently identified with one encountered on the 18th by other ships of the Armada in the Bay of Biscay, where it was already vigorous.

The cyclone centre was clearly moving NE from the region of the Azores to lie off the Hebrides on the 21st and off western to northwestern Norway on the 23rd.

There seems to be a possible analogy with the severe storm, a former Atlantic tropical hurricane 'Debbie' which passed along a similar track off the west coast of Ireland on 16 September 1961, causing enormous damage in the west of Ireland and devastating woodlands. It is possible, therefore, that the storm of 21 September 1588 also originated as a tropical hurricane.

*Maximum wind strengths*

The jet stream over the British Isles (region around this date seems from the measured 24-hour movements of frontal waves and young cyclone centres to have been of very great strength, such as would now be exceptional for the time of year, viz. up to 50–67 m/sec (97–130 knots) on 15th–17th and 55 m/sec (107 knots) around 22 September 1588.

The corresponding surface pressure gradient winds and order of magnitude of the strongest gusts observed at the surface were likely to be up to 18 to 23 m/sec (37–44 knots) in the warm air and 32–38 m/sec (62–75 knots) in the cold W'ly air stream.

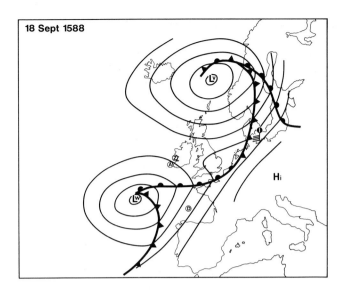

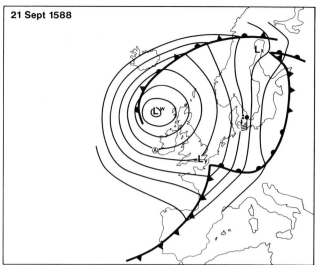

## Appendix A.   Note on the analysis of the storms in 1588

These synoptic weather maps are taken from K.S. Douglas *et al.* (1978, 1979). (The sources and archives used are listed in the 1978 publication.)

Douglas used the weather observations in the logs of Spanish ships which survived the disaster to the Armada in that year and in correspondence related to the expedition, supplemented by a smaller number of reports of the weather in England and Ireland, to construct a preliminary series of daily meteorological analyses for about 60 days between May and October 1588. Thirty-four of these maps, on which S.G. Aston and S.J.G. Partington of the Meteorological Office had also worked, were initially passed to H.H. Lamb for comment.

Luckily, a test of these preliminary maps was possible, since none of those who had taken part in the analysis, were aware that the Danish astronomer, Tycho Brahe, had recorded a series of his daily observations of wind and weather on the island Hven (55°55′ N 12°45′ E) in the Sound between Denmark and Sweden. These observations were reprinted by La Cour (1876).

Entering the Danish observations on the preliminary weather maps for 1588 showed reasonable agreement with the analysis on 72% of the days concerned. The entire series of maps for 60 days between 12 May and 26 October (New Style) 1588 was then thoroughly re-analysed by H.H. Lamb with the greatest care for logical continuity and consistency of the series. It is thought that the resulting analysis must be an acceptable approximation to the real meteorological situations on at least 80% to 90% of the days.

The analysis is clear enough to show the movements of individual low pressure centres and fronts from day to day, and measurement of the distances travelled provides an indicator of the strength of the atmospheric circulation systems involved. This is especially valued as one thing which can be measured in this case, at a time before the invention of the main meteorological instruments.

Relationships between the movement of cyclones and the wind systems prevailing at the time were investigated already in the 1880s by W.J. van Bebber and by Palmén in 1928. The latter provides the firmest basis for interpreting our map analyses of the summer of 1588.

Rates of movement of cyclonic features at the following different stages of development have to be considered as different cases:

(a) open frontal waves – i.e. the earliest stage of development

(b) young warm-sector cyclones up to the early stages of occlusion of the warm sector

(c) occluded cyclones

Occluded systems tend to be centred right outside the path of the strongest upper wind flow, away to the cold side of the jet stream. Systems in the middle category are usually closer to the core of the upper windstream, but the situation is liable to be complicated by developing curvature of the upper flow which steers the situation. The open frontal waves are commonly close to a straight jet stream aloft, and as a class they therefore tend to move fastest.

Results of Palmén's investigations showed:

(i) The speed $V_f$ of advance of an open frontal wave increases with the gradient wind speed $v_G$ in the warm sector and the surface temperature constrast $T_0$ at the front in such a way that

$$V_f = 4.2 + 0.7 v_G + 3.0 \sqrt{\Delta T_0}$$

where speeds are in m/sec and temperatures in °C.

The speed of displacement of the wave is generally greater than the surface wind measured in the warm air over land, often by 50% to 80%. This implies, however, that $V_f$ may be typically just 75% to 90% of the wind speeds observed in the warm air over the open sea.

(ii) The speed $V_y$ of a young cyclone, typically less than that of an open frontal wave, is best given by

$$V_y = 0.8 + 0.6 v_G + 2.6 \sqrt{\Delta T_0}$$

42

Table 8. *24-hour movements of cyclones (low pressure centres) and frontal waves and corresponding wind speeds July–September 1588*

| Dates | Identification of feature on the maps | Category: FW, frontal wave YC, young cyclone OC, occluded cyclone | Measured 24-hour advance (km) | Equivalent average speed (m/sec) | Implied wind speed (in round figures) (a) surface wind in warm air over the sea (m/sec) | (b) jet-stream (m/sec) |
|---|---|---|---|---|---|---|
| 22–25 July | Low a | OC | 240 | 2.8 | | |
| 25–26 July | b | YC | 1000 | 11.7 | 13 to 15.5 | 47 |
| 26–27 July | b | YC | 750 | 8.6 | 9.5 to 11.5 | 35 |
| 27–28 July | b | YC-OC | 850 | 9.7 | 10.5 to 13 | 40 |
| 28–31 July | b | OC | 340 | 4.4 | | |
| 28/29–30/31 July | c | FW-YC | 850 | 9.7 | 10.5 to 13 | 40 |
| 30/31 July–1 Aug. | c | YC | 800 | 9.2 | 10 to 12.5 | 37 |
| 28/29–30/31 July | d | YC | 750 | 8.6 | 9.5 to 11.5 | 35 |
| 30/31 July–1 Aug. | d | YC | 900 | 10.5 | 11.5 to 14 | 42 |
| 6/7–8 August | e | YC-OC | 900 | 10.5 | 11.5 to 14 | 42 |
| 8–9 August | e′ | OC | 850 | 9.7 | | |
| 8–9 August | e | FW | 1700 | 19.7 | 22 to 26 | 80 |
| 8–9 August | f | YC | 850 | 9.7 | 10.5 to 13 | 40 |
| 9–11 August | f | YC | 300 | 3.5 | 4 to 5 | 14 |
| 11–13 August | g | YC | 1300 | 15.0 | 16.5 to 20 | 60 |
| 13–16 August | h | YC-OC | 700 | 8.1 | 9 to 11 | 32 |
| 16–18/19 August | h | OC | 660 | 7.7 | 8.5 to 10.5 | 30 |
| 16–18/19 August | i | YC | 900 | 10.5 | 11.5 to 14 | 42 |
| 16–18/19 August | j | FW-YC | 1300 | 15.0 | 16.5 to 20 | 60 |
| 25–28 August | l | OC rejuvenated? | 850 | 9.7 | 10.5 to 13 | 40 |
| 24–25 August | m | FW-YC | 1400 | 16.1 | 18 to 21.5 | 65 |
| 25–28 August | n | YC | 380 | 4.4 | 5 to 6 | 17 |
| 28 Aug.–1 Sept. | o | OC | 350 | 4.2 | | |
| 1–2/3 September | n + o | OC rejuvenated? | 1000 | 11.7 | 13 to 15.5 | 47 |
| 2/3 September | p + q | YC-OC | 700 | 8.1 | 9 to 11 | 32 |
| 6–8 September | s | YC? | 500 | 5.8 | 6.5 to 8 | 23 |
| 8–10 September | s | YC rejuvenated? | 900 | 10.5 | 11.5 to 14 | 42 |
| 13–16 September | t | YC-OC | 800 | 9.2 | 10 to 12.5 | 37 |
| 13–16 September | u | FW-YC | 1450 | 16.7 | 18.5 to 22.5 | 67 |
| 16–18 September | v | YC | 1050 | 12.2 | 13.5 to 16.5 | 50 |
| 18–21 September | w | YC-OC | 500 | 5.8 | 6.5 to 8 | 23 |
| 21–23 September | w | OC rejuvenated? | 750 | 8.6 | 9.5 to 11.5 | 35 |
| 21–23 September | y | FW-YC | 1200 | 13.9 | 15.5 to 18.5 | 55 |
| 23–26 September | a | YC-OC | 600 | 6.9 | 7.5 to 9 | 27 |
| 26–28 September | d | YC | 1000 | 11.7 | 13 to 15.5 | 47 |
| 28 Sept.–1 Oct. | c + d | YC-OC | over 600 | over 6.9 | | |

This turns out to be on average about half the gradient wind speed $v_G$ given by the surface isobars in the warm sector. It is typically a great deal less than the wind speed observable over the Low at cirrus cloud levels, which we may now take as jetstream level. Chromow (1942) takes the wind at cirrus levels as about twice the geostrophic wind at near-surface level, which implies that the speed of the cirrus $v_{Ci}$ is given by

$$v_{Ci} \approx 4V_y$$

The measurement data were most abundant for Lows in this category, and therefore the results in these cases should be the most reliable. Although the investigation which Mr Loader was able to conduct with the chart material and observations at his disposal showed more scatter than might have been obtained with a more adequate map series, his result confirmed this last equation as closely as the data permitted.

(iii) The speed of old occluded cyclones $V_0$ decreases rapidly as the system ages, unless a fresh cold or warm airstream is drawn into the circulation, effectively rejuvenating the system. No rule connecting $V_0$ to $v_G$ or $v_{js}$ could be derived.

Table 8 lists the 24-hour movements of all the Lows in each category between 22 July and 1 October 1588 which could be measured and the equivalent wind speeds implied by the expressions above.

There is no ground for surprise in the fact that we sometimes derive different values of jetstream (or surface wind) strength from our measurements of the movements of different cyclonic features or frontal waves on the same dates. They are related to different parts of the map and of the upper wind-stream.

Table 8 indicates that there were jetstreams on six occasions reaching or exceeding 55 m/sec (200 km/hr) during these months, the strongest (on 8–9 August 1588) being about 80 m/sec (288 km/hr or 155 knots). Three other times in August jetstream strengths more than three quarters of this value are indicated, and we derived again in mid-September a speed of 67 m/sec (240 km/hr or 130 knots). The strongest mean surface winds in the warm air are estimated to have reached between 22 and 26 m/sec (80–94 km/hr or 43–51 knots): the greatest speeds in gusts and squalls near the cold fronts and in troughs in the cold air may have been over 50% stronger than this.

Although there is some margin of error or uncertainty attaching to the figures derived in this way, and this margin cannot be precisely gauged, it is unlikely that our one or two samples of moving surface weather systems controlled by the jetstreams of July–September 1588 register the effect of precisely the strongest part of the upper wind flow. It may be concluded therefore that winds of at least the strengths mentioned above occurred at some point within the wind circulation over the area analysed.

This is believed to be the first opportunity to glimpse the strength of the individual wind circulation systems at such an early date and during the Little Ice Age climate.

The repeated occurrence of jetstreams of the strengths noted could probably not be matched in July–September in modern experience (up to 1976) over the region studied.* It was presumably related to a very strong thermal gradient over the region associated with the anomalous forward (60° to 63° N?) limit of the polar water in the ocean surface, transported south by a vigorous East Greenland and East Iceland Current.

In the summer of 1586, in mid-June, and possibly (but with inadequate reporting of longitudes) again in 1587, expeditions under the command of Captain John Davis, exploring for a North-West Passage to Asia and the Moluccas, had found the Denmark Strait between Iceland and Greenland blocked by ice at latitude 66° N (Richard Hakluyt: *Voyages and discoveries* (1598–1600)). In June 1586 the East Greenland ice also extended south to latitude 57° N near 47° W, south of Cape Farewell. Little or no ice was found, however, at those latitudes in the Labrador Current. Although no reports for 1588 in those sea areas are known to the authors, it seems likely from the analysis here presented that the situation was quite similar to these reports of the two preceding years.

It should not be concluded that such wind conditions as here derived in the analysis of 1588 could not occur in the present century, since the polar water with surface temperatures about 3 °C was observed in the East Iceland Current advancing to near the Faeroe Islands briefly in the late spring of 1968.

*As far as can be estimated from the sample upper wind statistics kindly supplied by the Meteorological Office, the winds at jetstream level in July–September 1588 on at least the six occasions mentioned exceeded the maximum winds observed in the corresponding months in the years 1961–70 over northern Britain, in one case by up to 50%. And from the more limited sample of years 1949–55, for which the statistics have been mapped (Upper winds over the world, parts I and II, by H. Heastie and P.M. Stephenson, *Geophysical Memoirs*, No. 103, HMSO for Met. Office, London, 1960 and *Ibid*, Part III by G.B. Tucker, *Geophysical Memoirs*, No. 105, HMSO for Met. Office, London, 1960) over the whole area of our study, the same six occasions in the one season in 1588 indicated winds close to or somewhat beyond the probable limit of the modern distribution of upper wind strengths for those months.

## 23 October (Old Style)/2 November (New Style) 1592 and later occasions

*Area*

Denmark and neighbouring sea areas (most of the evidence from Skagen, the northernmost point of Jutland).

*Observations*

Long-continued S'ly gale 30 October to 7 November (New Style), confirmed by Tycho Brahe's observations on the island Hven in the sound between Denmark and Sweden, interrupted only on 2 November by a strong west wind which drove a great deal of water past Skagen into the Kattegat, producing an exceptionally high sea flood which did great damage, destroying many houses in the market town of Skagen and a number of farms. A first intimation of trouble had in fact occurred already in September 1591 when a NE'ly storm which raged for 3 to 4 days overlaid 25 farms with sand near Skagen. There had been an impression in the area about Skagen that storminess was increasing, and with it a tendency for more sand drift, from the time of a storm in 1568, if not indeed from as early as 1546.

This was the beginning of a series of misfortunes which then befell the hitherto prosperous town of Skagen, which began to be abandoned in favour of safer positions. The fishery was also failing at this time.

Gottschalk (1975, p. 790) has found records of a series of severe storms at approximately monthly intervals 'at the end of 1592': in Holland in September (Old Style), in Groningen in the north of the Netherlands in October and November, in north Germany in November, and in Belgium in December and January (all Old Style months). In Groningen and north Germany there was a storm surge flood of the North Sea in November 1592.

A further sea flood destroyed 14 farms and houses besides on 25 January (New Style) 1593: it was produced by a fairly hard SW gale, followed by a great storm from the NW. Over the days that followed there were strong SE and E winds.

Another sea flood afflicted the same coast for 4 days in late January or early February (New Style) 1595, apparently with a SW gale that was followed by a NW and later N'ly or NE'ly storm.

There followed a period of continually drifting sand, through repeated storminess, that lasted through the seventeenth and eighteenth centuries, and to some extent since, and completely altered the landscape, the heights above water level, and the position of the land-spit in the area of Skagen. The later village of Skagen was built on a new site. By the end of the seventeenth century the old market town lay deep amongst the sand-dunes, a condition registered in a painting by the landscape artist Vilhelm Melbye in 1848 (reproduced here as Plate 10).

The information here summarized comes from Hauerbach, Hansen and Nielsen (1983) and Tycho Brahe's original weather observations reprinted by La Cour (1876) of the Meteorological Committee of the Royal Danish Scientific Society. Hauerbach, a surveyor using his own measurements and the maps of various dates of the Geodetic Institute, Copenhagen, as well as older surveys, e.g. the *Atlas Danicus* of 1677 and a Videnskabernes

Selskab survey of 1795 (Hauerbach *et al.* 1983), give positions of the point at Skagen which show that the coast has been advancing to the northeast as a result of the transport of sand by wind, waves and current. From the earliest determination of the position, in AD 1355, to 1695 the average rate was 8 m a year. In the 100 years between 1695 and 1795 the advance was 2–3 km, an average rate of 23 m/year. The rate has since varied, averaging mostly between 5.5 m/year between 1795 and 1850 and 8 m/year in the twentieth century, but the average was 15 m/year between 1850 and 1887 and 25 m/year between 1954 and 1964.* This clearly points to a conclusion, mentioned by Hauerbach *et al.*, that 'in the cold climate periods the weather has been more blowy with frequent strong storms'. It may be that the nineteenth and twentieth century periods mentioned imply that strong storms and rapid advances to the northeast or east of the point Skagen Odde have been associated with periods of either colder or cooling climate.

*The advances mentioned have, as named, been period averages. There have, of course, been times when the point was broken and carried away by the waves and current. Such recessions (by up to 500 m) occurred between 1945 and 1954 and again between 1964 and 1974.

## From some time in the late sixteenth century or in the seventeenth century to 1725

### Area

Northern part of the island of Sjaelland (Zealand), Denmark – the island on which Copenhagen lies – most specifically Tibirke and Tisvilde Hegn, an area about 9 km in length from SW to NE along the north coast of the island by 4–5 km wide, facing the broadest part of the Kattegat.

### Observations

'A terrible sand-drift period', during which the whole area (of somewhat similar size to the Culbin Sands area on the coast of Scotland, which was overwhelmed by sand in the same period, especially by a storm in 1694) was turned into a wasteland by the blowing sand. Nine farms, the village of Tibirke (56°03′ N 12°06′ E) and its church, the parsonage, and other buildings were buried. The culmination came in 1725, when the village was abandoned in favour of a new site on fields to the north of the village. The church was by that time buried up to the roof in sand, but by order of the king it was preserved and restored, and was back

in use with a new pulpit and a new altar installed in 1739 and 1740. The area today is wooded.

The situation seems therefore to have become stabilized soon after 1725. That year itself is distinguished by the coldest summer in central England in 330 years of record, which must have been due to a great predominance of N'ly winds in the Norwegian Sea and the North Sea. The winds over Denmark at the same time were probably mostly W'ly and SW'ly; often in the central region of the, sometimes stormy, cyclonic activity.

This history runs parallel with that of the greatest period of extension towards the northeast of the point of Skagen in northernmost Jutland and was, at least roughly, contemporaneous with the moving sands inland in the Breckland of East Anglia and in the Veluwe in the eastern part of the Netherlands, mentioned elsewhere in this collection of storm reports. Another, broadly contemporary, instance seems to be the sanding up of the church and churchyard at Llandanwg on the west coast of Wales (52°50′ N 4°07′ W) near Harlech.

## Circa 1600 to 1720

### Area

Northern North Sea, especially Scottish coast.

### Observations

Walton (1956) notes: 'As early as 1654 . . . the small inlet of Strathbeg [was] by then almost obliterated with sand. Near it were the remains of the town of Rattray.' The areas referred to here were close to (just north and west of) Rattray Head (57°37′ N 1°50′ W), near the northeast corner

of the Aberdeenshire coast. The former inlet of the sea has since become a lagoon, the Loch of Strathbeg.

Walton continues: 'It is probable that this sand "inundation" was a temporary blockage . . . but caused a shallowing which seriously impeded the movement of small coastal craft . . . The bar may have been completely removed in time . . .' since an account (by one Alexander Hepburn) in 1721 mentions 'the river Rattray, at the mouth of which is situated the village of Rattray, famous for codfish which the inhabitants take in great plenty and have the best way of drying and curing them'. But Walton suspects that this account was written a considerable time before it was published in 1721. Another account, by Alexander Keith in his *View of the Diocese of Aberdeen* in 1732 reports that Rattray once had a good harbour but was then choked with sand and the town reduced to nine or ten houses.

*The New Statistical Account* (Edinburgh 1845) reports (vol. XII, p. 709) on Crimond that the lake, the Loch of Strathbeg, was much smaller in about 1700 than it subsequently became and had a small communication with the sea so that vessels of small burden could enter it. 'People born about 1700 well remember the overflowing (i.e. creation by flooding) of the western part of the loch; but the particular year is not known, though it must have been about 1720. Previous to that there was a hill of sand between the Castlehill of Rattray and the sea and still higher than the latter. A furious east wind blew away this sandhill in one night, which stopped the communication between the loch and the sea.' So sudden was the final sealing of the outlet that a small vessel is reported to have been trapped within the harbour and its cargo of slates used to roof the nearby house, Mains of Haddo.

It is astonishing that in the times around 1700 such a disastrous storm could happen without anybody writing down the exact date of it. Without the exact date, there is no possibility of collecting simultaneous observations from other places to perform a meteorological analysis. But the storm which finally cut off the sea connection to the Loch of Strathbeg must rank as a major storm.

Walton further reports (1956): 'The Royal Burgh of Rattray must have ceased to exist before the middle of the eighteenth century . . . some hardy fishermen continued to operate from the open shore . . . they probably gained a greater livelihood from plundering the wrecks which came ashore on Rattray Head than from the fishing itself . . . Eventually they moved away in the 1830s when six or seven families went to the new fishing village of Burnhaven' (near Peterhead).

This is another case, as in the Breckland storms, where we find evidence that storminess was increasing in and around the seventeenth century and continued for a long time. At other points around the coasts of northwestern Europe, there is evidence (see, for instance, Lamb, 1977, 1982) that blowing sand and the formation of drifting sand-dunes was on the increase from about AD 1300 onwards, but a more stable period seems to have occurred after 1450 until around 1550.

## 23 January (Old Style)/2 February (New Style) 1610 or 1611

*Area*

North Sea.

*Observations*

Netherlands and Oldenburg (northwest Germany) records report a very severe flood. Gottschalk (1977, p. 157) des-

cribes this flood as a catastrophe like that of the 8 March 1625, causing the loss of newly reclaimed land.

*Note.* The year 1610 to which this flood is commonly assigned may be a misunderstanding arising in the same way as in the 1552 case. All the reports come from Dutch and north German chronicles – i.e. areas where the Old Style ways of keeping the calendar were still in use. (See footnote to the 1552 storm, p. 38.)

## 26 February (Old Style)/8 March (New Style) 1625

*Area*

North Sea/German Bight.

*Observations*

An exceptionally great storm flood on this date is mentioned in several chronicles listed by Hennig (1904). Petersen and Rohde (1977) report a mark recording the highest level reached by this flood in Tönning on the Schleswig-Holstein coast at 3.90 m above the German sea level datum. The flood was highest in the afternoon. In places the effective height of the protective dykes was raised by the ice floes which the waters piled up, but elsewhere the flood overwhelmed the dykes, doing great damage. People, animals, and buildings were carried away, and sea-going ships were deposited in the midst of the cornlands.

Petersen and Rohde (1977, p. 104) further mention that the flood was driven by a storm which veered from W to NW.

According to Hennig, this sea flood also affected Friesland, Holland and France (but see below: it would be very unusual for a storm from the reported direction to bring a North Sea flood far enough south to affect the French coast). Gottschalk (1977, p. 157) agrees that the storm surge of 8 March 1625 was catastrophic in the Netherlands, a serious economic calamity because it flooded again land recently reclaimed by draining lakes and polders – a great loss of capital. The village of Petten in North Holland, which had suffered from earlier storm surges, was destroyed by this flood and rebuilt farther inland.

According to Gottschalk, this event did not affect places farther south than North Holland. Flanders and Dutch Zeeland were not affected.

After this flood the Dutch coast enjoyed a longer period of relative immunity from serious sea floods (Gottschalk, 1977).

## 11–12 October (Old Style)/21–22 October (New Style) 1634

*Area*

North Sea, especially the northeastern half, and Denmark and north Germany.

*Observations*

Hennig (1904) reports: 'Enormous storm flood of the North Sea, one of the severest which has ever occurred, in which 6000 people and 5000 cattle were lost and 1000 houses destroyed.'

Gottschalk (1977, pp. 121–2) reports: 'The North Sea coasts of Germany and Denmark were struck by a major storm surge on 21 October 1634, but this surge did not affect the Netherlands.' (Evidently it was more a WNW'ly storm, affecting the northern and central parts of the North Sea, not the South.)

Petersen and Rohde (1977, p. 40) report: 'A coloured mark on the north wall of the church at Klixbüll in North Friesland shows that the peak flood level was 4.30 m above normal sea level, compared with 4.16 m in the 1532 flood. (Minor errors may have been introduced by restoration of the marks several times since the event.) As the church stood already behind the protective dykes of the marsh-land, the storm flood may have been higher on the seaward side of the dykes.' And on pp. 44, 47: 'Peak water levels are indicated as +4.45 m and +4.61 or 4.68 m above normal German sea level in North Friesland and Tönning respectively.'

These observations suggest a weather pattern resembling the maps of the Danish 'storm of the century' on 23–25 November 1981 (q.v.). Judging by the exceptional height of the flood on the coasts of the German Bight, the 1634 event must have been centred slightly farther south than the 1981 storm. (The 1981 storm produced an exceptional flood level of 5.02 m above datum on the Danish coast at Ribe.) The intensities of the two storms seem to have been similar.

*Note.* Petersen and Rohde (1977) mention two other seventeenth century storms which produced very high flood levels in the German Bight: on 26 February (presumably 8 March New Style) 1625, when the water reached 3.9 m above the German normal sea level datum at Tönning, and 19–20 October 1663, when the water reached 4.74–4.81 m above datum at Hamburg. No further details of the situations accompanying these floods are known to the writer.

## November (Old Style) 1654 and February 1655

*Area*

Firth of Forth and eastern Scotland.

*Observations*

Great winds felled trees in Fife and blew ships out of harbour.

Further severe and protracted storms in the same area in February 1655 were followed by a long frost.

## 10 December (Old Style)/20 December (New Style) 1655

*Area*

East coast of Scotland, Fife and Lothians, Firth of Forth.

*Observations*

'All that day ... it did snow, but at night ther fell extraordinar mutch snow, and all that night ther blew a great wynde, which occasioned great losse and damage to the shyre of Fyfe, both by sea and land. As for the sea, it did flow far above ... its banks ... [There] were many small barkes and other vessels that perished, lying in harbours as in ... Dysart 18, Crail 30 ... Also peirs were dung downe in severall places, as in St Andraes, Eastor, Craill, Wemys, Leith.' Part of the stone granary at Leven was broken down and many ships in several places were borne over by the snow and destroyed. Some of the smaller houses were blown down and many trees blown over 'by the violence of this storme'; also 'several salt pans wronged (flooded by the sea?) both in Fyfe and Louthian (Lothian) side' (i.e. the north and south sides of the Firth of Forth).

From the distribution of damage it is reasonable to suppose that the storm wind blew from an easterly direction.

*Data source*

From the diary of John Lamont of Newton, the factor on an estate in the southeastern part of Fifeshire, near the coast of the Firth of Forth between Leven and Fife Ness. The diary is now held in the library of the University of St Andrews.

## 16 December (Old Style)/26 December (New Style) 1658

*Area*

East coast of Scotland, Fife and the Firth of Forth.

*Observations*

John Lamont of Newton, Fife, in his farm diary, reported that 'a great tempest of wind and rain' on this date followed a long period with much snow lying. There came a sea flood and people were drowned on the coasts of the Firth of Forth.

*Interpretation*

It is hard to judge the wind directions involved in this incident; possibly a gale from between NE and SE with temperatures no higher than +1 to 5 °C is likeliest, if general water levels in the North Sea had been raised by N'ly and NW'ly winds in latitudes between Iceland and Scotland.

## 18 February 1661 (Old Style)/28 February 1662 (New Style)

*Area*

England, especially the southern part, and at least the western part of the southern North Sea.

*Observations*

Defoe (1704) reports this as 'a very great storm of wind' which began 'in the early morning' and 'continued with unusual violence till almost night'. In London five or six people were killed by falling houses and chimneys. People were also killed in at least eight other places across the country from Hereford to Ipswich, by falling bricks, tiles and whole buildings, others by falling trees, and several in collapsing windmills. Churches were blown down in some places and three cathedrals were damaged. Steeples fell or were damaged at places between Berkshire, Bedfordshire and Ipswich – all in southeastern England.

The storm was accompanied by rain and hail, thunder and lightning.

Defoe remarks that the storm wind blew from the same quarter as in the great storm in 1703.

*Interpretation*

The storm was probably associated with disturbances on an active cold front, with very strong SW to W winds delaying the front's progress, but ultimately veering to NW and continuing strong for some time. The weather reported indicates the intrusion of thermally unstable cold air, presumably from the NW or N.

This storm, like that in 1703, was felt much less, if at all, in northern parts of Britain. Its effects were much less widespread than those of the 1703 storm, and no reports are mentioned of corresponding damage on the continent. This storm, though intense, seems to have been a much smaller system than that in 1703. (There may be some analogy with the storms that crossed England near latitude 53° N in 1972 and 1973.)

So far as the North Sea is concerned, there was a greater storm reported a month earlier on 16–17 January (New Style) 1662, when the coasts of north Friesland (Netherlands) and the German Bight, including Hamburg, were flooded. People were swept away and drowned by this flood, though it does not rank as one of the great North Sea flood disasters.

A Fifeshire diary kept by John Lamont of Newton near the Firth of Forth reports that the month of March 1661 (Old Style) – it is possible that this too refers to 1662 of the modern calendar – saw great winds and rains, culminating on 24 March (3 April, New Style) when flooding and torrents damaged mills and bridges and ruined crops.

The reporting of the storm in February of this year by Daniel Defoe, as remembered 40 years later, is valuable for what it tells us of the perceptions of intelligent observers around the times here concerned. The storm here described, and remembered so long afterwards, was plainly severe, causing many deaths and much damage across southern England along a belt whose axis was close to that of the much greater storm in 1703 (q.v.). The storm in February 1661/2 in southern England seems to have produced more rain, and hail and thunder besides, yet Defoe exaggerates in treating it as comparable with the 1703 case. He has reports of damage only from a narrower belt across England than in the latter case, and he mentions no reports of this storm from the continent, though it is hard to believe that nowhere in the nearer parts of the Netherlands and north Germany suffered damage. Although intense, this storm system was clearly of less horizontal extent than the great storm in 1703.

We may also notice that Defoe was already learning that there may be 'preferred' wind directions for severe gales. He was aware by 1703–4 that NW'ly or N'ly winds usually accompany storm surges and tidal floods of the North Sea.

*Note.* The storm which is the subject of this account should almost certainly be attributed to 1662 on the modern calendar.

## 2–4 December (Old Style)/12–14 December (New Style) 1662

*Area*

Firth of Forth and eastern Scotland.

*Observations*

A great storm destroyed the partly built new harbour at Dysart on the Fife coast and dispersed the heavy building materials. A large ship was blown out into the Firth and destroyed.

## 1663 (precise date or dates unknown)

*Area*

Northeast Scotland (Moray Firth).

*Observations*

Alexander (Lord) Brodie of Brodie Castle next to the Culbin estate records that 'the town of Nairn was in danger to be quite lost by the sand and water' in this year but gives no exact date. (The nearest sources of sand were the shoreline of the Moray Firth, running east and west immediately to the north of the town, and the lower reaches of the river Nairn just south of the town. The likeliest wind direction for this storm, from the most abundant source of sand, is therefore from some northerly point, especially as the language suggests a threatened sea flood besides.)

The only known North Sea storms in this year are:

(i) A note of a 'great wynde and raine' on 29 and 30 April 1663 (New Style) and on the same dates two ships from Newcastle wrecked, with the loss of 36 lives, on the coast of Fife near St Andrews (diary of John Lamont of Newton). Judging from the shape of the coast, this was probably a NE'ly storm.

(ii) The storm of 29–30 October 1663 (New Style) which produced a sea flood at Hamburg. This seems to have been a W to NW'ly storm.

If either of these storms gave rise to the anxiety at Nairn, it seems likely to have been the April one.

## 19–20 October (Old Style)/29–30 October (New Style) 1663

*Area*

North Sea/German Bight.

*Observations*

This flood is mentioned by Petersen and Rohde (1977, pp. 47–8) as having reached the unusually high level of flood marks at 4.74–4.81 m above the sea level datum at Hamburg. It is included in Hennig's (1904) compilation of weather events. It seems not to have affected the Dutch or English coasts to any great extent, since no references to it have been found.

*Interpretation*

A severe storm of winds from between W and NW must be presumed.

## 24 January 1666–7 (Old Style)/3 February (New Style) 1667

*Area*

Eastern Scotland, Edinburgh, Fife and the Firth of Forth.

*Observations*

John Lamont of Newton, Fife reported in his farm diary: 'A great speate of water at Largo' with thawing snow. 'Which night also, ther was a great tempest in divers places, as upon the water of Leven severall bridges incurred great damage.' At Wemyss a new timber bridge was wrecked. Several barns and two houses in Edinburgh had their roofs blown off.

*Interpretation*

There can be no clear diagnosis of the wind direction in this case, but from the locations of wind damage a S'ly seems likely: a gale from some point between SE and W can almost certainly be deduced.

## 13 October (Old Style)/23 October (New Style) 1669

*Area*

East coast of Scotland.

*Observations*

A great storm of wind, rain and thunder arose in the night and caused great losses both on land and sea. 'In divers harbours ... vessels were broken and clattered; as in Dundee, where they sustained, as some affirm, above ten thousand marks worth of losse; St. Andraes, Crail, Enster, Pittenwyme, Ferry, Wemyss, where a vessel of Kirkcaldie brake loose out of the harbour and spitted herself on the rocks.' In the Firth of Tay some of the islands used for grazing cattle were submerged by the sea and all the beasts were drowned. Trees were uprooted in many places.

*Interpretation*

This was probably an easterly or northeasterly gale, which raised the water level in the Firths of Forth and Tay. The sea flood seems to have been up to a metre or two above normal high tide level.

## Autumn 1676

*Area*

Northeast Scotland (Moray Firth).

*Observations*

The rich grain harvest on western farms of the Culbin estate, a few kilometres east of Nairn, near the southern shore of the Moray Firth, was waiting to be reaped when a NW'ly gale carrying clouds of sand before it covered the fields to a depth of over 2 feet (60–70 cm).

This was recorded by Alexander Lord Brodie in his diary covering the years from about 1650 to 1693. The source of the sand was the dunes at the coast, where the marram grass holding the sands had been eaten by sheep and other cattle.

A much greater storm disaster followed in 1694 (q.v.).

Another very violent windstorm in Easter Ross, north of the Moray Firth, in December 1674 (probably on 31 December, New Style) is mentioned by Sir George Mackenzy in *Philosophical Transactions of the Royal Society*, 1675, Vol. X (issue No. 114), p. 307, and this also may have shifted sand on the southern coast of the Firth near Nairn and Culbin.

## '1681' – almost certainly meaning the winter of 1681–2 (New Style)

*Area*

Scotland (Edinburgh).

*Observations*

R.C. Mossman notes in an Appendix to his great three part work on the *Meteorology of Edinburgh* (1898, p. 477) that in the winter of 1681 there were very severe storms of wind.

Mossman's note does not mention the direction of the storms, but there is synoptic evidence that they were predominantly W'ly – as were the winds in London that winter (90%). This is implied also by the report of a mild, humid winter in Denmark and a serious flood in Holland on 26 January 1682.

That the winter concerned was not 1680–1 is confirmed by indications that that winter was predominantly anticyclonic and cold in all three winter months in England.

## 20 October (Old Style)/30 October (New Style) 1688

*Area*

Southern North Sea, England and Holland.

*Observations*

Worsening weather had delayed the fleet loaded in Hellevetsluis to carry William of Orange and his supplies for England from 16 October (New Style). Brisk NW'ly winds turned to ENE, then a NW'ly storm on 25 and 26 October.

The fleet crept out into the North Sea in quieter weather on the 30th, but soon a SW'ly gale, which quickly veered to NW, sprang up and in a heavy sea the fleet returned to port, having to throw the horses overboard because of the wild sea. Violent cold winds and rain continued for about nine days while the fleet was re-stocked before sailing successfully to England with an E'ly wind, which carried William and his ships to the (English) Channel where they chose to land in Devon at Torbay, surprising the forces ready to welcome them in eastern England.

## Approximately 16–26 November 1689 (New Style)

*Area*

North Sea.

*Observations*

A fleet of 30 ships of the Danish–Norwegian navy, sent from Ribe (Jutland) to the aid of the English king in his wars in Ireland and against France, was beset by storms 16–19 November (New Style) and lay sheltering for some time in the area of Lister Deep (near the south tip of Norway). At first the storm had been a headwind (W to NW), but then the sea became really wild, apparently with N'ly weather. By the 22nd the now increased fleet had managed to cross to Hull, but 58 of the transport ships remained out at sea off the Yorkshire coast of northeast

England for some days and were obliged to throw a hundred horses overboard and some men who had died in the rough conditions on board.

## 10–11 September 1690 (uncertain whether Old Style or New Style date, considered probably Old Style)

*Area*

Skagerrak and at least northern Denmark and parts of the North Sea. Very stormy period in the North Sea generally.

*Observations*

Local records made by the priest at Skagen in northernmost Jutland record a 'gruesome' WSW'ly storm and a sea flood.

This was surely a continuing part of the stormy period recorded in the history of the Danish–Norwegian navy frigate *Lossen* on a crossing, which took 10 days, from Hull to Flekkerøy in southwest Norway in late August to September (probably Old Style) 1690. During the crossing, in which the *Lossen* was convoying 11 Norwegian and six English ships, they were in danger of being driven on land first on the Dutch and then on the English coast, although the voyage ended safely.

*Sources*

Danish local records, Skagen (quoted by P. Hauerbach) and Norwegian naval records, Norsk Sjøfartsmuseum, Oslo (article in 1976 yearbook).

## 1694. The Culbin Sands disaster in northeast Scotland

*Area*

No precise reports of the range of the storm or its date are known.

*Observations*

At some date, allegedly in the autumn of 1694, 16 fertile farms and farmland with a total area of 20–30 km² on the Culbin estate (57°38′ N 3°42′ W), near the north-facing coast of Moray Firth, close to the towns of Findhorn and Forres, were reportedly overwhelmed, in a single violent storm, by moving sand. The whole area and the buildings, including the mansion house, were buried, with depths of up to 30 m of loose sand.

Despite this emphasis on a single storm, it seems that trouble with drifting sand may have begun as early as the fifteenth century or even somewhat before. (There was a quite similar disaster at Forvie on the river Ythan estuary, on the eastern coast of Aberdeenshire, on or about 19 August (New Style) 1413, which buried an important district centre and township, covering it with sand-dunes (Plate 9 herein) which now stand up to 30 m high.) Conditions probably stabilized for some time afterwards, but later worsened again. There was another severe episode of blowing sand in the Culbin area on or about 21 April 1663 (New Style). Nairn, the neighbouring town to the west, was also threatened by drifting sand from its environs in 1663. And in the autumn of 1676 a NW'ly gale buried the harvest on the westernmost Culbin farms with up to two feet (60 cm) of sand. There was marram grass growing wild on the sand-dunes bordering the old coast at Culbin, helping to stabilize loose sand. And, as the marram grass provided the raw material for thatching and a variety of local crafts, it may have been overused, leaving the sandhills more vulnerable. Also, there may have been a series of violent storms, several of which shifted noteworthy amounts of sand, in the autumn of 1694 and for some years before and after. Nevertheless, this background

history should not blind us to the evidence that most of the damage, the end of the Culbin farmlands, was indeed brought about in the course of a single storm.

For ten years from 1694 to 1704 there were frequent periods of blowing sand, and there had been a history of severe coastal erosion at the mouth of the river Nairn, between 10 and 20 km west-south-west of the Culbin area, for well over 100 years before that (Edlin, 1976, and S.M. Ross of Forres, personal communication, 12 June 1975).

The sand and gravel carried by the Nairn and Findhorn rivers on either side of the Culbin estate comes from erosion of the glacial deposits (moraines etc.) beside the upper reaches of the rivers Nairn and Findhorn inland. There is also sand in the area derived from marine erosion of the soft sandstone coastline and cliffs of the Black Isle, across the Firth.

The area remained a desert of wandering sand for 230 years until its successful afforestation with pines by the Scottish Forestry Commission from the 1920s onwards.

Before the disaster, Culbin was shown on a seventeenth century map (see Edlin, 1976) as being on a peninsula between two bays. This prosperous area was at that time known as 'the Garden', or alternatively 'the Granary', of the county of Moray. The peninsula extended in an eastward curve around a much broadened, but presumably shallow estuary of the river Findhorn. And there was another bay to the northwest, beyond the present bar, possibly as an extension of the river Nairn's estuary. There would therefore have been great expanses of bare sand at low tide. The aftermath of the storm is described by Edlin (1976) as one of the greatest wind-borne deposits formed anywhere in Britain in recent geological time. Reports of the disaster tell that it came during the barley harvest, which in the cold summers of the 1690s was probably late: so sometime between late September and late October 1694 is the most probable date. In upland parts of Scotland, the largely failed harvests of that decade (including the year 1694) were cut very much later than that, but the fertile lowlands along the south shore of Moray Firth between Inverness and around Elgin are, and were, a more favoured district.

Edlin (1976) cites the following details from local traditions and original accounts of the event: 'At first only fields were invaded [by the sand]. A ploughman had to leave his plough, while reapers left their stooks of barley. When they returned, both plough and barley were buried for ever. The drift then advanced upon the village, engulfing cottages and the laird's mansion. The storm continued through the night, and next morning some of the cottars had to break through the backs of their houses to get out. On the second day of the storm, the people freed their cattle and fled with their belongings to safer ground. Their flight (southeastwards or eastwards) was obstructed by the river Findhorn: since its mouth had been blocked by the drifting sand, its waters rose until it could force a new passage to the sea.' This in the end resulted in sweeping away the old village of Findhorn on the east bank at the mouth of the river, the village having been abandoned in time by its inhabitants. When the population of Culbin returned after the storm no trace of their houses was to be seen.

In the course of the following two centuries the wind continued to shift the sand, and from time to time parts of a building or buildings would appear, only to disappear again, but there has been little further encroachment of the sand upon neighbouring farms to the south.

*Interpretation*

The above account supports the tradition that the disaster was essentially the work of a single storm.

Edlin supposes, doubtless rightly, that the prevailing nearly SW'ly or WSW'ly winds had over a long period built sand-dune ridges, in that orientation, along the coast behind the shore between the Nairn and Findhorn rivers. If then, perhaps after initially blowing from between SW and W, the violent storm wind came from the NW or N, the sand of the coastal dunes may have been brusquely carried over the farmland behind them along their whole length by the wind tearing at their slopes. It is reported by local tradition that the sea had been gaining on the land before the storm, so the windward – i.e. in this case also seaward – slope of the dunes may have become very steep, exposing the core of the loose sand of which the dunes were composed. The distance covered by the spreading sand lateral to the dunes was not very great, only 2–3 km (perhaps 4 km at one point). Even so, the wind must have been extremely strong – unequalled at that place in the time between 1694 and when the area was successfully afforested, 230 to 250 years later.

Other references consulted were Bain (1922), Dyke W.R.I. (1966) and Steers (1937).

Mr D.P. Willis of Fortrose Academy, Ross-shire, who has independently studied the Culbin Sands problem (Willis, 1986) adds 'Given that there may have been substantial dune alignment along the coast . . . the potential for sand movement may have been very considerable indeed if extremely strong winds did, in fact, come straight in to the Moray coast.' And he notes that the recorded rental of the barony of Culbin in 1693 confirms 'the undoubted productivity of the estate' up to that time and the magnitude of the 1694 disaster due to sand movement. He further writes 'there seems to be no evidence to suggest that the 1693 rental production was less than it had been in the past, nor . . . that the productivity of the estate had been diminishing due to environmental change – a fact that would surely have been commented on, bearing in mind the estate's reputation as the granary of Moray.'

That the disaster may have been due to the destabilizing of the dunes by excessive plucking of the marram grass for local industries is indicated by an Act of the Scottish parliament in 1695 (the year after the disaster) forbidding the pulling of marram grass for thatching.

Mr Willis's view on this is expressed in these words: 'Could the Culbin situation then [in 1694] have been one where deterioration of the bins (dunes) due to human intervention had indeed occurred, rendering them likely to blow on a broad front, given the right set of climatic circumstances?' It even seems possible that a modest amount of sand blown over the farmlands had come to be accepted as a common fact of life without being taken as a dire warning of the potential for disaster from a severe storm.

*Meteorology*

The storm, which seems to have continued for about 30 hours or even more, can be safely deduced to have been from NW to N. The most effective direction, as regards moving the sand, would have been at right angles to the coastal dunes, i.e. from about NNW. This is likely to have been the actual wind direction, since such extraordinary quantities of sand were moved across the 2–3 km-wide belt.

The wind must also have been exceptionally strong. The strongest gusts, coming in straight off the sea, may have been comparable with the strongest observed in storms in the Northern Isles and at sea: i.e. rivalling the other severest storms in this compilation, perhaps 100–130 knots (up to 50–65 m/sec). The mean wind speed may therefore, at a reasonable estimate, have reached 50–60 knots (25–30 m/sec) or rather more for some hours during the worst periods of this storm, especially during the night.

Since the precise date is not reported – astonishing to modern ideas as that may seem – no collecting of reports from a network of places for synoptic meteorological analysis is possible. However, since the wind responsible seems to have been so nearly N'ly, it is likely that some influence of such an outstanding storm would be felt all down the east side of Britain, including in the London area, where there are some series of daily weather observations extant. There seem to be three storm periods reported in London in the autumn of 1694 which might be connected with the Culbin Sands storm on the Moray Firth: 11–13 and (22–)23 October or 1–2 November 1694 (all dates converted to the modern calendar).

The London observations read as follows (all dates New Style):

> 9 and 10 October   Cloudy mornings, pleasant afternoons. Wind W

11 October   Frost early, then rain, with great winds.
Wind NW

12 October   Frost, then pleasant till rain at noon.
Wind N

and in the second period:

21 October   Mist early, cloudy later. Some rain.
Wind SW

22 October   Frost early, glorious day. Wind W

23 October   Frost, snow, cold and stormy. Wind N

and in the third period:

31 October   Mist early, cloudy then rain and snow.
Wind NW

1 November   Cloudy. Some rain and snow. Cold
wind. Wind NW

2 November   Tempestuous. Some rain and snow.
Wind NW

The last of these dates, which was followed by a long period of quieter, fair, but sometimes foggy, often frosty weather during the rest of November, may appear as the likeliest to have been involved in the Culbin Sands storm, since it alone among the three occasions listed coincided with the spring tide at Aberdeen (according to computations kindly supplied by Dr J. Vassie of the Tidal Institute at Bidston Observatory, Birkenhead (Institute of Oceanographic Sciences); personal communication, 12 December 1977).

*Synoptic interpretation*

The information available hardly permits a sketch map of the general weather situation. The N'ly windstream in all the cases listed that autumn presumably reached south along the whole extent of the coast of Britain and, with the conditions reported in London, it may have been continuous from much farther north in the Arctic. Since no report of any North Sea tidal surge flooding the low coasts of the North Sea is known from 1694, the situation seems to have been rather complex. It might be explained if contrary winds, from southerly or easterly points, dominated the eastern North Sea and its continental coasts.

There is one further point known about the background to the season in which this storm occurred. It was the time when the Arctic sea-ice was advancing most rapidly from the north, to surround Iceland completely by the end of the year.[1] Although the polar waters had long extended south, far beyond their present limit, the swift advance of the polar ice-pack between October and December 1694 must have required continual N'ly winds over the Norwegian Sea and the regions north of Iceland. And it can be presumed that these winds were often notably strong.

If 1–2 November was the real date of the storm, it had been preceded by an unbroken 10-day period of N and NW winds in London with frequent frost, snow and sleet. The likeliest synoptic wind and weather pattern to explain the

development seems to be that a cyclonic centre was stationary over, or near, the central North Sea during most of the period. The fronts of any warmer air steered from the southeast by such a centre were probably unable to progress farther west, being blocked by the extremely cold Arctic airstream from the north. This windstream may well have been intensified as some secondary system coming from the southeast caused the warmer air to make its greatest advance towards the western part of the northern North Sea.

*Addendum*

After the foregoing account and suggested analysis were written, a surprising wealth of observation reports from near, or bearing upon, the regions of concern to the study of this storm and covering the whole month of October 1694 on the Old Style calendar (11 October to 11 November New Style), was received. The reports were from the logs of ships of the Danish–Norwegian navy of that time from the archives of the Danish Meteorological Institute, made available through the kind assistance of Dr Knud Frydendahl of the Climatological Division: these included:

(1) the frigate S/S *St Johannes* near the south coast of Norway and sometimes in Kristiansand
(2) the warship S/S *Lindormen* in or near Kristiansand
(3) the warship S/S *Gyldenløve* visiting Portsmouth and London
(4) the S/S *Flyvende Fisk* in Danish home waters, sometimes in the Great Belt and sometimes near Copenhagen
(5) a frigate S/S *Packan*, probably farther south but whose positions it unfortunately seems impossible to resolve. The log sheets are variously headed 'Oosterend' (Ostend?), 'St. Martin Pas de Calais', and 'Portland Bill', but the latitude and longitude positions seem garbled, some suggesting the middle of the northern North Sea but also a position south of Brittany near Nantes has been suggested by some investigators of the log.

The reports indicate that a very severe storm – force 11 is suggested – reached the position of the S/S *Packan* about sunset on the 29th and there was some rain that night. The storm wind veered NW during the 30th under overcast skies and blew at an estimated force 11 throughout the following night and all day on the 31st, but there was no further report of rain.

It seems impossible to reconcile the sequence of weather reported by S/S *Packan*'s log with a position near Brittany. The ship was probably even farther south, on the Biscay coast of France, at or near La Rochelle. This confusion was caused by the number of places along the coasts of France called St Martin,[2] but the probable solution has emerged in a letter from the Captain dated 4 November 1694 (Old

---

[1] This should be a prime example of the development of a large scale abnormality predisposing the atmosphere and ocean surface to favour singular events as suggested by Namias and referred to in this compilation under 15 December 1986.

[2] The only description of the ship's position on a number of days.

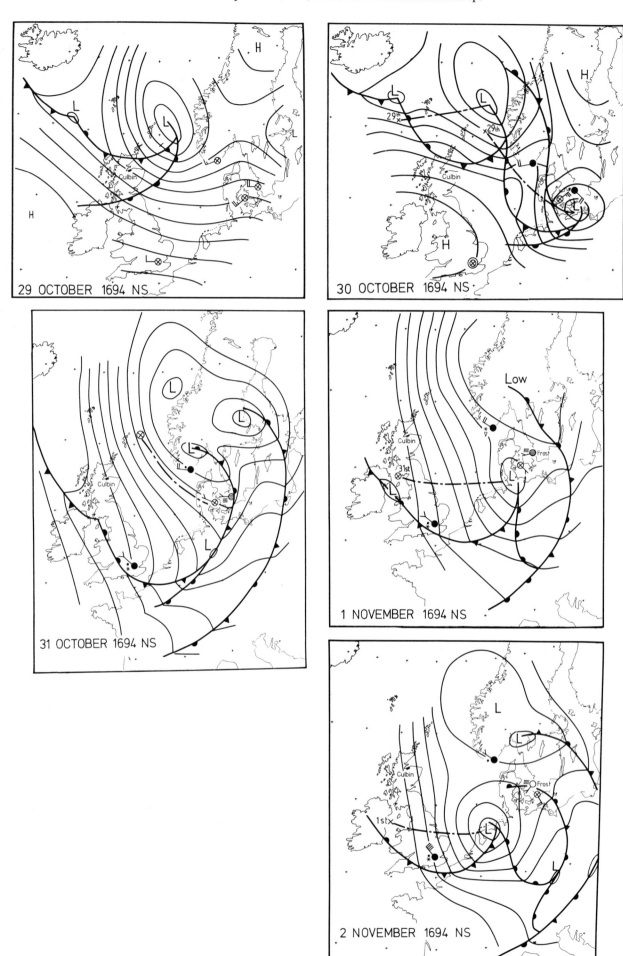

29 OCTOBER 1694 NS

30 OCTOBER 1694 NS

31 OCTOBER 1694 NS

1 NOVEMBER 1694 NS

2 NOVEMBER 1694 NS

Style), recently discovered by K. Frydendahl among the papers concerned with the voyage.

The other Danish ships, reporting in Danish and Norwegian home-waters, supply observations of winds and weather that fit the proposed analysis very well. They also confirm the suggestion that no severe N'ly or NW'ly storm lasting more than 4 hours without a break was experienced at the same time on the Danish side of the North Sea. The winds were sometimes S'ly there.

The approximate course of the storm development over these dates is illustrated here by once daily miniature maps referring to the midday or afternoon period of each day from 29 October to 2 November 1694 on the New Style calendar.

These Danish ships' logs show that the light to moderate winds in the days preceding the storm had been mainly from points between NW and about NNE or NE on 13 of the preceding 18 days. Winds from between NW and about ENE were reported on 21 of the 31 days of that month of October on the old calendar (11 October to 10 November 1694 New Style). The wind pattern over London was somewhat similar, with winds between W and N, but on the whole stronger than on the Danish side. It is a reasonable deduction that there were usually anticyclones dominating either the Atlantic between about 50° N and Iceland or the Norwegian–Greenland Sea or north Norway–Finland or all these regions.

In amplification of the data on the maps, the following significant changes reported later in the days mentioned should be noted:

| | |
|---|---|
| 30 October | Wind at Kristiansand, south Norway NNW force 4 in the evening, with rain. Wind in the Great Belt, Denmark became NW force 4 in the evening. |
| 31 October | Colder air reached Kristiansand in the afternoon and evening with snow and hail showers. Copenhagen had frost and fog with the light wind in the evening. |
| 1 November | The light wind in the Great Belt, Denmark turned gradually to SW in the evening and night. |
| 2 November | The light wind at Kristiansand turned to SE by S in the afternoon and then SSE force 10 in the following night, presumably indicating the depression shown over the southern North Sea coming nearer. Winds over Denmark on the 3rd were moderate to fresh SE to ESE. |

The features of the observation reports here shown are believed strongly enough marked to indicate the sequence reliably in its broad features, whatever uncertainties remain about the ability of the seafarers of that time to follow what must have been still an embryonic system of estimating wind force and weather reporting.

It may be that this analysis at last comes near to settling the date of the main Culbin Sands disaster and at least in its broad framework the nature of the meteorological situation that produced it.

The critical features of the occurrence seem to have been:

(1) a great NW'ly or NNW'ly storm of exceptional strength and duration, lasting perhaps 36 hours, and preceded by

(2) about 24 hours of a very strong W'ly storm wind with already rough seas;

(3) the NW to N wind blew over a very long fetch of open sea from the Arctic; and

(4) the sequence coincided with spring tides which probably caused the wild seas to tear open the face of the coastal sand-dunes, thus exposing much fresh dry sand to wind scour. An exceptional extent of other sand, moved by wave action, was probably also newly exposed to the wind at low water.

## 12 September (Old Style)/22 September (New Style) 1695

*Area*

Southern North Sea and Channel, Belgium, northeast France, southeastern areas and coasts of England.

*Observations*

Neumann (1981), extracting data from the special article in *Theatrum Europaeum* (1702) on the exceptional weather of the year 1695, reports:

'On September 12, at night, a violent storm sprang up along the English coast. Many ships were torn from their anchor and driven to the west coast of the English Channel. The Dutch warships that sailed from Duyns (the Downs, off Deal, Kent) at the beginning of the previous week were forced to return. It is reported in letters from Yarmouth [apparently Great Yarmouth] that out of a fleet of 200 coal ships 70 were dashed on the beaches; in 50 of these ships all the people perished. However, the king's ships *Mitford* and *Assistance* remained unhurt in port. This storm was so severe in the London area that trees were felled in St. James's Park and a market ship overturned in the Thames. Further, 13 ships were stranded between Dunkirk and Boulogne. One of them was a Portuguese ship named *Maria*. Only 3 people could be saved from it.'

The logs of HMS *Mitford* and *Assistance* in port at Sheerness in the Thames show that they experienced winds from between NNW and N by E on the 12th. The winds had blown mostly from about N since the 9th and continued in that quarter until the 16th, when they became E'ly (Old Style dates).

Part of this description tallies so closely with Daniel Defoe's account (1724) of a storm, which he no longer remembered the exact date of, that it seems safe to conclude he was writing about the same storm and gave some further details, from the east coast of England:

'About the year 1692, (I think it was that year) there was a melancholy example of what I have said of this place [Winterton Ness, Norfolk]: a fleet of 200 sail of light colliers (so they call the ships bound northward empty to fetch coals from Newcastle to London) went out of Yarmouth Roads with a fair wind [presumably from a southerly point], to pursue their voyage, and were taken short with a storm of wind at NE after they were past Winterton Ness a few leagues. Some of them, whose masters were more wary than the rest . . . tack'd, and put back in time, and got safe into the roads. But the rest pushing on, in hopes to keep out to sea, and weather it, were by the violence of the storm driven back when they were too far embayed to weather Winterton Ness . . . and so were forced to run west, every one shifting for themsleves as best they could. Some run away for Lyn Deeps but few of them, (the night being so dark) cou'd find their way in there. Some but very few rid it out, at a distance. The rest, being above 140 sail were all driven on shore, and dash'd to pieces, and very few of the people on board were sav'd. At the very same unhappy juncture, a fleet of loaden ships were coming from the north, and being just crossing the same bay [the Wash?], were forcibly driven into it, and not able to weather the Ness. Also some coasting vessels loaden with corn from Lyn, and Wells, and bound for Holland, were with the same unhappy luck just come out to begin their voyage, and some of them lay at anchor. These also met with the same misfortune, so that in the whole, above 200 sail of ships, and above a thousand people perished in the disaster of that one miserable night . . .'

There are no surge reports on the coasts of the continent anywhere farther north than Belgium. Petersen and Rohde make no mention of any flood on the coasts of Schleswig-Holstein or in the Elbe. It seems almost certain that the winds at the time were easterly there. It is clear from the shipping movements in and out of the port of Dundee in eastern Scotland (56.5° N) that the influence of this storm did not affect areas so far north.

*Inference*

This seems to have been a rather fast travelling intense storm of relatively small diameter, presumably developing from an open wave on the main polar front and steered east or east-north-eastwards in or near the latitude of the English Channel (near 50° N) by a long run of strong W or WSW winds from the Atlantic, although ahead of the storm centre a S'ly breeze seems to have developed for a time as far north as 52° to 53° N (Yarmouth and the east coast of southern England), while a northern high pressure system blocked the storm cyclone's progress farther north and brought cold air from the N and NE. It may well have been in the wake of this storm that 15 cm of snow fell in Bohemia between the 25 and 29 September (New Style).

The summer of 1695 was disastrously cold in the northern countries, including Scotland. The Arctic sea-ice surrounded Iceland all year, impeding shipping. The polar water seems to have extended south to the Faeroe Islands and to near Shetland as well as to the whole coast of Norway (Lamb, 1979). All summer, and indeed throughout

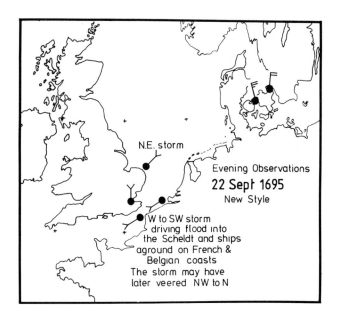

the preceding winter and spring, polar anticyclones seem to have dominated the Greenland–Iceland–Norwegian Sea region.

The implied long prevalence of easterly and northerly winds over most of the North Sea and the coast of England may mean that the Dutch fleet which sailed forth from the coast of Kent the previous week had been continuously held back in that area. The winds reported by ships HMS *Mitford* and *Assistance* between Suffolk and the Thames were continually between NW and NE to E from the 7th or earlier to the 19 September (Old Style) apart from apparent brief interruptions in the mornings of the 9th and 12th. The ships in Danish waters reported winds always between NW and about ENE at least from the 7th to the 15th (Old Style). The N'ly winds in these areas were reported as gentle to moderate apart from the storm near southeast England and the eastern Channel late on the 12th. So any surge of water into the Scheldt could not be a North Sea surge and must have come up the Channel from the west.

Gottschalk (1977, pp. 399, 402) reports that 'at an unknown date in 1695 Belgium was struck by a rather serious storm surge. The dikes gave way at Ostend on the North Sea Coast, causing flooding of a large area with numerous villages south of this place. The flood waters rose to the 5 metres contour.

'Relatively little damage seems to have been caused east of Ostend. In Antwerp, on the other hand, a contemporary (witness) speaks of the Scheldt being pounded up to a greater height by a gale . . . than during the All Saints Flood of 1570 . . .'

Since the *Theatrum Europaeum* report indicates that ships were driven onto the eastern shore of the Channel near Dunkirk and Boulogne, the winds must have been W'ly there. So these reports may well relate to the same storm, with a strong westerly gale driving water against the French and Belgian coasts and as far as into the Scheldt, during the later half of the 12 September (Old Style) when strong NE winds were raging farther north on the coasts of England. Gottschalk notes that a polder on the west coast north of Antwerp must have been breached by the sea flood, but concludes that this storm surge had little effect on the coast of Holland.

## 21–22 September (Old Style) 1–2 October (New Style) 1697

*Area*

North Sea, especially the northern and northeastern parts, probably also Hebrides and northern parts of the British Isles.

*Observations*

Reports of a storm flood reaching notably high levels on 1–2 October 1697 (New Style) on the coasts around the German Bight from Denmark to north Holland come from Gottschalk (1977), Petersen and Rohde (1977) and Weikinn (1961). Sources quoted by Weikinn refer both to a strong storm on 1st October from SW driving more water into the German Bight and the Elbe and then to a strong NW'ly storm which brought the water over the dykes between the Weser and the Elbe.

Gottschalk reports that the highest level of the flood at Petten (52°46′ N 4°39′ E) in north Holland was 3 m above normal high tide, 2.5 m at Den Helder, and 2.4 m off the island of Wieringen. The northern third of the island of Texel was flooded. These levels in Holland were well below the 1570 and 1953 floods. Petersen and Rohde give the highest tide level in this flood at Hamburg as 4.86 or 4.93 m above the German sea level datum. This is higher than in 1634 or 1717 and was certainly the highest flood in the 222 years between 1570 and 1792.

Little or nothing is said in these sources about the weather situation which produced the high water level. It seems to have the characteristics of a flood produced by storm winds from about WNW over the North Sea, since Gottschalk reports that there is no trace of this storm surge in the southwest Netherlands.

There seems to have been no interruption to the movements of short-haul local shipping in the Firths of Forth and Tay or along the east coast of Scotland between latitudes 56° and 56°35′ N at this date as noted in the records of the port of Dundee. But these would be relatively sheltered coastal areas in a generally strong NW'ly wind situation.

It may be significant that a very great gale 'in the autumn of 1697' is reported to have buried in sand a site in the Outer Hebrides – Udal on the island of North Uist (57½° N, 7½° W) – that shows archaeological evidence of more or less continuous habitation for almost 4000 years until that time. Trouble with drifting sand, which has left great bands of the nearly white, loose sand from the shore burying the earlier cultural levels, seems to have begun about AD 1400, presumably due to increasing storminess but probably also through the greater tidal range about that time than for many hundreds of years before that or at any time since. There was a return to more stable conditions around 1500, but by 1542 the storms were increasing again. And the depth of sand that covered the place in the gale of autumn 1697 caused the site to be abandoned (Crawford, 1967; Crawford and Switsur, 1977; Morrison, 1967–8). The sand depth on the site reached 6 m.

The storminess in the northern North Sea in late September 1697 on the modern calendar caused a ship to strand on the south coast of Norway in Aust or Vest Agder on 22 September.

*Meteorology and inferences*

There is no complete certainty that the Hebridean (Udal) storm was the same storm that went on to cause the highest North Sea tidal surge of the seventeenth century in the German Bight on 1–2 October 1697 (New Style), but this seems rather likely to have been the case. If so, it tells us something about the meteorological situation, suggesting a cyclonic system which having produced an unusually severe gale from a westerly or northwesterly direction in North Uist in the Hebrides, travelled (or developed) southeastwards towards southern Norway and Denmark (see map).

The probability that it was the same storm which the Hebrides experienced in a severe form is heightened by the discovery that a weather diary kept for 12 months, daily from 1 March 1697 to 28 February 1698 (Old Style), by one Thomas Evans, vicar of Llanberis, among the mountains of North Wales, reported stormy weather on the 28 and 30 September (New Style), with hail showers on the 30th, followed by snow on the mountains on the 1 and 2 October. There was hail again on the 3rd. (All these dates have been converted to the modern New Style calendar.)

The 1 and 2 October were rainy in the valley at Llanberis. The ocean surface near southeast Iceland was probably still almost 5 °C colder than in the present century, as it is thought to have been already for about 20 years (Lamb, 1979). Nevertheless, snow at such an early date covering the mountains of North Wales implies an extremely strong northerly windstream bringing air rapidly south from the Arctic. This would have passed over Scotland and the Hebrides on its way to North Wales. And it was, no doubt, this N'ly gale that produced the great surge of water in the North Sea which seems to have produced the highest water levels on the German and Danish coast since 1570. There was, moreover, no other such cold outbreak reported in the Llanberis record that year and no more snow until 16 December (New Style).

Series of daily weather observations are available for about three different sites in and around London in 1697, mostly without instrument measurements. As in North Wales, the summer had been windy, often wet, and sometimes cold, since late July of the modern calendar. But in London there were some warm spells: 22–25 July and 6–19 September (New Style). Then it turned to being frequently rainy (20–21, 25 and 28 September to 1 October) with strong winds: the 23rd and 24th described as 'stormy', 25th–27th 'cold with great winds'; 28th 'much rain and great winds', on 1 October 'much rain and great wind' (all these dates New Style). Comparison with North Wales becomes closer when we notice frosts (evidently at the first observation of the day) mentioned in the London records on 22, 26, 27 and 30 September and on 2, 3, 4 and 5 October (New Style). Rain and hail fell on 3 October. The winds in London had been generally from the W in late September, but were reported as NW on 26th–27th and SW on 28th, then W again till 3 October, but NE on 4 and 5 October. (The site concerned in London is generally supposed to have been underexposed to NW winds, which seem consistently under-reported: so the wind, reported as W, may, on some of the dates concerned, really have been NW.) It is evident that London, like North Wales, experienced the remarkable cold outbreak at the end of September and was affected indirectly, when not directly, by the storm in the North Sea. It is also clear that after 19 September there was cyclonic activity affecting the south of England, which had more frequent rain and was less cold than North Wales, where the 29 September in particular was cold and dry in the first part and cold and wet in the afternoon.

The general weather situation becomes clear enough when the reported observations are put together on the map. The winds on 1 and 2 October seem to have been mostly from between NW and N over the western parts of the British Isles, W'ly in the Channel and over the south of England, where a small secondary depression may have passed, about WNW over the central North Sea and the German Bight, and probably from between W and SW near southernmost Norway, where a cyclonic centre may have passed close. The data mapped suggest that the centre probably passed over southern Denmark on the 1st and then became slow moving, with mostly light winds near the centre over the following days.

But, without data from a wider area to permit a more complete synoptic analysis, there is nothing to give any close indication of the wind strengths ruling. It seems likely, but cannot be proved, that the feature which brought the sharply colder air to North Wales on 1

October, marking an intensification of the northerly wind-stream, passed close to the Hebridean site mentioned earlier on that day or late on 30 September and reached the eastern North Sea within 24 hours. That is, a progress of roughly 600 nautical miles in at most 24 hours, an average rate of advance of the storm of possibly no more than 25 knots. The surface winds which produced the high tidal surge in Hamburg, however, must have been of exceptional strength, as must those that produced the coastal changes in the Hebrides, and it can be taken as certain that they were in cold post-frontal air. This airstream is also presumed to have travelled very fast from the Arctic to produce the snow in North Wales. These observations mean that this was a mature cyclonic storm system, itself moving at a rather modest speed of this order, and not an open frontal wave depression.

There is no statistical basis here for calculating the probable strength of the west to northwesterly jetstream associated with this mature storm cyclone, such as it was possible to use in the case of some of the Spanish Armada storms in 1588 which could be identified as open wave depressions. This storm in 1697 was probably more strongly developed than any of the mature cyclones in 1588.

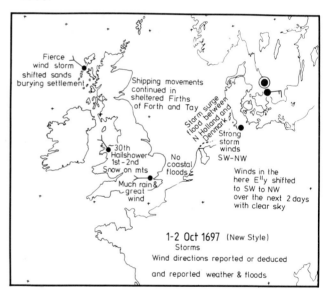

17 December (Old Style)/27 December (New Style) 1698

*Area*

North Sea, Skagerrak and southern Norway coast.

*Observations*

A great storm (probably S'ly or SE'ly) carried off the sails of a Scottish ship bound from Riga to Montrose and wrecked many other ships on the coast of south Norway, while others lost their masts etc.

*Meteorology*

This was a very mild wet winter with a great prevalence of SW'ly winds in January across England to Denmark and most of the North Sea, the winds apparently more S'ly near Norway and in the Baltic.

## 18 January 1700–1 (Old Style)/29 January 1701 (New Style)

*Area*

Southern England, Channel and southern North Sea.

*Observations*

Southerly gale caused many shipwrecks and great loss of life at sea, damaged buildings and brought down many trees inland (report in Lowestoft diary, Revd Mr Say).

Earlier in the month there had been severe frost, continuing since late December, until by 19th January (New Style) ice was 50 cm thick. The remainder of the month after the 19th was changeable with rain and fine days and more snow and frost. The 28th as well as the 29th was windy with much rain and flooding (Lowestoft report).

## 22 October 1702 (New Style)

*Area*

Northern Scotland, Moray Firth and North Sea coast

*Observation*

A storm, apparently from about WSW, but the wind perhaps again shifting later to NW, caused further severe drift of the loose sands that had covered the Culbin estates in 1694 (see the account of the 1694 storm), extending the area devastated until the River Findhorn was forced to change course to the east, to reach its present mouth.

In this storm the old town of Findhorn at the mouth of the river, was destroyed by the sea. (This suggests battering of the outer coast at some stage by a heavy north-westerly or northerly sea.)

## 26–27 November (Old Style)/7–8 December (New Style) 1703

*Area*

This storm's passage was marked by a 300 nautical miles (about 500 km) wide belt of exceptional destruction across southern England and Wales, the southern North Sea, Netherlands, north Germany, Denmark and, according to Defoe (1704) writing soon after the event, also parts of France, Sweden, 'The Baltick Sea . . . Finland, Muscovy and part of Tartary . . .' until 'at last it must lose itself in the vast Northern Ocean where Man never came and Ship never sailed'.* The map indicates this great swathe of destruction.

Hennig (1904), with less attempt at precision, reports this storm as exceptionally severe 'over practically all Europe'.

*Observations*

Many have suggested that this was the severest storm of which we have any good account. In fact, many details are known.

Short (1749, Vol. 1, p. 428) reports that 'England lost more ships in this storm than ever were lost in any encounter with an enemy.' One estimate at the time reckoned that 'on the English coasts 1500 men perished'. (This estimate is from the *Lincoln Date Book*, published as late as 1866, which was kindly supplied by the Grimsby Central Library.) Other reports specified that a fleet of 160 to 200 ships of the English navy, with their victuallers and supply ships sheltering 'in the Downes', off Deal on the Kent coast, was scattered by the storm and one-third of all the seamen in the navy were lost in the night of 7–8 December (New Style), some 10 000 men in all. Thirteen navy ships from Portsmouth were scattered and destroyed.

Buildings were blown down in at least 35 different places in southern Britain, including the Eddystone lighthouse, many windmills and parts of a city wall. A sea wall along the Severn was destroyed by the battering of wind and water. A church was unroofed in southern England and three cathedrals damaged.

In the Netherlands, Utrecht cathedral was severely damaged and partly blown down. Many houses and churches, and some mills, were badly damaged in Jutland and on the Danish islands of Fyn and Sjaelland. Many ships ran aground on the Jutland coast and at Ålborg in northern Jutland few houses were left undamaged. In north Germany the biggest churches in Wismar, Rostock and Stralsund had their spires blown down.

There was evidence of whirlwinds or tornadoes in a few places – probably near a cold front, since they were far from the centre of lowest pressure. At Whitstable, Kent a ship was lifted out of the water and deposited 250 metres from the water's edge, on rising ground. And in the same part of Kent a cow was lifted into the uppermost branches of a tree.

Whirlwinds were observed at Delft in Holland some time in the night of 7–8th, described as a night of 'dreadful storm', and at 8–8.30 a.m. the observer at Delft noted 'the barometer never seen so low'. At Utrecht, too, there is a suggestion of whirlwinds in that the storm crusted the windows on the northeast as well as the southwest side of the buildings with salt carried far inland from the sea.

There was a sea flood in the river Avon and Severn estuaries, evidently extra water flooding into the Bristol Channel on the SW'ly storm winds. The tide in the Avon at Bristol on the 8th was reported as reaching 8 feet (about 2.5 m) higher than ever remembered.

There was a very high tide in the Thames two days later, when strong NW'ly winds were blowing over the North Sea. But there the tide was not so excessive.

---

*This colourful writing (perhaps not surprising for its time) displays ignorance of the exploits of Ottar (or Othere) with a fleet which explored far to the north from north Norway between AD 870 and 880 and which he himself reported to King Alfred in England.

Ships were blown from their moorings, and many of them wrecked as far apart as England's south coast, both sides of the Bristol Channel, and at Grimsby and all about the river Humber and Spurn Head. Fifty of Grimsby's hundred ships were lost.

There had been strong winds for a good many days before. Defoe (1704) says 'It had been blowing exceeding hard . . . for about fourteen days past.' Brooks (1954) says they were generally westerly winds, but there seems no warrant for that statement. Our maps give a better perspective in this regard. Certainly, it is noticeable in the port records at Dundee on Scotland's coast, as at the southern ports, that shipping movements had for some time come more or less to a standstill, as vessels had sought shelter. But the winds were, of course, not entirely persistent and the directions varied. When the great storm came, 'the ports were full of ships which had come in for refuge from the previous gales' (Defoe, 1704, p. 65).

According to Brooks (1954), the north winds behind the storm depression were also of unusual violence. He opined that there were no well-marked fronts in the storm cyclone: like true hurricanes, 'it had a rather calm centre' and the wind direction veered gradually. These remarks are, however, not altogether supported by the sequence of observations analysed in this study. Defoe, quoting Derham's account from his observations at Upminster, east of London, imagined the wind to have blown from about SW by S, or nearer S, in the beginning and to have veered to W, or at least WSW, towards the end of the storm. The night was too dark to see, and people were afraid to go out of doors. Innumerable trees fell too.

There was little mention of rain. Derham noted only that 'at four [in the afternoon of the 8th] it blew an extreme Storm, with sudden gusts as violent as any time [in the preceding night] . . . it came with a great black cloud and some Thunder, it brought a hasty shower of Rain, which alayed the Storm so that in a quarter of an hour it went off and only continued blowing as before' [presumably as before 'the extreme Storm'].

The duration of the most violent stage of the storm was reported as 5–6 hours at Helston, Cornwall and about 4–5 hours (at successively later times) in London, Amsterdam, and Ålborg in Denmark. A contemporary report, kindly supplied by the Danish Meteorological Institute, stressed that the storm and the damage caused were gruesome in southern Jutland and on the Danish islands, less serious in Ålborg diocese (i.e. over northern Jutland) although few houses escaped damage even in Ålborg. There were many shipwrecks on the Danish coasts.

*Meteorology: instruments and scales*

We know something of the barometric pressures ruling from correspondence about the storm published in a letter from 'The Revd Mr William Derham FRS' in the *Philosophical Transactions of the Royal Society* (1704, Vol. 24, pp. 1530–5) and from the observations made at the Observatory of Paris (kindly supplied in manuscript for this report). Derham compared his three or more times daily observations at Upminster, Essex (near London) with those

made by his friend Richard Towneley, Esq. at Towneley, Lancashire in northern England between 6 and 10 December (New Style). The barograms reconstructed from these atmospheric pressure measurements are depicted in the diagram as pressures reduced to sea level. Barometric corrections to sea level have been derived from such general information about height of the observation point above sea level as could be gained from topographic maps and the probable height of buildings, together with reported temperatures, assuming unheated rooms. The results of such estimation have been compared in the case of the Paris Observatory with what is known from fuller information about the site towards the end of the eighteenth century. Although the barometer corrections and adjustment to sea level cannot be certain to within 2 or 3 millibars (mb), the changes with time which each of these keenly scientific observers noted are doubtless trustworthy. All three register the passage of a deep low pressure system. Of these three places, London experienced the sharpest pressure changes, although the depression centre which produced the main storm seems to have passed no nearer than 100 nautical miles (185 km) farther north. Derham's readings indicate that pressure near London fell 21–27 mb in the last 12 hours of the fall and rose quickly afterwards: over 12.7 mb in the first three hours. The rise continued until the pressure was 56–60 mb higher 48 hours later. The central pressure of the cyclone must have been below 960 mb: both Brooks' * and my analyses have suggested independently a central value around 950 mb over the English midlands.

Thomas Short (1749) noted that the storm was 'followed by a length of dry weather', a 'happy (circumstance) for all those whose roofs had been stripped'. This confirms the indications of a much higher pressure situation following the storm, although the maps analysed indicate that the winds continued W'ly over England at least until 14 December. Defoe says the winds continued strong till afternoon on the 12th.

*Meteorology: synoptic analysis*

The synoptic meteorological analysis here presented (pp. 65–9) is based on a series of once daily weather maps for 12–15 h each day from 1–14 December. Much additional information for other hours of the day was entered on the original maps. Daily observation reports from London, Lancashire, and Paris were supplemented by reports from many other places at the height of the storm on the 8th. Ships' observation reports kindly made available from the Danish, Dutch, and English navies have added valuable coverage to the maps, extending at times to the northern North Sea and the western part of the English Channel. Regular reports from ships in or near harbour in Ports-

---

*Brooks (1954, p. 30) only ever published a single sketch map of the weather situation on 8 December 1703 with the storm cyclone (about 950 mb) shown centred over Nottingham (near 53° N 1° W) at 4h on the 8th, as an intense secondary to an older Low with a central pressure below 950 mb near 61° N 1° E somewhat below 950 mb. He never indicated how much information he had been able to use to arrive at this solution.

mouth, Grimsby, Amsterdam and Copenhagen completed the coverage obtained.

This is the earliest case for which it has been possible to gather together a comparable network of reports. The plotting conventions and symbols used for presenting the meteorological observation data on the synoptic maps in this study may need explanation for some readers. A brief appendix following this report of the great storm in 1703 is devoted to this (pp. 70–2).

The lack of a standardized hour of observation must widen the error margins affecting the barometric pressure and temperature values used on the maps and the applicability of wind directions and strengths reported, especially at times when rapid changes were going on. Thus, on the map for 8 December, during the quick rise of pressure at the rear of the storm it had to be assumed that the pressure reported at Towneley, Lancashire in northern England was as much as 4 mb too high for the time of the map. The ships' observation reports were made at more regular times than some of the reports by diarists and others inland. Weather reports were written for each (generally 4-hour) watch, but tended to be written at the end of the watch and the times at which the events reported occurred had a characteristic margin of uncertainty which probably averaged almost two hours.

*1 December*

The series opens with a small depression centred over the south of England, near the coast and apparently moving northeast. The temperatures reported round the depression indicate no strong airmass differences, but the wind directions suggest quite sharp frontal shifts. The winds experienced by HMS *Antelope* at Portsmouth, closest to the low pressure centre, were squally and shifted round the compass during the day. Ships at Plymouth and near Lands End also reported squally, mostly southwesterly winds, but at the Dutch ship S/S *Schieland*'s position the wind veered to NW during the afternoon. Rain was reported everywhere near the active frontal zones, heavy rain at the nearest point to the cyclone centre.

Another Dutch ship, S/S *Callenburgh*, coming into the North Sea from the Skagerrak, reported a strong wind from about SW, but improving conditions. This is interpreted as an easing of the wind and sea as the vessel left behind the narrower waters of the Skagerrak (which acts as a funnel for SW winds). There may also have been some slackening of the pressure gradient as a small ridge approached from the south.

*2 December*

The small depression has turned north since yesterday, as another small, but quite deep, depression advanced about 500 sea miles from the southwest to a position over southern England, giving further rain. The winds backed to S ahead of the new Low, and were strengthening from the S over the North Sea generally as the low pressure systems seem to have coalesced over the British Isles. S/S *Schieland*, now sailing east up the Channel, reported a wind veer to WSW, confirming the passage of a front, by the time she passed Dover in the evening. Skies were clearer everywhere west of the cold front concerned.

At the same time the high pressure system seen over south Sweden on the 1st has moved south and the winds reported near Copenhagen changed from E to W'ly by the morning of the 2nd.

*3 December*

Another small Low has moved some 600 sea miles from the southsouthwest, from near Spain, to the southernmost North Sea. Its progress is marked by wind changes, but no excessive strength, and little rain was reported. Farther north in the North Sea S/S *Callenburgh* reports almost calm conditions, apparently in the little ridge of high pressure dividing this system from its forerunner of yesterday, the fronts of which have passed to the north while the low pressure centres have coalesced somewhere in the region of the northern British Isles. The depth of this complex Low and its exact position cannot be determined at this stage.

A squally situation with a sharp veer of wind from S to NW near the middle of the south coast of England (Dorset) cannot be confidently interpreted, but is treated as an active trough south of the main low pressure centre. With the rather heavy rain reported near it, it is taken as an indication of considerable instability of the airmass in that area – evidently originally cold air, although the surface temperatures were not low for December.

*4 December*

This map indicates a continuance of the same sequence of developments, with yet another small frontal wave depression apparently having passed (a bit more quickly, some 700 sea miles in 24 hours or about 30 knots) nearly northwards, from Spain or southern France to the North Sea. Winds backed to SE or E before it, became S'ly, and then veered to SW as it passed. The general south to north movement seems to have produced a small northward shift of the main British Isles Low (depth still indeterminate) and represented a generally S'ly wind situation over the North Sea.

The strongest winds reported were over southern England and must indicate the approach of another, deeper cyclone from the southwest or south to near Cornwall.

*5 December*

The new Low that has come in from the south seems now centred over, or near, Ireland. It had probably largely absorbed the older systems, whose fronts are still indicated over the northern North Sea and had not yet reached Copenhagen at the (uncertain) time when the wind was still given as S by the ships in that area.

The new Low clearly brought a sharp front which veered the winds over southern England as it passed. HMS *Antelope* near Portsmouth reported a sharp wind veer from S by E to WSW, with some rain, at 11h. Nevertheless, there was little change in the surface temperatures reported in the generally windy and rather cloudy weather. The general level of barometric pressure has fallen in the region of the British Isles and areas to the north.

The winds over southern England increased again later, apparently from a more S'ly point, and a deepening frontal wave disturbance is presumed to have approached from the Bay of Biscay. By evening the wind in London was so strong that part of a house fell so near Defoe that he felt lucky to have escaped injury.

### 6 December

The frontal wave disturbance detected yesterday has by midday reached northern England, and we can be sure from the pressures reported at Towneley, Lancashire that the central pressure was 980 mb or below. The wind reported by the Dutch ship S/S *Callenburgh* in the northern North Sea indicates that the older low pressure centre still has a separate circulation of the winds around it. So we have a deepening low pressure complex over or near the north of the British Isles.

The winds continue very strong to hard gale over southern and eastern England. Probably another frontal wave disturbance developed over Biscay and moved north, contributing to the increasingly vigorous cyclonic situation focussed over Britain. Defoe described the winds in London as unusually violent all day. Pressure fell again in the later part of the day, from 998 mb at noon to 989 mb in the evening with violent squalls, lightning, and some rain and hail. Rain reached Amsterdam in the evening and the wind became a S'ly storm which veered to WSW later in the night.

Defoe remarked that this day of violent winds 'would have passed for a great wind, had not the great storm followed so soon.'

And in southwest England the storm which had been blowing up since 5 December 'became so redoubled' on the 6th that no boats dared to go out to relieve the keeper of Eddystone lighthouse, off Plymouth.

### 7 December

The main centre of the cyclonic activity that has concerned us until now seems to have passed away to the northeast or east, evidently as a very deep system. S/S *Callenburgh*, near 58° N in the northern North Sea had a SW gale on the 6th, increasing to storm in the night, but now reports wind and weather improving and 'the scattered [Dutch] fleet returning' and regrouping around this ship's position.

The accelerated movement and eastward turn of the old low pressure system again suggests influence of a westerly jetstream farther north (probably near latitude 65° N or the Arctic circle).

The change to westerly winds at Copenhagen indicates a further fall of pressure over the northern part of the map. A westerly jetstream is suspected near latitude 60–65° N, and a fast-moving warm front breakaway depression has probably been carried forward by it from near Shetland.

Over eastern and southeastern England, the winds continue from strong to hard gale, generally SW'ly, with rain and drizzle, although pressures are higher than yesterday evening. There seems to be a secondary low pressure centre, probably associated with a frontal wave disturbance which brought rain and higher temperatures at Paris. This disturbance evidently moved northeast, and the winds eased somewhat for a time and veered to WSW or

W. There was a tornado reported in Oxfordshire at 16h, about the time this system would have passed. In the S'ly winds, on the eastern side of this small secondary low pressure centre, a gradient wind strength of about 100 knots is suggested by the map (cyclonic curvature allowed for).

Brooks (1954) states, presumably from original, contemporary reports, that the main storm centre which is now about to come on the scene and which was an obviously separate system, announced itself 'off the west coast of Ireland' on the morning of the 7th, travelling east and crossed the coast of Wales before midnight. At Milford Haven in southwest Wales the wind had been increasing from S by E from early afternoon.

Defoe (1704, pp. 24–6) records that towards night the wind in London, which was still strong, increased again and 'about 10 o'clock our barometers informed us that the night would be very tempestuous. The mercury sunk lower than ever I had observed it.'

Defoe (1704, p. 60) adds that 'we are told they felt upon that coast [Florida and Virginia] an unusual Tempest a few days before the fatal 27th of November' (Old Style). The respected climatologist C.E.P. Brooks (1954, p. 29) wrote that 'a few days before the 26th a West Indian hurricane travelled up the coast of Florida and out into the Atlantic', but gave no hint of what this amplified statement was based on. Nor do we know the source of Defoe's statement. No report of the alleged hurricane appears in Ludlum's list of *Early American Hurricanes* (1963), which does include one 'a month and a week earlier'. Neither is it listed in Tannehill (1956). The date is almost exceptionally late in the season: out of over 1300 hurricanes listed by the two authors between the years 1500 and 1955, only four or five touched the American Atlantic seaboard after mid-November, but this included three in December. A further 15 or 16, several of them in December, originating well south in the Caribbean region and moved on tracks far out into the Atlantic, sometimes later approaching the United States or Canadian seaboard, more often passing near Bermuda or the Azores. Storms on such tracks were little likely to be reported before the mid-nineteenth century and may therefore be less unusual than they seem from the available listings.

The progress of the great storm in early December (New Style) 1703 across Europe is consistent with an origin near Florida about four days earlier (see map). It also appears remarkable how this storm continued its steady march, seemingly uninfluenced either in direction or speed by the proximity near the coast of Europe of an older, and still deep, depression. These observations probably support the suggestion that it originated as a tropical storm, possibly of hurricane strength. At least, it seems easiest to suppose that tropical air was brought close to the front of very cold North American continental air close to the coast of Florida and Virginia, producing a strong jetstream. The storm, moving steadily across the map at a speed of about 40 knots (c. 20 m/sec), displayed a progress which is notably rapid for a frontless and later a partly occluded system. This hints at a notably strong jetstream and would be consistent with the proposed tropical origin.

*8 December*

Barometric pressure in London was lowest in the early hours, sea level value about 975 mb at 00.30 and 4h, possibly as low as 965 mb around 3h. Winds veered to WNW or NW over southern England and Wales, with squalls and lulls, but generally continuing at storm force for up to five or six hours. With the Low centred near Nottingham at 3–4h these values suggest a central pressure about 950 mb.

For the drawing of the isobars over England on this map of the situation at about 12h on the 8th barometric pressure measurements at a third place, Townley in Lancashire (near 53.8° N 2.2° W), were available, but no value has been plotted on the map because the time of observation did not fit. A mean sea level pressure of 979 mb was indicated at 7h on the 8th and 990 mb at 15h. Later, at 21h, the pressure had risen to 994 mb. The appropriate value for 12h has been taken as about 986 mb.

Defoe reports that 'a tin ship' from Cornwall, with only one man and two boys on board, was forced by the NW wind out of Helford Haven, near Land's End, about midnight, and was blown (presumably under bare poles, with no sail) a distance of about 150 sea miles to the Isle of Wight in just eight hours. The party's safe arrival, blown ashore between two rocks, can reasonably be called miraculous. During this time the Eddystone lighthouse fell, with the loss of everyone there. The storm had been blowing in west Cornwall since about 21h and houses were blown down there as in so many other places. The SW'ly gale produced a record high tide in the Severn estuary region.

The gradient winds implied by these barometric pressure maps are most easily measured in the sector of the depression with nearly straight isobars over Denmark and northernmost Germany on the 8th; and, although there are no barometer readings to support the drawing there, the isobar spacing is thought to have been nearly the same over southern England between midnight and about 6h on the 8th: indeed this was the basis of the drawing of this map near the low pressure centred over Denmark around 12–15h. This reasoning indicates gradient winds in that part of the system of about 150 knots.

We must suppose that the ship blown from west Cornwall to the Isle of Wight between midnight and 8h, representing an average speed made good of 18 or 19 knots if the report is accurate, was carried along with a great deal of wind-driven surface water at a speed which could probably be between a fifth and a tenth of the prevailing surface wind. This reckoning hints at surface winds certainly over 60 knots and perhaps approaching 80 to 90 knots.

The northernmost place in Britain which reported violent winds was Spurn Head and the area of the river Humber. It is unfortunate that HMS *Dartmouth* in Grimsby roads, where she reported 'violent storm' gave no report of the wind direction. From the way local ships from Grimsby were torn from their moorings and grounded about the Humber and wrecked near Spurn Head, it seems likely that the wind direction changed a good deal during the passage of the storm. There may have been a period when it was E'ly or SE'ly and then SW'ly. The cyclone centre seems to have passed very near. A report from Leeds, which lies only a few kilometres farther north but by being much farther west was probably 30 to 50 km north of the path of the low pressure centre, says the storm was not felt there. Clearly, the strong winds were at this stage almost confined to places south of the track of the low pressure centre.

*9 December*

Over the British Isles and North Sea the W to NW winds of the rear side of the depression are now established and the barometric pressures are much higher. The winds are still strong: in London 'still blowing furiously', while HMS *Dartmouth* in Grimsby reports 'fresh gales [equivalent to Beaufort force 8], but much more moderate than yesterday'. The Dutch ship S/S *Gouda* in Amsterdam reported a 'stiff sailing breeze'. Skies were clearing, but S/S *Callenburgh* in the central North Sea reported a heavy shower.

The main depression centre is believed to have continued on its path across southern Sweden to Finland. But continuing very strong wind from NW in the early part of the day at Copenhagen, turning N by midday, suggests that a deep trough of low pressure or a secondary centre developed over the southern Baltic before moving away east. By evening the weather was calm at Copenhagen. The main cold front of the storm depression, which lay over northern France at noon on the 8th, can be presumed to have reached the Mediterranean. The temperature at Paris on the 9th was 5 °C lower than on the 8th.

The later history of the winds at Copenhagen and over other northern parts of the map suggests that a complex low pressure area continued over northern Scandinavia, possibly extending beyond the northwest coast of Norway.

*10 December*

The NW winds over England early in the day were, by the afternoon, backing to W or SW and strengthening again. In the central North Sea S/S *Callenburgh* reported NW wind at noon, becoming lighter and then variable: by 17h it was again blowing a 'gale' from the SW. The weather became fair for a time before this next strong wind. This new gale is most easily attributable to an occluding depression advancing quickly east from the Atlantic, passing north of Scotland on the 10th and reaching Sweden by the 11th. A speed of advance between 45 and 60 knots is indicated, implying that there was again a strong westerly jetstream between latitude 60° N and the Arctic circle.

The temperature of 7 °C at Paris at noon, with 1031 mb, indicated that this was a ridge of high pressure formed in cold air behind the storm depression. Amsterdam had SW, estimated Beaufort force 5, by evening. At Copenhagen, too, the wind had become SW'ly. The extra water driven into the North Sea by the NW winds of the early part of the day, which had also prevailed on the 8th and 9th over the northern North Sea, produced a more than ordinarily high tide in the Thames on the 10th, but not as excessive as the record tide in the Severn estuary produced by the SW'ly storm on the 8th and not such as to cause further damage in the Thames after the storm.

### 11–14 December

Daily weather map analysis was continued for four more days, but with the return of the S/S *Callenburgh* to port in Holland – she was already south of 55° N on the 10th – and the withdrawal of HMS *Dartmouth* from Grimsby to more southern bases in England the synoptic observation network no longer existed over the northern part of the map. It seems likely that the Dutch navy was still respecting the Christmas holiday on the old calendar (abandoned there in 1700). And the generally less stormy weather led to much more restricted information for posterity about the days which followed.

Strong gale to storm force winds were again reported in the Copenhagen area in the night and morning of the 11th, and are attributed here to the new depression from the west deduced from the data for the 10th.

It seems that yet another fast-travelling disturbance, probably another warm-sector depression centre rather further north than the earlier cyclones of this month, was approaching Ireland (and Scotland?) from the west on the 10th and caused the S to SW winds over England to strengthen on the 11th. We are told that the wind in London on the evening of the 11th was 'so furious that people feared to go to bed', but this probably was a fear exaggerated by their experience just four nights earlier. The navy ships in Portsmouth agreed in reporting 'hard gales' (equivalent to Beaufort force 8 or 9) around the middle of the day but moderating later on the 11th. HMS *Dartmouth* in Grimsby also reported hard gales but in the night of the 10th to 11th. The wind was easier there during the 11th. In London, according to Defoe, the wind did not moderate until noon on Wednesday the 12th. There was some rain about that time, and the weather improved in the afternoon so that by 16th it was calm. At Grimsby there was one more chapter to the story, the wind increasing to 'hard gale' from the NW in the afternoon of the 12th and some rain at 6 o'clock in the evening before the weather finally moderated. This may have been due to a secondary front from the north.

S/S *Callenburgh*, now off the north coast of the Netherlands, was struck by this new gale in the night of 11–12 December and driven willy-nilly inside the belt of islands. Presumably the gale-force wind was from the NW or N.

The winds continued generally W'ly and moderate to fresh over southern England and Holland at least until the 14th. Evidently cyclonic sequences were still active farther north, but with barometric pressure holding steadily at about 1012 mb in London these areas were under some influence of the fringe of a higher pressure regime to the south.

### General inference

This analysis broadly confirms Brooks' earlier diagnosis. A long-continued cyclonic regime dominated the half-month studied and is reported to have affected the British Isles for two weeks before that. There seems, however, to have been an anticyclone over Scandinavia until 1 December. And after 9 December pressure was high over France and exerted some influence over southern England.

The cyclonic activity mentioned became concentrated over the British Isles in the first six days of December, and had moved north or northeast to a position near 60–62° N off the coast of Norway on the 7th, with lowest pressure probably about 950 or 955 mb.

Another system, responsible for the historically severe storm, arrived from the west or westsouthwest over the British Isles on the 7th and seems to have marched on steadily towards the eastnortheast across northern Europe during the 8th and 9th, at something like a constant speed of about 40 knots. Its course seems to have been more or less unaffected by the proximity of the former low pressure centre, which may, however, have been losing energy at this time.

The lowest pressure attributed to the storm depression centre, about 950 mb over central England on the 8th and over Denmark on the 9th, is the same as Brooks' estimate although independently arrived at.

There is some support for the hint already made by Defoe in 1704 that the great storm may have originated as a West Indian hurricane which (he reported) affected Florida perhaps four days earlier.

This analysis has produced an estimate of the strongest gradient winds in the system as about 150 knots. The strongest surface winds may have been 90 knots or rather more and the gusts and squalls probably much stronger. The margin of error of a pressure gradient wind estimate based on these early instruments and charts makes it impossible to say with certainty whether the winds in this case exceeded all others in this compilation, much as the evidence of surface destruction suggests that. So it may be that the strongest gradient wind exceeded 150 knots.

### Acknowledgements

I am greatly indebted to the Public Records Office, London and the National Maritime Museum, Greenwich for the English navy ships' logs used in this analysis; also to Dr A.J. Clark, Deputy Librarian, The Royal Society, London for photocopied excerpts from the *Philosophical Transactions*; to Dr K. Frydendahl of the Danish Meteorological Institute, Copenhagen for photocopied Danish ships' data and much advice about early developments of the meteorological observations on Danish ships and in Danish waters; to the Algemeen Rijksarchief, s-Gravenhage for photocopied material from ships' logs of the Dutch navy in the North Sea and Channel; to the Observatory of Paris for photocopied original manuscript observations, and to the Directeur de la Météorologie Nationale for information about early meteorological observations in Paris; to Mr J.S. Williams, Acting City Archivist, Bristol for photocopied material from the British Records Office; and to Mr Cullen, Assistant City Archivist, Dundee. I also acknowledge my indebtedness to Dr Kenneth C. Spengler, Director of the American Meteorological Society, for the valuable information about early American hurricanes.

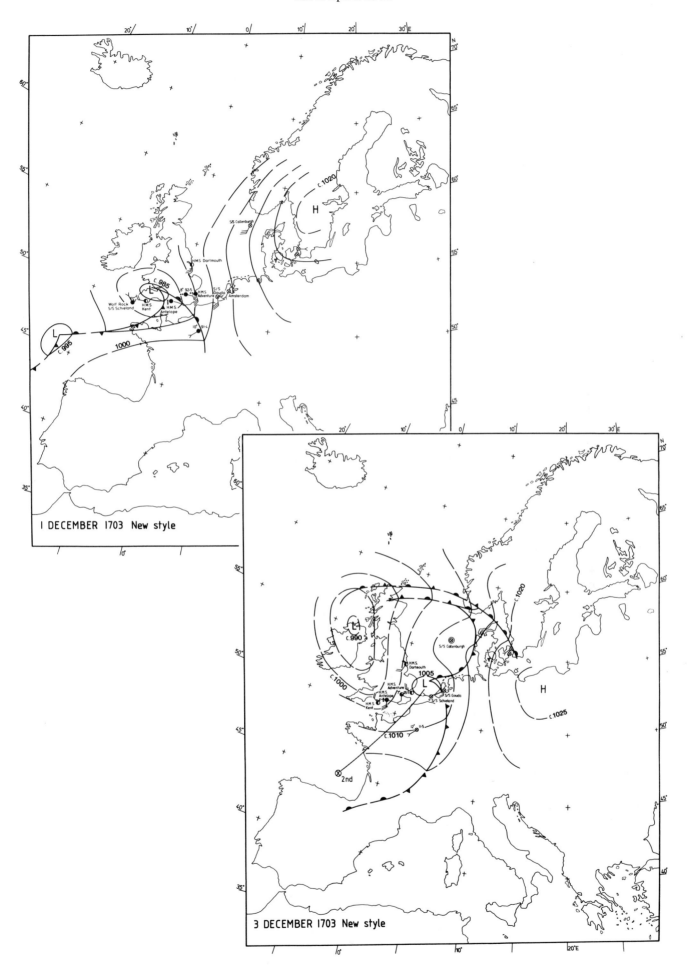

1 DECEMBER 1703 New style

3 DECEMBER 1703 New style

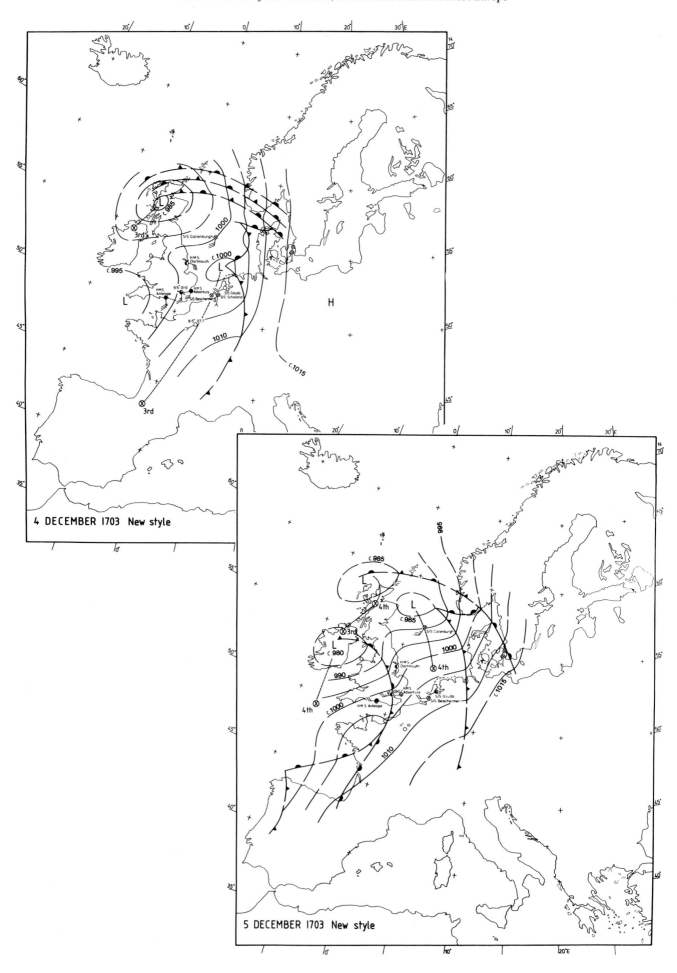

4 DECEMBER 1703  New style

5 DECEMBER 1703  New style

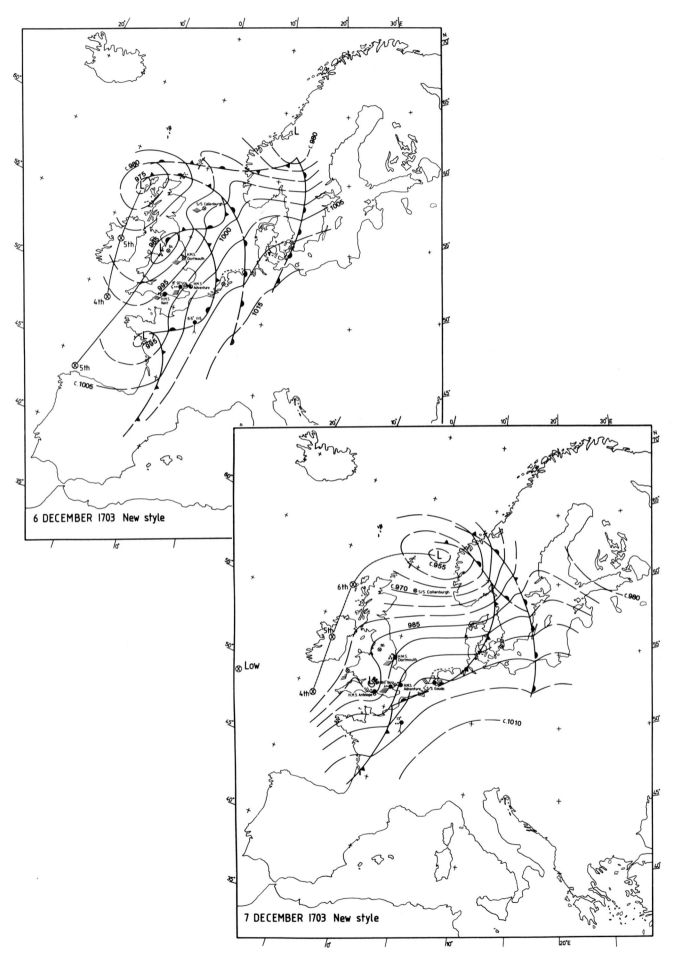

6 DECEMBER 1703  New style

7 DECEMBER 1703  New style

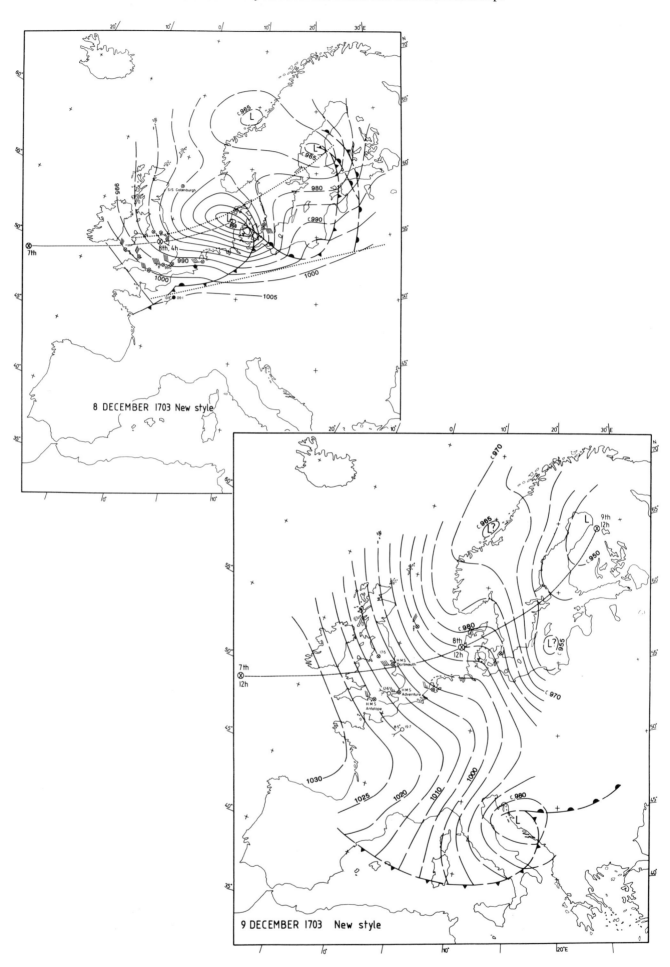

8 DECEMBER 1703 New style

9 DECEMBER 1703 New style

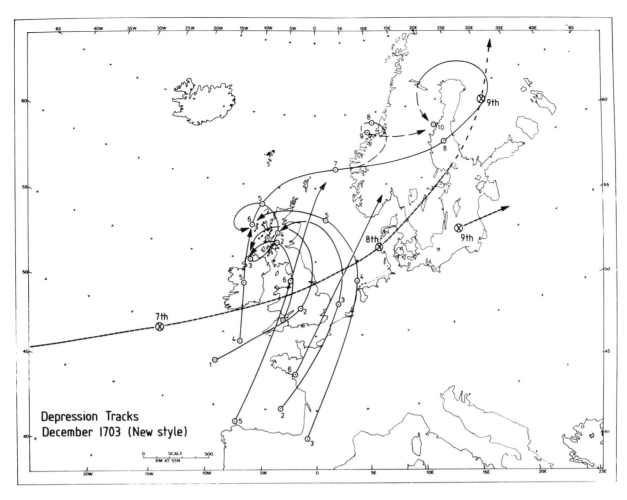

Depression Tracks
December 1703 (New style)

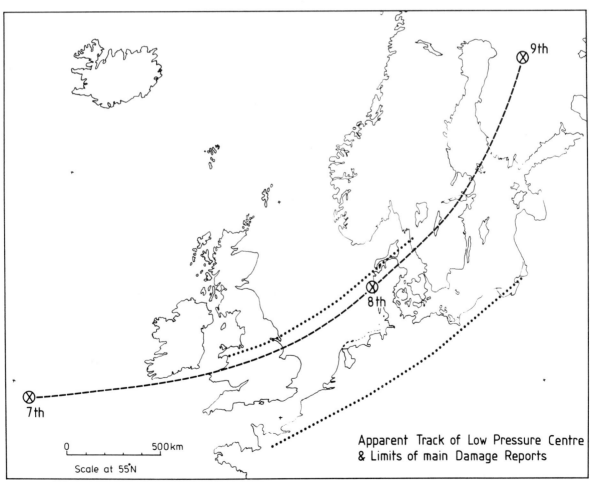

Apparent Track of Low Pressure Centre
& Limits of main Damage Reports

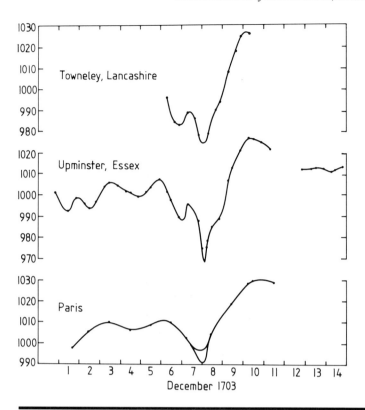

December 1703

## Appendix B. Note on the reports on the earliest full synoptic meteorological maps of the storm sequences studied

The plotting model used for presenting the data for each place is as close as may be to that long established in international practice in meteorology:

The circle marks the position of the observing station or ship. The blacked in portion of the circle indicates the proportion of the sky covered with cloud, in this case three quarters. An X in the circle is used when the cloud amount was not or could not be observed, presumably because the sky was obscured by fog or mist (or perhaps darkness). The tail attached to the circle shows the direction from which the wind was blowing, and the flecks on the tail indicate the estimated force of the wind, each half fleck corresponding to one point on the Beaufort scale of wind force. When only the wind direction and not its force is known, the direction is shown by a twin-tailed arrow. When the wind is still (calm), an outer circle surrounds the station position circle.

The figures to the upper left of the circle give air temperature in degrees Celsius, obtained by conversion from the various early scales used by the original observers. The figures to the upper right of the circle represent barometric pressure in millibars with the hundreds figure(s) omitted, adjusted in accordance with the best information available to the equivalent value at sea level and standard gravity (as at 45° N). On some of the later eighteenth century maps

observations made at high level stations in the Alpine passes or elsewhere on high ground in central Europe begin to make their appearance: in these cases the pressure entered on the map is as observed at the station height, in millibars with the hundreds figure included. To the left of the circle are the conventional symbols for any weather reported.

Examples of the common symbols used are:

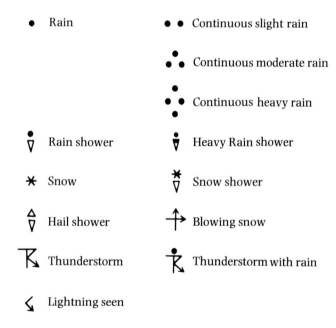

Any of these symbols placed to the lower right of the circle refers to recent past weather.

Conversion tables for ancient units of atmospheric pressure and temperature have been given in Lamb (1986),

Table 9. *Observers Scales of Wind Force*

| The bald terms used by English sailors' about 1700 (Defoe, 1704) | Admiral M. Bille's scale codifying Danish seamen's usage since 1690 (published 1836) | Admiral Beaufort's wind scale (originated 1806) | | | | | |
|---|---|---|---|---|---|---|---|
| | | Beaufort number | Standard names | Specifications for use at sea by mid-nineteenth century sail-rigged warship | Inland | Beaufort number | Equivalent wind speeds (knots) |
| 1 Stark calm | | 0 | Calm | | Smoke rises vertically | 0 | Less than 1 |
| 2 Calm weather | 1 Little motion | 1 | Light air | Just gives steerage way | Wind direction shown by smoke, not by vanes | 1 | 1–3 |
| 3 Little wind | 2 Light top-gallant sailing breeze | 2 | Light breeze | Speeds of man-of-war under full sail in smooth water Bft. 2 1–2 knots | Wind felt on face | 2 | 4–6 |
| | 3 Top-gallant sailing breeze | 3 | Gentle breeze | Bft. 3 3–4 knots | Leaves and small twigs in constant motion. Small flags extended | 3 | 7–10 |
| 4 Fine breeze | 4 Topsail breeze | 4 | Moderate breeze | Bft. 4 5–6 knots | Raises dust, leaves and loose paper. Small branches moved | 4 | 11–16 |
| 5 Small gale | 5 Reefed topsail breeze | 5 | Fresh breeze | Ship can just carry all sails | Small trees in leaf begin to sway | 5 | 17–21 |
| 6 Fresh gale | 6 Double-reefed topsail breeze | 6 | Strong breeze | Ships can just carry topsail | Large branches in motion. Whistling heard in overhead wires. | 6 | 22–27 |
| 7 Topsail gale | 7 Three-times reefed topsail | 7 | Moderate gale | Reefed topsail | Whole trees in motion. Difficulty in walking against the wind. | 7 | 28–33 |
| 8 Blows fresh | 8 Close-reefed topsail gale | 8 | Fresh gale | Double-reefed topsail gale | Breaks off twigs. Generally impedes progress | 8 | 34–40 |
| 9 Hard gale of wind | 9 Mainsail gale | 9 | Strong gale | Closer reefed topsail | Slight damage to buildings (roof slates, chimney pots) | 9 | 41–47 |
| 10 Fret of wind | 10 Stiff mainsail gale | 10 | Whole gale | Ship can barely carry reefed lower mainsail | Trees uprooted. Considerable damage to buildings | 10 | 48–55 |
| 11 Storm | 11 Flying storm | 11 | Storm | Reduces ship to storm stay-sails | Widespread damage | 11 | 56–63 |
| 12 Tempest | 12 Hurricane | 12 | Hurricane | No canvas could withstand | | 12 | Over 63 |

Lamb and Johnson (1966). For accounts of early meteorological instruments and their exposures, the reader is referred to Lamb (1977, pp. 22–4), Manley (1953, 1962), and especially Middleton (1969).

Derham's letter containing his observations concerning the 1703 storm provided an opportunity to decipher the temperature scale he used, which seems to have differed from Newton's and Hawksbee's scales. In it he stated (*Philosophical Transactions of the Royal Society*, Vol. 24 (1704), p. 1531) that the freezing point of his thermometer was about 84. This agrees nicely with the conclusions previously derived by the present writer from examination of his observation series that the freezing point was 84.8° on Derham's scale and the average temperatures of the summer months suggested that 20 °C was about 154 on Derham's thermometer. Some uncertainty arises because the thermometer may have been exposed indoors in an unheated north room (as was the practice in some cases at that time). If we assume a linear conversion scale, the Celsius temperature derived – at least in winter time – must be a close enough approximation (within about 1° below 10 °C) for the identification of air masses and fronts on a daily map sequence.

Defoe details (1704, pp. 21–2) a twelve-point scale of wind strength descriptions, 'bald terms used by our sailors', as early as 1703–4, which apparently already had some equivalents in other countries' practice and from which evidently the Beaufort scale was later derived. This and the other scales used in the observations employed in the synoptic maps of eighteenth and nineteenth century storms in this study are detailed in Tables 9 and 10 in this appendix. It will be seen that a degree of common currency was developing in the assessments of wind strength. Most observers used twelve-point scales (Frydendahl, 1986), but it is plain to see that the numbers on the different scales were not exactly equivalent (as with the different units then used for length, pressure, temperature and weights). All the observations reported have been converted, as best possible, to Beaufort scale wind strengths before entry on the maps here presented.

A simpler 4–5 point wind scale was used by some of the early observers on land. One such was specified by the *Societas Meteorologica Palatina* of Mannheim in 1781 as in Table 10.

Table 10. *Mannheim network observers' scale*

| | |
|---|---|
| 0 | Quite light wind |
| 1 | Moderate air motion, which brings the leaves of trees into movement |
| 2 | Stronger air motion, causing the branches of trees to move |
| 3 | Very strong wind in which heavy branches of trees move, and dust swirls up from the ground |
| 4 | Storm wind |

## 24–25 December 1717 (New Style)

*Area*

Whole North Sea and much of Europe, especially south Norway and Sweden, Denmark and north Germany, also the British Isles.

*Observations*

This was one of the greatest historically recorded storm disasters on the coasts of the North Sea in terms of loss of life – in this sense the greatest since the beginning of effective dyke building in early modern times. About 11 000 people are reported to have died, particularly many in the region of Emden at the coast of northwest Germany, besides many thousands of animals. There was storm damage and flooding all along the coasts of the southern North Sea, on the English as well as the continental side, and the damage extended to the French coast. The Danish-Norwegian fleet, which was concerned in the war against Sweden at that time, was at least partly scattered by the storm in the eastern Skagerrak, and a famous frigate, the S/S *Lossen*, was wrecked on the northwest shore of Hvaløyen (an island near the eastern side of the mouth of Oslofjord) towards midnight on 24 December, probably about the time of passage of a cold front over the region.

The associated tidal surge carried the sea into the streets of Gothenburg in southwest Sweden. And all along the North Sea coast of the then Danish provinces of Slesvig and Holsten (today's Schleswig-Holstein) and of northwest Germany and Holland the sea flood caused a great tragedy of lost lives, drowned cattle and destruction of the dykes that had been built for protection against the sea. In the Netherlands 2000 people were reported drowned and 15 000 houses destroyed, especially in the northern province around Groningen. Parts of England were also flooded and there was damage on the northern coasts of France.

Harrowing accounts of the suffering from the sea flood on the northwest German coast about Emden were collected by a pastor in the town, Gerhardus Outhoff (*Verhaal van alle hoogen Watervloeden* (1720)) and reprinted in an article by W. Schweckendieck (1876). The SW and W gales of the previous days had eased when the wind veered NW about sundown on the 24th, and many were drowned in their beds when the storm sprang up again 'with unheard-of fury' between 1 and 2 o'clock on Christmas morning, the 25th, with a rapidly rising flood, as the sea invaded the streets. Others escaped hurriedly in their night clothing or in none and many of them perished and drowned after exposure to the bitter wind. 90 000 cattle are estimated to have died. And it took 8 years, until 1725, before the sea dykes could be fully repaired.

This flood seems to have ranked no higher than fifth highest in the eighteenth century on the German and Danish coasts: in the region of Hamburg and the Elbe, the highest water level recorded was 5 cm below that in the flood of 11 September 1751, about 10 cm below the flood of 7 October 1756, 30 cm below that of 22 March 1791, and about 40 cm less than that reached on 11 December 1792. It was not a spring tide; and had it been, no doubt the flood would have been higher. Farther north on the coast, at Tönning, the flood mark for Christmas day 1717 was 26 cm lower than in 1756 (Petersen and Rohde, 1977; Rohde, 1964). Nevertheless, although the surge of water into the southeastern part of the North Sea in 1717 was not quite as great as on some other occasions, clearly the combined effect of the wind and waves battered the dykes until they were breached in a number of places. The damage and deaths in 1717 were much greater than in the later storms here mentioned, by which time the coastal defences had been renewed and greatly improved as a result of the experiences in 1717.

*Meteorology: data available*

The analysis of the Christmas 1717 storm and the series of storms in that month of December is in some ways more difficult, because of data problems, than in the case of some of the earlier storms analysed. There is, of course, much more information by this date, enough to attempt some detail in the analysis. But the reports from several key points – London, Paris and Amsterdam – turn out to be of doubtful quality, in some cases almost certainly inaccurate about wind direction and other details. Nevertheless, the broad outlines of the weather development become clearly established. And one crucial feature, the progress of the major cold front that introduced an outbreak of Arctic cold air from the north and was accompanied by the direst wind, weather and sea flood events as it approached and crossed Europe, can be traced and timed with something near precision. Only one barometric pressure record, at Paris, as against three in the case of the 1703 storm, has been found.

In these circumstances, no pressure values can be assigned to the 'isobar' lines on the map analysis, which can, however, give an indication of the probable pressure and wind pattern and, in a general way, of the nature of the pattern of fronts.

The observations at Paris, for which we are indeed grateful to the Paris Observatory and the French meteorological service, were made – apparently when time and other duties permitted – by the astronomical staff of the observatory, being entered in the same handwriting (though somewhat untidily) as in the 1703 record. The hours of observation varied but were usually described as 'midy' or 'deux heures' (2h), which description was in any case misleading to a twentieth century reader, because in accordance with astronomical practice of that time (and for long afterwards) the days were counted as beginning at noon: hence 'midy' in this record actually means 'midnight', as the temperature observations testify. The temperatures were entered in °F. The barometer readings in 1717 (reported in Paris inches and lines) can be reasonably adjusted to sea level and standard gravity by adding about 8–9 mb. This indicates that the lowest pressure there, in the night of 24–25 December, may have been about 996 mb and by the 30th pressure apparently had risen to 1020 mb. (In the 1790s, based on a strict knowledge of the height of the barometer station, an overall correction of 6 mb was used.) A plot of the successive, more or less 24-hourly barometric pressure readings indicates clearly enough the main features of the sequence and at

least broadly confirms the diagnosed method of reporting times because of the pressure minimum which accompanied the cold front in the night of 24–25 December.

The only observations in the London area available at this date are found in the diary of one, George Smith, 'Procurator in all Causes . . . Maritime, Foreign Civil and Ecclesiastical to Queen Anne,' and are believed to have been made in his garden at Richmond, Surrey within the grounds of the Old Palace of Richmond. The record was maintained from January 1712 to June 1745. No instrument measurements were reported. And some of the weather reported is so obviously at variance with the indications from surrounding points on the map that it seems they may have been entered in the record some time afterwards, on the basis of already failing recollection, particularly since for a number of days the main part of the entry is 'Ditto'. On some days entries seem, moreover, to be exaggerated impressions of beautiful weather in contrast to the storm on the 24th–26th. (From the 27th to the 31st the weather was described as 'fine', sometimes 'very fine and warm' or even 'warm like April'.)

At Amsterdam the winds and weather were reported one or more times in the day, at somewhat varying hours, by the keeper of the tide-gauge. The weather elements were clearly regarded as of only secondary interest and the wind direction, particularly when the wind was light, seems often to have been wrongly reported. (Over the period of the 1703 storm the wind directions reported here showed clear discrepancies with the reports from a Dutch naval ship in Amsterdam harbour, often differing by 45° or 90° and occasionally more.) Nevertheless, we are very grateful to the Dutch archives for these reports which in general strengthen our knowledge of the overall storm sequence.

The best covered part of the maps of this storm sequence is in the waters where the Danish and Norwegian fleet was operating, with up to 20 ships at sea reporting the weather in their logs.

There was also a very useful network of observations over Germany and central Europe maintained by an organization centred on Breslau (now Wrocław, Poland) and published in Breslau in a series entitled *Annalium Physico Medicorum oder Geschichte der Natur und Kunst*. *Classis I* of these annals is the weather history comprising daily observations of wind and weather. *Classis IV* included an essay account of the storm and its impact over Europe. An account of how the reported temperature observations were handled is the subject of a note appended at the end of the report on this storm.

*Meteorology: analysis*

As in 1792 (q.v.), December 1717 was a very stormy month in Europe generally. There are many reports of a continuing sequence of storms from about 28 November to 28 December (New Style). Reports from western Europe, including the British Isles, speak of the winds, nearly always strong, coming mainly from S and from NW by alternations. It was a mild month on the whole (Manley gives a figure 5.0 °C for the mean temperature in central England, +0.8 °C above the 1921–50 average). At

Breslau/Wrocław temperatures as high as $+7\,°C$ to $+8\,°C$ were observed on the 22nd and perhaps occurred again in the early part of the stormy night of 24–25 December. Another indicator of the prevailing mildness is that there had been no ice on the Danube in Hungary up to mid-month.

A report from Norway says 'it began to freeze a little in early December, then became mild again'. right up to the 30th there had been 'no settled winter weather, a little frost, but always soon rain again'. Much snow fell between whiles and there were many storms, the one of the 24th being particularly notable in south Norway.

There is, as the above reports imply, evidence of abundant moisture and heavy precipitation, but also of occasional intrusions of very cold air from the Arctic. There was particularly heavy snow in the Italian Alps on 5 December which made the roads impassable in the Turin area. The same storm wrecked many ships and flooded the coastland of Italy about Leghorn (Livorno). Reports from Rome and southern Italy indicate storminess as notable in the western and central Mediterranean as farther north. There were many ships wrecked in and near Palermo, Sicily between the 14th and 18th of the month; on the 25th a storm tide there produced 'a higher flood than anyone could remember', and on the 31 December–1 January another storm uprooted trees and wrecked many houses in Palermo and Messina. Another index of the coldness of the occasional Arctic air intrusions may be seen in the rapidity with which ice appeared on the harbours at Arendal and Stavern in south Norway and temporarily blocked the harbour at Ålborg in north Denmark (Jutland) on the 28th, following the cold front which passed over the region on the 25th.

The passage of this major cold front southeastwards across Europe on the 24th and 25th can be timed by the often rather precise reporting of the thunderstorms which accompanied it almost everywhere:

> Mandal, near the south tip of Norway, circa 21h, 24th
> Oslofjord, 22–24h, 24th
> Emden, near the German North Sea coast, 1–2h, 25th
> Pritschwald, north Germany (Pritzwalk, Brandenburg? 53.2° N 12.2° E) 5h, 25th
> Breslau, 8–12h, 25th
> Vienna, afternoon 25th (probably between 13 and 15h)
> Strassburg (now Strasbourg, France) 19h, 25th
> Paris, in or about the early hours of the 25th.

This timing indicates a steady progress at about 50 knots between south Norway and Vienna, but there were 8 to 10-hour delays in the progress of the front from northern Iceland to the places mentioned in south Norway and again in its arrival in Strassburg, where it might have been expected before reaching Breslau. These delays indicate the occurrence of wave disturbances on the front (a) over the North Sea on the 24th, and (b) over France and Belgium on the 25th. It is thought that further waves, forming over France at about 24-hour intervals on the subsequent days, developed and travelled northeastwards, explaining the prolonged storminess reported to have affected Emden where the greatest disaster occurred. The winds at Neuchatel in western Switzerland are reported to have continued strong southerly until the 31st.

It is clear that the ingredients of the severe storm of 24–25 December 1717, as of many of the other historic storms here analysed, were a developing meridionality which injected warm, very humid air from the south – in this case passing northward over the British Isles on 22–23 December and over France on subsequent days – at a time when very cold Arctic air was advancing behind a cold front from the Iceland region.

(In England, after a brief quieter period, another southerly storm was reported on 7 January 1718.)

*Maximum wind strengths*

No precise measure of the strongest winds in this storm is likely to be possible. As in the case of the Spanish Armada storms in 1588, however, some fairly reliable measurements of the distances travelled by the cyclone centres can be made. An occluding centre travelled 700 nautical miles in 24 hours southeast from the Iceland region to Norway on 23–24 December and apparently a similar distance east from there in the following 24 hours. A frontal wave then travelled 800 nautical miles eastnortheast from Belgium to the eastern Baltic on 25–26 December. These speeds are statistically consistent (Douglas *et al.*, 1978) with jet streams of the order of $150 \pm 25$ knots or even more. They probably mean that the strongest surface gradient winds on the southern and western sides of the depression were of the order of 100 knots and strongest gusts of wind at the surface could well have been 100 to 110 knots over the sea and coasts.

*Note*

A good deal more weather information was available from the various places reporting on land, and from the 20 Danish-Norwegian ships at sea, than it has been possible to indicate by symbols entered on the maps of this series. Some of the ships were also too close together for every one to be entered at all on the maps. Some 'station circles' will be noticed on the maps indicating places from which there was useful observed weather information which, however, could not be indicated by recognized symbols. Some of this information was for earlier or later hours than the time of the map, including very useful wind observations which also gave an indication of the approximate times of wind changes. Present weather is entered, where known, using the now conventional symbols placed to the left of the station (or ship) circle. Past weather is indicated by symbols to the lower right of the circle, sometimes in the form of another circle showing the prevailing amount of cloud. Past weather up to 12 hours before the time of these maps is included.

On the map here included (p. 78) to show the logical continuity of the proposed frontal pattern sequence in the analysis of this storm, the patterns shown as very thin lines mark the boundary of the warmest air in the storm system – air which gave temperatures up to $+8\,°C$ in eastern Germany – on the dates concerned.

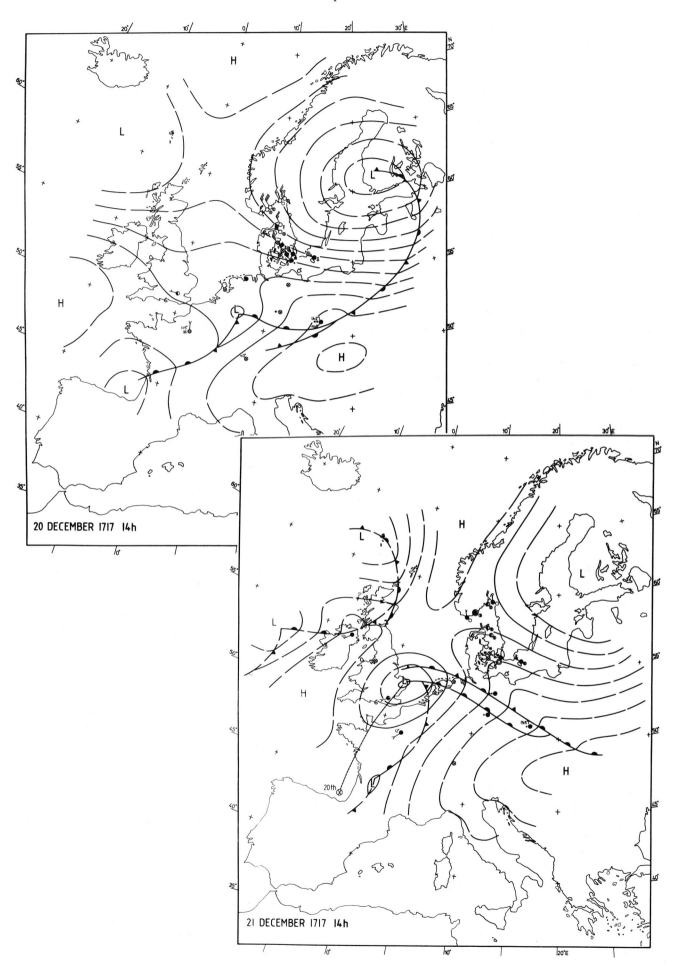

20 DECEMBER 1717 14h

21 DECEMBER 1717 14h

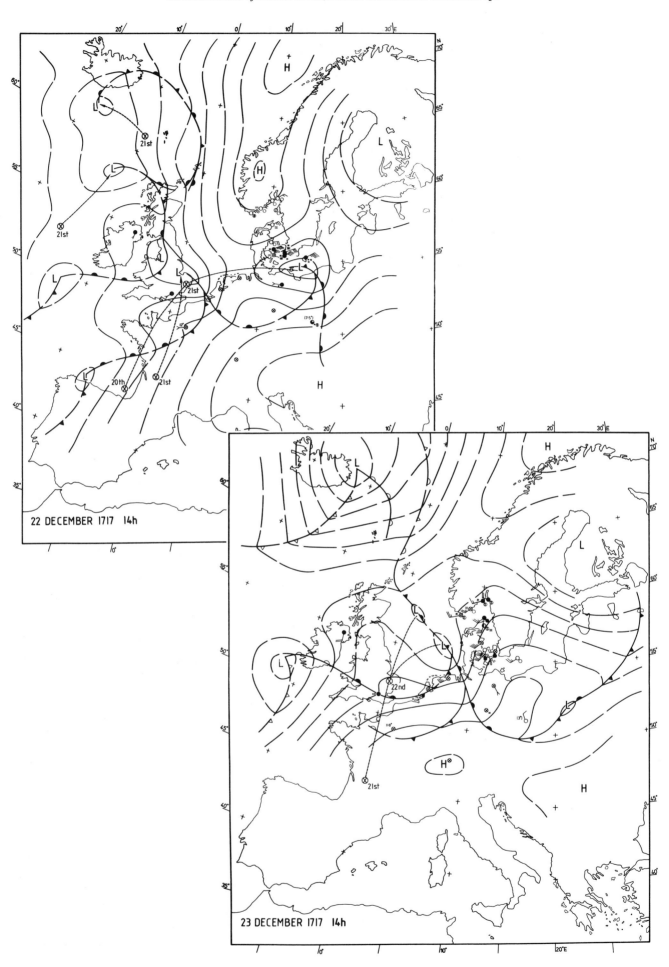

22 DECEMBER 1717 14h

23 DECEMBER 1717 14h

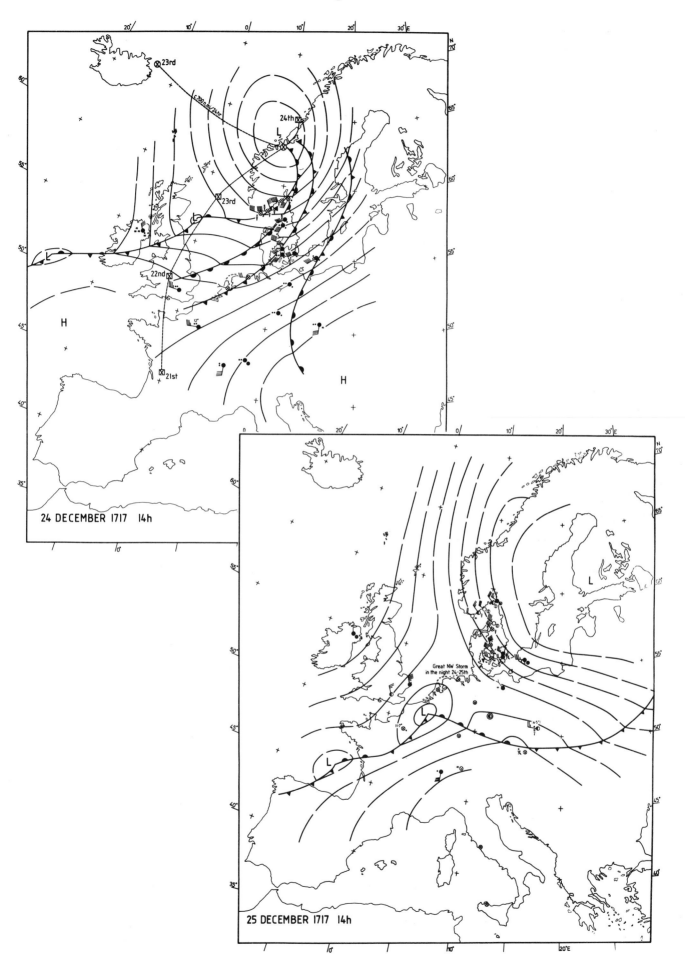

24 DECEMBER 1717 14h

25 DECEMBER 1717 14h

Great NW Storm
in the night 24-25th

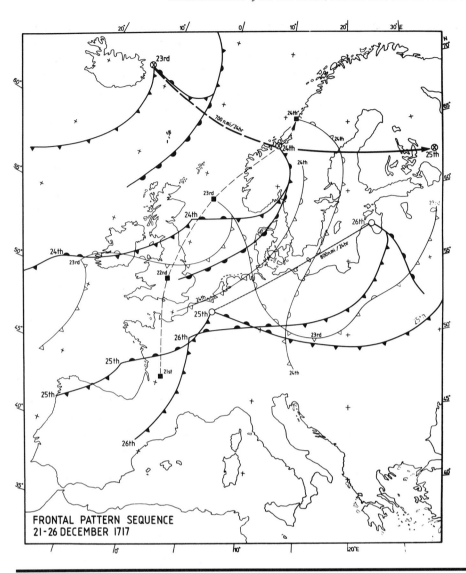

FRONTAL PATTERN SEQUENCE
21-26 DECEMBER 1717

## Appendix C.    Notes on the Breslau temperature measurements in 1717.

The *scale* used was clearly of the (inverted) Hawksbee type, with freezing point around 75°, and 0° represents about the highest temperatures that were presumably considered possible, or perhaps likely to occur some time, in summer.

Treating it as a Hawksbee thermometer shows:

(i)   All the months from July to the end of 1717 appear colder than the 1850–1932 averages, except December which would be 1 °C above the said period's average. (It *was* a mostly mild December.)

(ii)   However, the September 1717 mean, equivalent to 7.7 °C, is unbelievably low. And the July value of 13.2 °C is also very unlikely.

(iii)   The overall 6-months average of the readings, when converted from °H to °C, is 3.6 °C below the 1850–1932 average, which is also unbelievably low. 1717 is reported to have been a good summer, which produced good though not exceptional wine, in central Europe. So it may be that the true 1717 average for July–December should be close to the 1850–1932 mean.

If we apply a correction of + 3.5 °C to the reported observations, the monthly means at Breslau become:

| | |
|---|---|
| July (1717) | 16.7 °C |
| August | 20.0 °C |
| September | 11.2 °C |
| October | 8.6 °C |
| November | 4.5 °C |
| December | 4.7 °C |

It may be supposed that the thermometer was reading low, particularly in the summer, partly because of its exposure – likely to have been in an unheated North room (indoors), as was quite common practice at that date. This type of exposure seems to be confirmed by the reading at 8 a.m. on 23 December 1717, which on conversion appears (*with the above correction*) to be +1 °C, on a morning when the observer reported sharp 'window frost'. It is also noteworthy that the extreme range of the temperatures reported in each month, with the exception of September 1717, when a rapid cooling of the season took place, is rather small for any outdoor exposure. Thus the range between highest and lowest temperature reports at any hour on any day each month in 1717:

| | |
|---|---|
| July | 8.4 °C |
| August | 6.7 °C |
| September | 16.0 °C |
| October | 9.2 °C |
| November | 8.4 °C |
| December | 8.4 °c |

This suggests the conclusion that the instrument was a Hawksbee thermometer, with a fault that made it read about 3.5 °C too low, perhaps partly – but only some small part – due to the nature of its exposure. It seems to have been exposed indoors, presumably in an unheated room, or possibly in an excessively enclosed shelter in a garden. Its readings will be useful in indicating day-to-day changes of temperature, but cannot be treated as accurate.

## 8 January (Old Style)/19 January (New Style) 1735

### Area

Southeast England and a zone across Europe from Biscay to Denmark and south Sweden, presumably also most of the North Sea on the 19th–21st.

### Observations

Westerly gale in London 'the most violent since the great storm of November 1703': it did great damage in London (according to the weather register kept by G. Smith at Richmond, Surrey from 1713–45; reported also by Lowe (1870)).

### Meteorology

It has been possible to map the weather daily from 16–19 January with reports from eight points in Europe between the Bay of Biscay, Scotland and central Europe.

Mild weather brought by SW'ly and W'ly winds culminated with temperatures as high as 13 °C in Edinburgh, where there was a SW'ly gale on the evening of the 17th. By midday on the 18th the temperature in Edinburgh had fallen to 1 °C and remained near the freezing point over the next two days. The barometric pressure there seems to have fallen precipitately from about 1025 mb on the 16th to 965 mb by the afternoon of the 19th. The SW, and sometimes S, winds continued to blow farther south and east until at least noon on the 19th, becoming increasingly stormy as the frontal system approached and the disturbances on it became more active. The depression which crossed England and moved more slowly across the central North Sea seems to have been exceptionally deep. Pressures well below 950 mb are indicated over central England on the 19th and perhaps as low as 935 mb. Strong gale to storm force winds were reported in the Danish

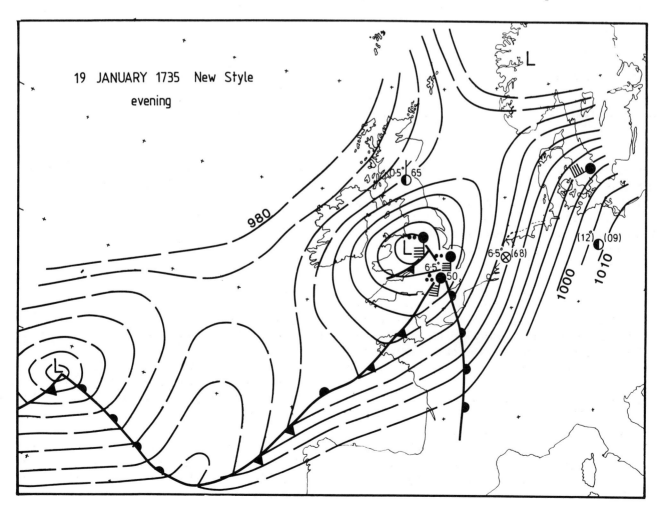

Sound as well as in London on the 19th and by a ship in the Bay of Biscay on the 19th and 20th. Gales and very strong winds were reported between midday and midnight on the 19th over an area that extended from the English midlands to Berlin.

Rainfall was also heavy and prolonged over most of this area on the 18th and 19th.

The pressure differences shown on our map for the 19th imply very strong winds over the southern and eastern parts of the North Sea and all the continental coasts from Biscay to Sweden. It seems that the winds from about NE

affecting northern England and the central North Sea on the northern side of the low pressure centre may have been even stronger: the gradient wind in part of that area seems to have exceeded 100 knots.

There seems little doubt that the intensity of this storm development owed much to the intensity of the temperature contrast between the typically mild humid SW'ly winds bringing maritime tropical air from the Atlantic Ocean and the Arctic airstream from the N, which came into close proximity over the British Isles.

## 16 February (Old Style)/27 February (New Style) 1736

### Area

Western North Sea and coast of Britain.

### Observations

Highest tide level in the River Thames for 50 years past. London experienced a brief W'ly gale which veered N.

### Meteorology

Our weather map analysis shows a strong direct N'ly outbreak passing due south from the Arctic over the British Isles and western North Sea. Presumably this wind current drove extra water south into the North Sea and particularly along its western side.

The pressures shown over the eastern part of Britain seem to indicate a gradient wind close to 100 knots.

The situation is what is known in meteorology as meridional, with prominent N'ly and S'ly windstreams and little sign of westerly winds in this sector. The deep depression approaching the Alps and central Europe has been travelling southeast or south-southeast. The winds in London on the 26th were S'ly before the brief episode of W'ly reaching gale force. The NW'ly and N'ly winds on the 27th had been strong to gale in eastern Scotland and East Anglia. Frost and heavy showers, including snow, also were, or had been earlier, reported on the 27th at various places

in England and Scotland. In Holland temperatures in the N'ly wind fell during the day from positive values to −2 °C in the evening.

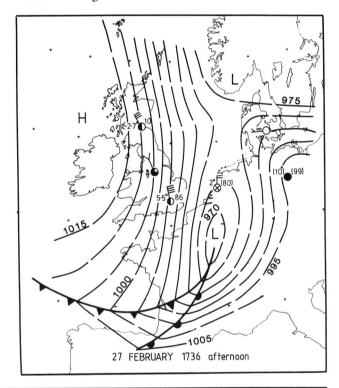

27 FEBRUARY 1736 afternoon

## 24 December 1736 (Old Style)/5 January 1737 (New Style)

### Area

North Sea (?) and Thames.

### Observations

Lowe (1870) reports that a very high tide in the Thames flooded Westminster Hall to a depth of 2 feet (60 cm).

### Interpretation

Flooding of Westminster Hall was a rare event in history and seems not to have happened in connection with the storm tide on 27 February 1736. If Lowe's report in the present instance is right, it seems to imply a very great storm surge in the North Sea on 5 January 1737. But it is possible that the surge entering the Thames encountered the river in a swollen condition. There had been alterna-

tions of heavy rain and heavy sleet and snow, with stormy winds from between S and W, from the 24–31 December, more snow and some frost, with N and NW winds, between 2 and 4 January. Barometric pressure in London had been 993 to 994 mb on the 29 to 31 December and is given as 996 mb on 3 January according to Manley's collected London records. Pressure then rose quickly to 1033 mb on the evening of 4 January and the wind at that time was described as 'N. high'.

It is reasonable to suppose that the swift rise of pressure with winds that had swung round to NW, and later N, marked an anticylone moving in from the west. And the swiftness of the changes suggests both that the high pressure system was moving rather fast and that the pressure gradient was strong enough to give a N'ly gale or storm over the North Sea for some hours. It is also possible that the westerly winds of previous days and weeks had driven extra water into the eastern Atlantic and North Sea. But the known facts do not suggest that the conditions existed for an exceptionally high storm surge. Nor are any other indications or reports of stormy winds known to this writer. (See, however, report on 1 March 1791, p. 91.) It is likely that the Westminster flood was partly due to the river being in a swollen state after the previous two weeks' weather over the Thames basin.

## 3 August (Old Style)/14 August (New Style) 1737

*Area*

Southeastern England and East Anglia, and the following day all Denmark and Danish waters south of latitude 56° N, by implication also southernmost Sweden and the southern North Sea and continental coast.

*Observations*

Lowe (1870) reports: 'Violent gale ... innumerable trees uprooted, chimneys blown down and ships sunk in the Thames'.

*Meteorology*

The general situation on the evening of the 14th is shown by our map with nine observation points entered but becomes clearer on the 15th when the storm had moved into the centre of our observation network. This was a small, but intense, depression with central pressure probably about 980 mb on the 15th, moving northeast across the map at about 30 knots in a rather unusually low lati-tude for August. The rather brief gale was easterly in the Thames and East Anglia. The system was probably also deepening and by the 16th a NW'ly to N'ly storm, with estimated winds of Beaufort force 9 to 10 was blowing over southern and eastern Danish waters.

Places in northern England, and points in the northern North Sea and the Skagerrak, experienced no strong winds.

That the storm had a complex structure somewhat as shown on our map for the 15th is indicated by the succession of changes of wind direction and strength reported from the ships (with observation reports for each of the six 4-hour watches of the day and night) in southern and eastern Denmark, where the SW'ly to S'ly gale increasing up to force 9 was followed by lighter winds backing around the compass through SE and ENE, and becoming much lighter for a time before the force 9–10 NW'ly to N'ly set in and blew for 8–12 hours in some places.

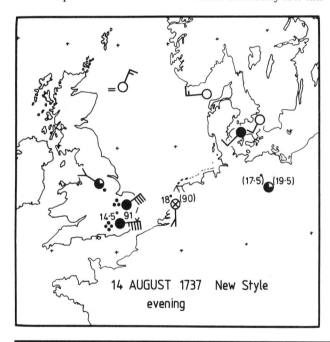

14 AUGUST 1737 New Style
evening

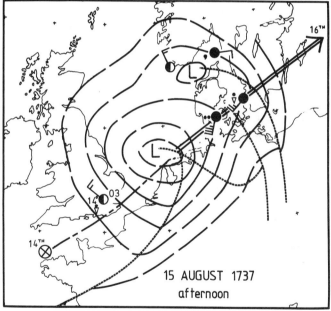

15 AUGUST 1737
afternoon

## Alleged date: 1 December (Old Style)/12 December (New Style) 1737

*Area*

Report by Lowe (1870) of 'A violent gale ... (which) did much damage' in London and the Thames.

*Comment*

It seems impossible to reconcile this report with Manley's records of the daily records of London's weather and instrument readings by observers in the London area at this date.

Possibly the report should be related to 12 December 1738, when Manley found records stating that the wind was 'S. stormy' and on the 13th 'S. very high'.

## 14 January (Old Style)/25 January (New Style) 1739

### Area

A belt across the British Isles between latitudes about 53° and 57° N, and presumably a similar belt across the North Sea, was swept by a violent WSW'ly gale.

### Observations

A newspaper report in the Caledonian Mercury (which was published on alternate weekdays) records that a devastating WSW'ly storm in the early hours of 25 January did enormous damage to buildings. One building on Parliament Square, Edinburgh had its lead roof lifted and carried away by the wind. Trees were uprooted, and many ships were torn from their anchors and wrecked on both the western and eastern coasts of Scotland. The belt affected stretched from Arran and Glasgow in the Clyde to Berwick and Dundee to Montrose on the east coast. A violent storm was also reported in Dublin on this or the previous day (probably the latter, since the same source mentions that Edinburgh was similarly affected 'on the 13th' (Old Style)). The WSW wind was reported as a strong gale in Sheffield also (on the 25th), but only moderate winds farther south.

### Meteorology

This storm was presumably associated with a depression moving quite quickly from about WSW to ENE.

The storm seems to have been strengthened by a funnelling effect through the Lowland valley of Scotland between the Highland region to the north and the Southern Uplands to the south. Convergence ahead of a cold front advancing from the north was probably part of the situation – a case somewhat analogous to the Tay Bridge disaster storm in 1879.

Thomas Short (1749, vol. 2, p. 248) reported that 'This whole winter was remarkable for Heat, Thunder, Lightning, Hurricanes, great stormy Winds, which caused general Health', also that all the roads were 'very deep and bad, yet no great Rains.'

This was the last winter of a decade that had been characterized by a great frequency of winds from between W and S affecting Britain and western and northern Europe, and in several cases the winters were very stormy. The unusually mild conditions seem generally to have extended even to Iceland.

## 31 December 1739 (Old Style)/11 January 1740 (New Style)

### Area

Southern and central North Sea and nearer parts of England, probably also the Channel.

### Observations

Easterly gales reported on the coast of England from Alnwick in Northumberland (55°25′ N) to the Thames. In 'the severe easterly gale' several vessels laden with corn and with coals were sunk by the ice in and off the Thames. Great shortages ensued because for about seven weeks thereafter the ice prevented ships entering port.

### Meteorology

The general easterly wind stream is illustrated by our skeleton map for 11 January and is attested by descriptive reports of the character of the weather of that January from all over Europe north of the Alps. The cold was severest in the winds over the European plain. Dieppe harbour on the north French coast was frozen in early February. But in Sweden the cold was not very notable. At Uppsala only three days produced temperatures below −20 °C and only seven days below −15 °C.

In Copenhagen on the 11th the wind later became ENE. The ice was already 9 cm thick on the harbour and the dry snow was blowing. The weather continued overcast for seven days.

The E'ly gale in England on the 11th was not exceptionally severe but is included in this compilation because of the historic difficulties with the sea-ice. Temperatures of −6 °C on the 11th to −9 °C on the afternoon of the 12th were observed in England with the gale.

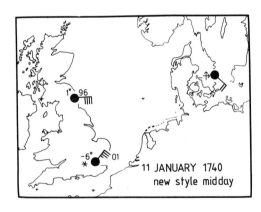

11 JANUARY 1740
new style midday

## 4–8 September (Old Style)/15–19 September (New Style) 1740

*Area*

England and North Sea.

*Observations*

Lowe (1870) reports 'S'ly gales in London on 4 and 7 September (Old Style): much damage to shipping on the 7th' (i.e. 18 September, New Style), though no support has been found for the gale on the earlier of these two dates.

The meteorological observations available to the writer (from the late Professor Gordon Manley's assemblage of London Weather) confirm a SW'ly gale through the day on the 18th increasing to S'ly storm in the evening, with thunder squalls.

A ship, S/S *Dronning af Danmark*, in the port of Arendal in south Norway (58°28′ N 8°46′ E) reported SSW force 10 for some hours in the middle of the day on the 19th, preceded by SSE force 5 in the night before and falling to light in the following evening.

*Meteorology*

The situation seems to have been becoming progressively more cyclonic over this period in the northern North Sea, with the pressure level falling over 20 mb at Alnwick in Northumberland. There were only smaller oscillations of the pressure level in London, but a great deal of rain – particularly from the 16th to 18th – over a belt through the middle of our observation network over England between Sheffield and London and apparently extending to Denmark (though less prolonged there).

The easiest interpretation seems to be that there was an active front, with rather closely spaced minor wave disturbances on it lying across the region, through most of these days. The 18th was the wettest day of the period in England, and it seems probable that the warm sector over England on that day extended farther north than its forerunners and was narrowing. With this occlusion process cold, clearer air reached London on the 19th though the wind continued generally southwesterly there, while the front approached and cleared south Norway.

There had been a weaker low pressure centre in the neighbourhood of the Skagerrak through much of the period 15–17 or 18 September, and it seems likely that the deeper depression on the 18th and 19th was centred not much farther north.

## 1 November (Old Style)/12 November (New Style) 1740

*Area*

North Sea and coasts of England.

*Observations*

Gale blew down one of the spires of Westminster Abbey. Thomas Short reported (1749, II, p. 258): 'November 1st was a memorable Day, with a N. wind, Snow, Sleet, Hail, Rain, and Floods, and a Hurricane from 6 at Night to 12, Wind N. and N.E. the terriblest of many Years; it did inestimable Damage on the N. E. and N. coasts of England in Shipping, Goods and People's Lives; its Effects continued at Sea to the 5th' (i.e. until 16 November on the modern calendar).

*Meteorology*

A deep low pressure centre moved northeast across England, evidently close to London, and on beyond to lie over the central southern North Sea on the 13th. Barometric pressure in London corrected to sea level was lowest about 6 a.m.: 972 mb, and had risen to 996 mb by afternoon.

Strong to gale NW'ly winds with low temperatures (5 °C at midday) had been reported in England on the 10th, associated with a previous cyclonic system, but pressures in London reported on the 10th and 11th were high (around 1020 mb). Gale force winds from between NE and NW were reported at Alnwick on the Northumberland coast on the 10th and 11th, and sleet fell there on the 12th. In London on the 12th the wind was S'ly up to force 7 before turning W to NW and becoming stormy. Winds were still strong on the 13th and snow showers reached London with the wind NW later turning to NE.

The low pressure system probably later moved east to southeast over the continent on the 13th. A key part of this sequence must be the presence of such cold air, apparently firmly established so far south at an early stage in the winter, towards the end of England's coldest year in a more than 300 years long record. The winds at Alnwick and on the south coast of Norway seem to have been persistently from between NW and E.

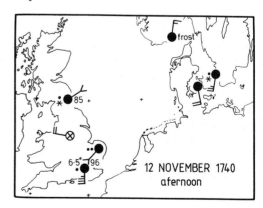

12 NOVEMBER 1740
afternoon

## 8 September (Old Style)/19 September (New Style) 1741

*Area*

England and presumably neighbouring part of the southern North Sea.

*Observations*

A great storm blew down the southwest tower and steeple of the great church (St Margaret and St Mary) at King's Lynn: the tower fell across the nave destroying much of the building. (The angle at which the steeple fell across the building seems to indicate a wind from WSW or SW by W.)

Thomas Short of Sheffield (1749, II, p. 262) (using Old Style dates) reported: September 8th was a terrible hurricane at Lyn Regis, St. Ives, Huntingdon &c. The Wind was neither S nor N, as was reported, but N.E. for during all the long Rain, it kept between N. and E. On some days were opposite Currents, as September 11th, when the lower was N.E. and the upper S.W.

Short's report must surely be a second-hand account from a correspondent in Lynn or thereabouts. It seems likely that the correspondent was not conversant with the (at that time not firmly established) convention of naming the wind by the direction from which it blows and not where it is going towards. There may be a hint in the report of tornadoes having been seen.

It also seems fair to deduce that the storm was not felt, or not much felt, near Sheffield.

The late Professor Gordon Manley's collection of reports of London weather indicates that the wind in London shifted from light SE'ly early to a SW'ly which became squally and increased to storm force, accompanied by periods of intermittent heavy rain.

The barometric pressure, given by Manley as 977 mb when reduced to sea level, was very low and 25 mb lower than the day before. It rose to 990 mb by the following day. The temperature on the 19th was about 15–16 °C; slightly higher than on the 18th.

*Meteorology*

Our network of observation reports is rather poor at this date, but what seems to be indicated is an intense cyclonic centre, with lowest pressure perhaps in the range 950–960 mb, passing rather quickly across central England during the 19th in a direction roughly parallel to the storm wind at King's Lynn.

The winds in London continued southwesterly for several days thereafter, so the general meteorological situation was probably steered by a large low pressure area near Iceland or between Britain and Iceland.

## 26 February (Old Style)/9 March (New Style) 1751

*Area*

England, mainly or perhaps only the South, southern North Sea, and on the next day Denmark.

*Observations*

Lowe (1870) reports that a severe gale did much damage in London and the Thames on 26 February (Old Style). Ships in Danish waters reported WSW gales force 9 to 10 on the 10th.

After a cold February over the entire area – the only cold month of that winter in England – March became mild and wet. On the 9 March there was still much ice in Danish waters: the Great Belt was at least more or less blocked and there was heavy drift ice in the Sound between Denmark and Sweden. When this gale struck, the ice began to move and open up. It became possible for the ships in the Great Belt to send smaller boats ashore, but a warship off Helsingør (Elsinore) reported on the 10th that the ice was still threatening: 'hard work to save the ship'.

*Meteorology*

Our coverage of reports from the British Isles is weak at this time, but this W'ly storm may have been concentrated

in a belt across the North Sea from southern England between latitude 50° and 53° N to Denmark.

Barometric pressures measured in London fell by over 20 mb from the 7th to the 9th, to below 990 mb at mean sea level (MSL), and had risen quite sharply to 1004 mb by the 10th. The wind was again SW'ly. Hail showers on the 10th, and a thunderstorm with the S'ly gale the previous evening, indicate that a cold front had probably passed from the west.

In Denmark, too, the wind became S or even SE increasing to force 9 or 10 on the 9th, and there were many hours of rain and drizzle, before the wind veered on the 10th to SSW and for much of the day WSW.

These Danish observations more particularly suggest a pattern with the cyclone centre passing close. The longer-continued WSW'ly gale may best be envisaged as nearly parallel to a cold front advancing slowly from the north or northwest across the region.

There seem to be no reports of any sea flood on any of the North Sea coasts with this storm.

## 31 August (Old Style)/11 September (New Style) 1751

*Area*

All North Sea and Denmark.

*Observations*

Winds from between SW and S, in some places SSE for a time in Denmark and along the continental seaboard on the 10th to early hours of the 11th, increased to about force 9 and in the Danish Sound to force 10, then veered sharply to WNW to NW and continued to blow at about these highest strengths for up to 12 hours or more.

Even as far inland as Berlin there was an 'unpleasant wind' as early as the 9th, although the barometric pressure was as yet high (1028 mb), and the Kirch family record describes the weather there on the 10th and 11th as 'very moist and stormy'.

At Hamburg the tide on the 11th brought a flood of about the same height as in the 1717 disaster: measurements indicated that this 1751 flood was perhaps 5 cm higher. But there is no record of loss of life or serious damage, presumably because the sea defences had been thoroughly restored, with improved technology, following the events of 1717. This 1751 flood was, however, one of the three or four highest storm surges of the century.

*Meteorology*

The weather situation can be confidently analysed as in our map for the 11th about midday from the network of observations available.

The pattern as drawn indicates gradient winds well over 100 knots (possibly up to 150 knots) over the central

North Sea. But the pressure values cannot be vouched for to the same extent as for the 1703 and 1790s studies, because far too little is known of the instruments and observing practices used at the few places where barometric readings were available to indicate the pressure differences.

On the 12th winds generally between W and N were no more than force 5 over the whole area here mapped. And on the 13th they had fallen light and were backing to SW before the approach of another, rain-bringing but relatively minor, disturbance.

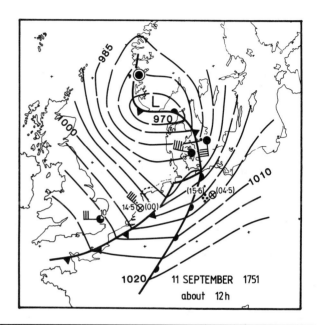

*Area*

Southern and central North Sea, and later on the 7th probably all the North Sea, also most of Britain, Denmark, the Netherlands and northern Germany.

*Observations*

The most violent winds in Britain with this storm were reported at Wigton near Carlisle, and from Penrith also in Cumberland, between 23h on the 6th and 3–5h on the 7th, with numerous trees blown down or twisted off their trunks by tornadoes (especially around 3h). Pieces were blown off stone buildings. Sea salt was blown inland and blasted all the vegetation about Carlisle – a circumstance reported only after exceptionally strong storms such as those here listed in 1703 and October 1987 near the south coast of England. At Newcastle-upon-Tyne, on the east coast of northern England, this storm was reportedly at its worst around 9h on the 7th; houses were blown down, ships sunk and overturned in the harbour and others blown out to sea.

On the Dutch coast the 'hurricane' on the 7th stranded and sank many ships.

Hamburg and Tönning, on the coast of Schleswig-

Holstein, were affected by a great North Sea storm surge which brought the highest tide of the century until it was surpassed by a margin of just 2 cm in December 1792 (Petersen and Rohde, 1977). This tide level at Tönning exceeded that in 1751 by about 7 cm and that which caused the great disaster in 1717 by 26 cm. As in 1751, there is no mention of damage or loss of life: the improved sea defences evidently held. (About half of the 26 cm rise in surge level may be attributable to the local rise of sea level, which had been particularly great in the German Bight.)

Storm winds from SW, later veering WNW, were reported from Danish waters in the Great Belt and later in the Sound (Øresund). By the late evening of the 7th the winds had veered to NW or NNW in the Sound and had dropped to force 7. On the 8th they were NW to NNW only force 4 to 5, while Stockholm experienced an E'ly gale, force 9.

As far inland as Berlin, the weather had been 'very windy' already on the 5th – 'a great wind' that evening – and then after two days with lighter winds and variable rain, showers and sunshine on the 6th and 7th, the experienced observers in Berlin (Kirch family record) reported 'a terrible storm' in the night of 7–8 October.

*Meteorology*

For three days, 5–7 October, an active frontal system with small wave disturbances developing on it, and giving rain, as they travelled eastnortheast seems to have lain over a zone from the Atlantic near 40° N 30° W to the southern Baltic, as seen on our map of the sequence. The rainfall seems to have been considerable. But at first the only indication of strong to gale force winds was reported from Germany, probably ahead (south and east) of the initial warm front of the sequence (seen in the frontal pattern for the 5th on our map below).

What introduced the first violent (damaging) winds into the situation seems to have been a new feature that announced itself in northwest England near the Scottish border around midnight on the 6th–7th. Temperatures fell a degree or two, and the winds became a W to NW storm, force 9, which spread as far south as the eastern midlands of England (Lyndon, Rutland) and perhaps to near London later. It was presumably this feature that later brought the strongest winds to the Dutch, German and Danish coasts and brought the North Sea surge. The heights attained by this tidal surge suggest that during much of the later part of the day on 7 October a strong NW to N'ly gale was also blowing over the northern North Sea. This would be likely in the rear of the deep low pressure system formed by merging of the two centres shown on our midday map. This depression had apparently moved to the Baltic or Swedish coast south of Stockholm, near the island of Gotland, by midday on the 8th.

If we apply in this case Palmén's (1928) statistical associations, which we used in analysis of the 1588 Spanish Armada storms (see Appendix A to our entries for 1588),

to the day to day advance of the biggest frontal wave across Britain to the eastern part of the North Sea from 6th to 7th October 1756, this indicates a WSW'ly gradient wind in the warm sector air of the order of 60 knots. The pressure gradient in the squally cold air stream coming in across England on the 7th was obviously much stronger, but we have too few barometric pressure measurements and too little information about them to indicate the pressure gradient directly. By analogy with other, comparable North Sea flood situations studied in this compilation, and with other cases of reported blasting of the vegetation by sea salt transported inland on the wind, it is probably safe to assume a gradient wind in excess of 100 knots.

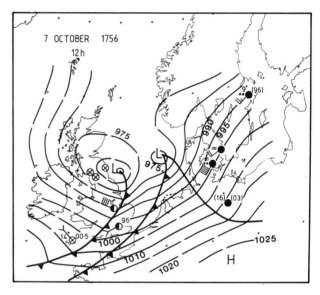

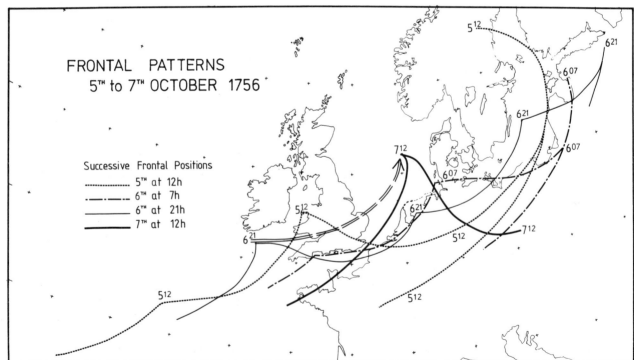

FRONTAL PATTERNS
5ᵀᴴ to 7ᵀᴴ OCTOBER 1756

Successive Frontal Positions
............... 5ᵀᴴ at 12h
.—.—.—. 6ᵀᴴ at 7h
——— 6ᵀᴴ at 21h
━━━ 7ᵀᴴ at 12h

## December 1761

*Area*

North Sea.

*Observations*

A Dutch ship, the S/S *De Liewde*, anchored in the mouth of the Scheldt, was driven loose by violent S'ly gales and driven helpless, without sails, to Egersund in southwest Norway. (Report in the *Jaeren og Dalene Tingbok* in Stavanger Museum, Norway.)

*Meteorology*

The winters of 1760–1 and 1761–2 seem to have been persistently mild in western, central and northern Europe (mean temperatures in central Europe +1.1 °C and +2.2 °C respectively) with high pressure over Spain, Biscay and France and a very high percentage of S'ly, SW'ly and W'ly winds across the British Isles and all northern Europe.

## 20 April 1773

*Area*

Eastern Scotland, probably also northern North Sea.

*Observations*

Unusually severe gale blew down 400 trees in the area of Aberdeenshire around Kemnay (on Donside). (From a Kemnay farm diary.) N'ly or NW'ly gale presumed.

*Note.* A great prevalence of N'ly winds was reported in central Europe (Bohemia) in the early summer of 1773 and on 18 June a gale brought many trees down in that region.

## 23–26 December 1783 and first days of January 1784

*Area*

Northern North Sea and eastern Scotland.

*Observations*

A violent E'ly storm on the 25 and 26 December wrecked three ships on the Scottish coast, one in the mouth of the River Don at Aberdeen and the others on the open coast between there and Stonehaven to the south. Gales had brought snow in Aberdeenshire from the 23rd and by the 31st the frost was the keenest experienced there for many years. On 2 January a violent SE'ly storm blew the snow into 5–6 m deep drifts inland in Aberdeenshire, imprisoning many people in their houses for 2 days before they could be dug out and closing all roads to wheeled traffic. On the 2 and 3 January houses in parts of eastern Scotland were unroofed, rocks were blown into the harbours on the east coast, and stacks of corn and hay were carried away. (From a farm diary at Kemnay, Aberdeenshire.)

A somewhat similar storm affecting these areas in 1740 was considered to have been not quite so severe.

*Meteorology*

These storms occurred in the early stages of a long spell of severe weather, with mainly E'ly winds across central Europe to the British Isles, that continued from early in December till well into January; and apart from rather unimportant breaks and changes of pattern the weather remained severe and winds from between NW and E dominated the scene until about the 20 February.

The storm situations are best illustrated by the maps for 24 December 1783 and 2–3 January in our series.

On 23 and 24 December a depression that had arrived over Scotland from the west or northwest deepened quickly and developed a small intense centre, luckily attested by

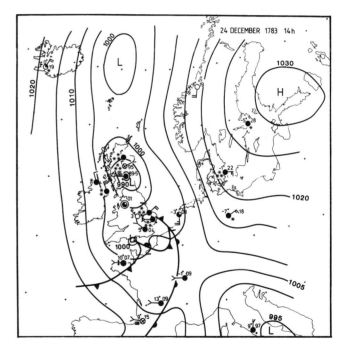

pressure measurements by observers believed to be reliable, although we have inadequate knowledge of the station and instrument details. The strong cyclonic curvature of the isobars, with some uncertainty about the radius of curvature (probably around 70–100 nautical miles in parts of the strong wind zone around the centre) makes calculation of the gradient wind uncertain, but values of 70–100 knots probably occurred on the western, northern and eastern sides of the centre. Gusts within this range probably occurred at the surface.

Two other centres also appeared farther south, attached to the frontal system which had proceeded south to the Channel and southern England. The quick development of these centres from a simpler situation on the 23rd with

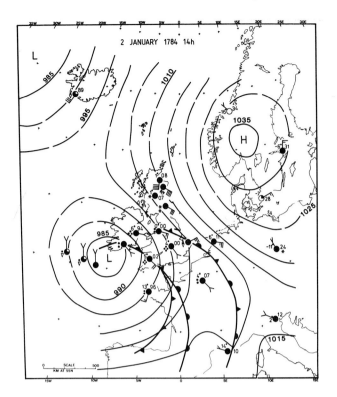

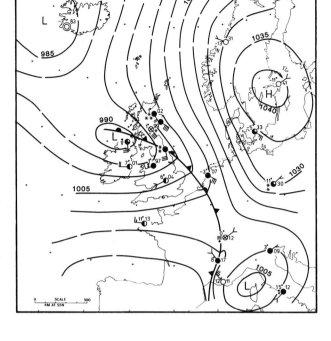

just one centre in this region, over Scotland, depth about 1001 mb suggests that thermal instability in the air that had recently arrived from the Arctic north of Iceland played a part.

On 2–3 January mild air was pushing in over France and England from the Atlantic; but the intense cyclonic activity gradually weakened and the mild air was excluded again.

## 14–15 September 1786

### Area

British Isles, especially southern half, also North Sea and its continental coasts.

### Observations

Information recorded from the correspondence of a farmer's wife living at Kemnay in Donside, 20 miles inland from Aberdeen: 'A hurricane of wind in many parts of England (but not Scotland) threw down houses, overturned coaches and waggons, and killed many people'.

There was much loss of life at sea.

This is confirmed by Gilbert White of Selborne's *Naturalists journal*, which mentions damage 'in particular about London' and by diarists at Stroud and Harwich, the latter mentioning ships driven onto the Dutch coast by the gales. Many trees in the New Forest in the south of England were 'torn up by the roots' during these two days of very heavy gales, according to a Portsmouth reporter.

A gale similar to that which swept across the English Midlands was also reported from Dublin in Ireland on the 14th and places at, and near, the coast of the Low Countries, where there were many ships stranded and one French ship wrecked on a rock, and from Germany, from Flanders to Hamburg. By deduction from the barometric pressure map on the 14th, the North Sea coasts of Denmark and southern Norway probably experienced S'ly gales of quite notable strength at some time on that day.

### Meteorology

This was a WSW'ly gale of damaging strength over a broad belt of country across the English Midlands. The barometric pressure distribution indicates the strongest gradient winds there, probably up to about 80 knots, in a cold, unstable air stream with many showers, frequently heavy. The wind was probably therefore unusually squally.

The wind had probably taken only 20–24 hours to travel on a circuitous track over the ocean from Iceland. In England it brought afternoon temperatures generally of 14–16 °C, but in northern Iceland it had arrived as a N'ly from the Arctic, bringing snow showers, hard frost and blowing snow, even at this early stage of the cooling season. It is known, that the Arctic sea-ice was unusually prominent, and may have largely covered the seas north of Iceland. The rapidity of the heating of this air over the ocean over the intervening hours suggests that great, possibly very great, thermal instability must have been induced in it by the time it reached England.

The approach of the cold front introducing this Arctic air into southern France, central Europe and the Alpine region on the 14th and 15th produced very widespread thunderstorms.

*Note.* Daily synoptic weather maps covering a slightly wider area of Europe than here shown have been analysed in the Climatic Research Unit, University of East Anglia,

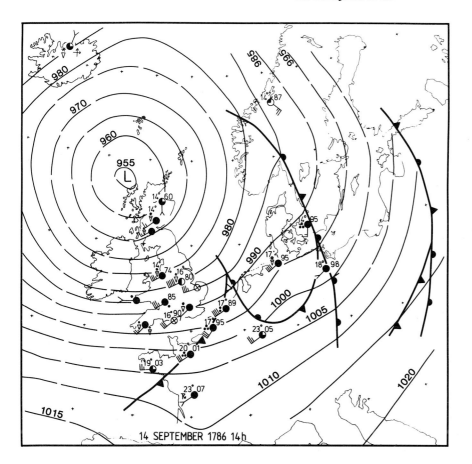

14 SEPTEMBER 1786 14h

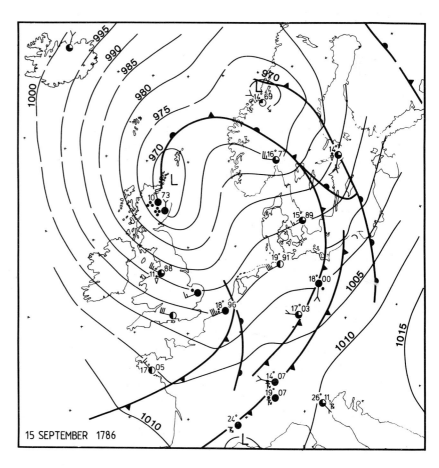

15 SEPTEMBER 1786

Norwich, by Mr J. Kington for every day of the years 1781 to 1786.

A three-day sequence 13–15 September 1786 was analysed in more detail for the study of this storm, here illustrated by simplified maps for the 14th and 15th, showing the observations at only a sample selection of the roughly 50 points entered on the original maps.

## 1791–9, especially 1791 to 1795

*Area*

North Sea – German Bight – River Elbe.

*Observations*

Petersen and Rohde (1977, pp. 52–3) remark that these years saw an exceptional sequence of high and very high storm surge sea floods, which by that time (starting in 1786) were monitored by regular observations of the water level in the Elbe mouth area at Cuxhaven and Hamburg and published twice weekly with daily meteorological observations in the newspaper *Hamburgische Adresse-Comptoir Nachrichten*.

After the floods recorded in 1751 and 1756 there had been a long lull. Not until 1777 and 1778 were there any of any note and these cases seem not to have been of much importance. But between 1791 and 1793 there was a very remarkable sequence. And between 16 November 1792 and 3–4 March 1793 there were four very high flood tides, reaching over 2.5 m above mean high water level, and four more that reached between 1.5 and 2.5 m above the normally expected high tide. These included floods on 29 January and 24 February 1793.

There were further storm surge flood levels in 1794, 1796 and 1797, though these seem to have been less noteworthy.

Petersen and Rohde remark (1977, p. 53) that no such frequency of storm floods occurred again until 1973 and then with not quite the same severity.

Since synoptic weather maps were analysed for every day of the months of March 1791, December 1792, and May 1795, which were particularly stormy in the North Sea, it has been possible to derive the maps of the monthly mean pressure distribution in each of those months as here printed. They all show patterns conducive to prevalent NW'ly to N'ly winds over the North Sea, particularly the northern part.

This notably stormy period also affected the British Isles and coasts and other neighbouring countries as far as Sweden. At one exposed point on the east coast of Scotland, Milton of Mathers a little south of Johnshaven in Kincardineshire, a small port and village known as Miltonhaven was destroyed by the sea. The disaster was prepared by the unwise quarrying for lime, for agriculture on the nearby Lauriston Castle estate, from the projecting ledge of rock, which ran across the bay and had formed a natural protection for the village. Three houses were carried away by a sea flood in 1792 (Watt, 1985) and then during an easterly gale in 1795 nearly all that remained of the village and harbour wall were carried away in one night (Lyell, 1830–3, *New Statistical Account* 1843): the destruction was finally completed by another storm in 1829. Unfortunately no record has been found of the precise dates of any of these stages in the destruction of Miltonhaven, so the individual storms involved cannot be identified.

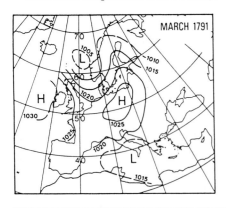
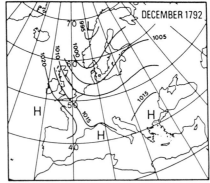
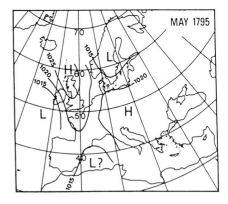

## 1 March 1791 and continuing part of next day

*Area*

Southernmost part of the North Sea, coast of Kent and East Anglia.

*Observations*

Report from Canterbury, evening of 1 March: 'A perfect hurricane from the NW'. On the 2nd, Maldon (Essex) report: 'Exceeding hard gale and flood tide'. As early as 27 February, Parson Woodforde at Weston Longville near Norwich had noted 'A very cold, wet, windy day almost as bad as any day this winter'. Though somewhat less on the 28th, there were snow and sleet showers at Hoveton, north of Norwich; and on the 1 March the Hoveton House diarist noted 'frequent storms of hail and snow'. No reports have been found of strong winds on the continental side. At Haarlem and Zwanenburg in the Netherlands and at Brussels only moderate winds were reported. Nevertheless,

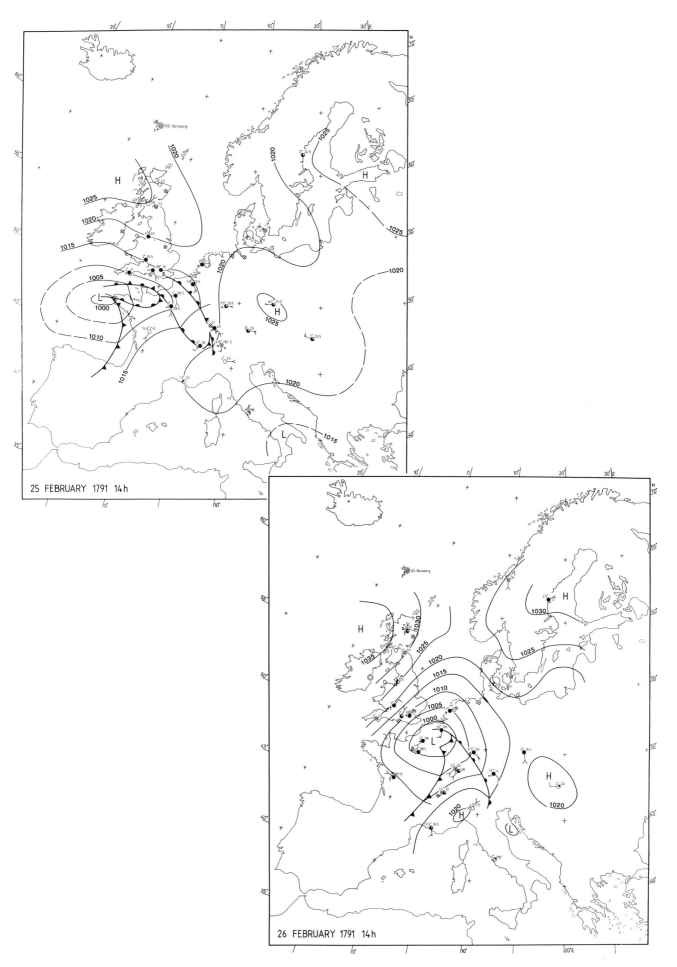

25 FEBRUARY 1791 14h

26 FEBRUARY 1791 14h

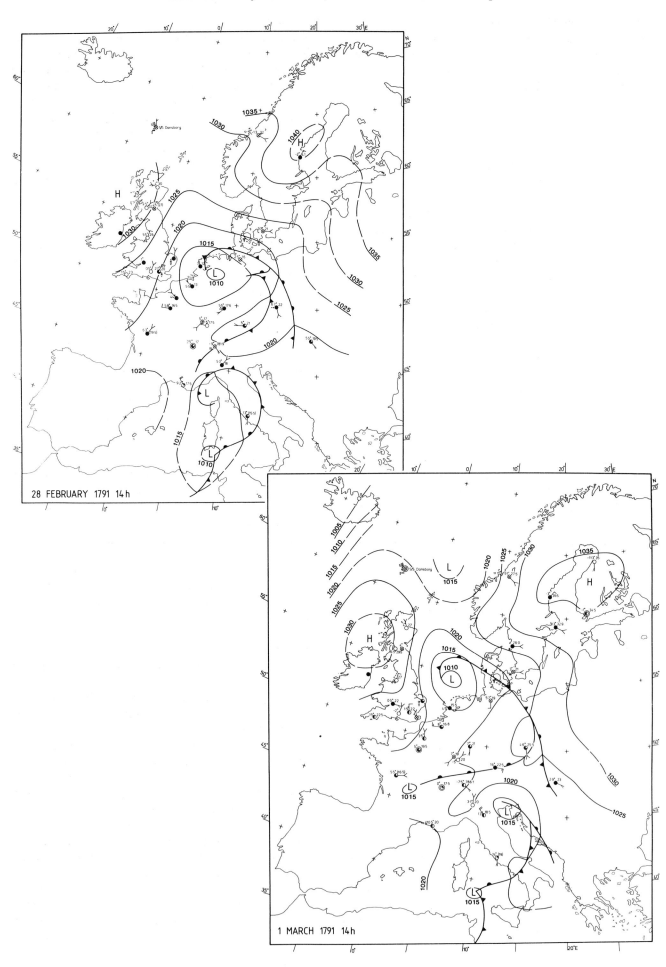

28 FEBRUARY 1791 14h

1 MARCH 1791 14h

on the morning of the 2 March: 'it was flood tide along the sea coast of Kent fully 2 hours before the usual time ... Sea walls and banks at Seasalter were broken down in several places and washed away and all the marshlands overflowed to a great depth.'

Similar flooding was reported at Whitstable, Deal, and other towns on the coast: 'This tide was considerably higher than the remarkable one some years ago on 1 January'.

On the Essex side we learn from Maldon that 'the unexpected high spring tide of this evening (1 March) has done immense hurt and on each side of the Black Water and Crouch Sands the tide ran more than 2 feet above the height of the sea walls and laid the whole land marshes under water. No man remembers so high a tide.'

On the 2nd, all the marshes in the neighbourhood of Queenborough and Sheerness were covered by the sea water. And in London on the 2nd the tide rose in the Thames 'to such an amazing height that in the neighbourhood of Whitehall most of the cellars were full of water. The parade in St James's Park was overflowed, as was Palace Yard and Westminster Hall as well as the Temple Garden ....' Damage to the cornlands beside the Thames was estimated at £20 000.

There was reference to 'a similar disaster to this' on 24 December 1736 (Old Style) q.v.

## Data sources

Most of the above reporting is what the Revd James Meek of Cambuslang near Glasgow copied from his correspondents. The flooding in London is, however, verified by official records and is referred to in recent literature about the Thames Barrier.

## Meteorology

There is some mystery about the severity of the flooding, but the synoptic weather maps indicate gradient winds over the area from Norfolk to the Straits of Dover of 60 knots from N or NNE already on 27 and 28 February, i.e. for up to 60 to 70 hours before the flooding, and on 1 March the gradient wind was about 75 knots from the N all down the coast of Britain from Aberdeenshire to the Wash. A feature of the N'ly windstream was that its course was nearly straight (great circle) over the distance from latitude 57° to 52° N so that the gradient wind and geostrophic wind were presumably of the same strength. The wind turned, to become NW, around the coast of East Anglia. The pressure gradient was of similar strength as far as 200 km out over the North Sea, near the centre of the depression, when this was approaching the central North Sea, though the gradient wind would be less strong there (perhaps 55–60 knots) on account of the cyclonic curvature of the wind's path.

This was a depression, probably already complex and with a long previous history, which had approached Britanny and the Channel from the SW on 25 February and moved slowly ENE passing over Normandy and Belgium to central Holland by the 28th, then curved N and later NW to the central North Sea by the night of 1–2 March before passing away more quickly N up the coast of Norway.

## Maximum wind strengths

From the measured pressure gradients, N'ly winds of about 45–50 knots were probably sustained at the surface over the westernmost parts of the North Sea for 3 days, with gusts on the 1 March to around 75 knots.

# 21–22 March 1791

## Area

North Sea, especially western, southern and (on the 22nd) central and eastern parts.

## Observations

'One of the severest floods of the century' in the River Elbe at Hamburg at 5h on 22 March, according to the report in the newspaper *Hamburgische Adresse-Comptoir Nachrichten* giving weather and water level observations for the week. Comparisons given with the extreme tides earlier in the century, including some details from M. Petersen and H. Rohde in *Sturmflut– die Grossen Fluten an den Küsten Schleswig-Holsteins und in der Elbe* (Neumünster, Karl Wachholtz Verlag, 1977), imply that this flood reached higher than the disastrous flood of 25 December 1717. The floods of 11 September 1751 and 7 October 1756 (all dates New Style) reached about the same height as this one, that in 1756 apparently 7 cm higher. (All these were overtopped by the flood of 11 December 1792, which exceeded this March 1791 case by 12 cm and reached 3.70 m above the mean high water level. (The December 1792 case came in the course of a remarkable sequence of storms and floods, the greatest such sequence in the century.)

Direct reporting of the outstanding gales of the 21–22 March 1791 extends as far south as Norfolk and Paris; the reported violence of the westerly wind was also noted in the journal of Deluc at Geneva.

## Meteorology

A quite small and weak low pressure centre began to develop on 18–19 March 1791 over the southernmost part of the North Sea, its central pressure falling from 1025 to 1015 mb, and a very slow movement towards the east was registered.

The first sign of a potentially serious development seems to have been a cold front which moved slowly south over Iceland during the same 24 hours, bringing very cold Arctic air from the north, and 'adding a great deal to the snow cover since yesterday' in north Iceland with a prolonged blizzard of blowing snow – here attributed to a wave on the cold front during the night of 18–19 March. The same front seems to have turned the winds to N over Scotland between 20 and 21 March, while the depression over the southern North Sea turned northwards to reach the central North Sea and a depth of 985 mb by 14h on the 20th. It seems to have had a depth of about 955 mb off the coast of Jutland by 14h on 21 March, having absorbed another low pressure centre previously associated with the cold air invasion coming from the north.

The depression lingered in the Skagerrak, filling up, during 22–24 March (central pressure c. 980 mb on the 22nd, 985 mb on the 23rd, 1015 mb on the 24th) and was ultimately replaced by another centre off Trondheim, which moved away northwards off the Norwegian coast.

*Maximum wind*

Gradient winds up to 150 knots from WSW, later W, are indicated over the German Bight near Hamburg and just south of the Bight, over Friesland, on the afternoon of 21 March. Strongest gusts may well have reached 120–130 knots. At the same time a great current of winds from about NNE was blowing into the northern North Sea, where gradient wind strengths 120–130 knots are indicated over the western half of the sea area all the way from around Shetland to the coast of Northumberland. As in the storm of 1 March 1791, the wind's path was nearly straight (great circle) over this stretch. The entire breadth of this northerly airstream bringing Arctic cold air southwards extended from Iceland to about 150 km off the coast of Norway. In the central North Sea near 53° N, gradient wind strength was about 100 knots from W and WNW, but nearer the centre of the depression the cyclonic curvature effect presumably reduced the gradient wind to about 80 knots and at the Danish coast at 14h on 21 March winds were much lighter.

By 14h on 22 March the winds over the German Bight had veered to WNW and decreased a good deal (gradient wind 70–75 knots), but farther north, off southwest Norway, gradient winds from NW of about 100 knots were still blowing towards the German Bight and the coast of Denmark.

At 14h on 23 March, the N'ly gradient winds were still 100 knots off the coast of Norway between Bergen and Ålesund and 80 to 90 knots around Shetland.

*Data source*

The main collection of daily weather reports used in the map analysis of the 1791 storms was the 1791 volume of the *Ephemerides* published by the *Societas Meteorologica Palatina* in Mannheim. (It should be noted that the dates given for observations at places in Russia in that collection are on the old Julian calendar and need to be corrected by adding eleven days.) Besides the data in the Mannheim *Ephemerides*, many diaries and registers of daily weather observations in other countries, notably Britain and Ireland, have been used, and I am particularly indebted to Dr K. Frydendahl of the Danish Meteorological Institute, Copenhagen and Dr E. Hovmøller of the Swedish Meteorological Service for collections supplied, as well as to Sjöfn Kristjánsdóttir of Reykjavik for obtaining the record from northern Iceland.

## Appendix D. Error tests: March 1791

A check upon the magnitude of the probable errors affecting the March 1791 map analyses in the northern parts of the maps such as the North Sea was made by adding subsequently Swedish and Finnish stations in the gaps in the observational network between:

| | |
|---|---|
| Stockholm–Trondheim | 600 km |
| Copenhagen–Trondheim | 830 km |
| Stockholm–St Petersburg | 720 km |

comparable with the breadth of the gap across the North Sea in the observational network:

| | |
|---|---|
| Norfolk–Hamburg | 600 km |
| Edinburgh–Hamburg | 870 km |
| NE Scotland–Trondheim | 1000 km |

Unfortunately, the addition of the Finnish stations revealed clearly that the St Petersburg observations were made on the old Julian calendar, so that the maps had to be altered in that regime and redrawn.

Tests in the western and central/northern Sweden area in or near the Copenhagen–Trondheim and Stockholm–Trondheim gaps revealed that:

(i) The *barometer at Gothenburg* was not one of the best and did not respond to the quicker/bigger changes of pressure fully, but eliminating the minority (eight out of 31) of days when this made the Gothenburg pressure readings obviously wrong, the *Standard Error of the pressure shown at Gothenburg appeared to be 1.24 mb* (though part of this may still be due to misbehaviour of the barometer). This would represent an error of 1 mb per 200 km from the nearest station, at Copenhagen, corresponding to between 5 and 10 knots in the gradient wind indicated.

(ii) The wind direction observations at Gothenburg fitted the isobar pattern excellently on about 25 days of the month, showed departures of up to 45° on five days probably attributable to light winds and local effects, and showed a bigger departure (100°) on one other day of very light wind near an anticyclone centre doubtless due to some quite localized detail of the pattern.

(iii) The weather and wind changes registered full agreement with the pattern of fronts and frontal passages indicated by the analysis throughout the month.

(iv) At Härnösand bigger errors might be expected because the station lay 250 km beyond the line joining the northernmost observation stations entered on the map. The position was also sometimes affected by the misleading data entered for St Petersburg when the first analysis was drawn.

The *observations at Härnösand* showed a standard 'error' of 3.72 mb, really a 'Standard Departure' from the first analysis. However, when the charts were altered and redrawn with full observation data including one station 540 km NE of Härnösand this station still showed a standard error of 1.45 mb which must be a measure of the observable unreliability (similar to the Gothenburg barometer) of the responses of the instrument to pressure changes. It seems therefore that the *standard error of the analysis was of the order of just 2 mb at this point.*

(v) The wind and weather observations showed virtually a 100% fit with the pattern of frontal analysis and frontal passages at Härnösand throughout the month.

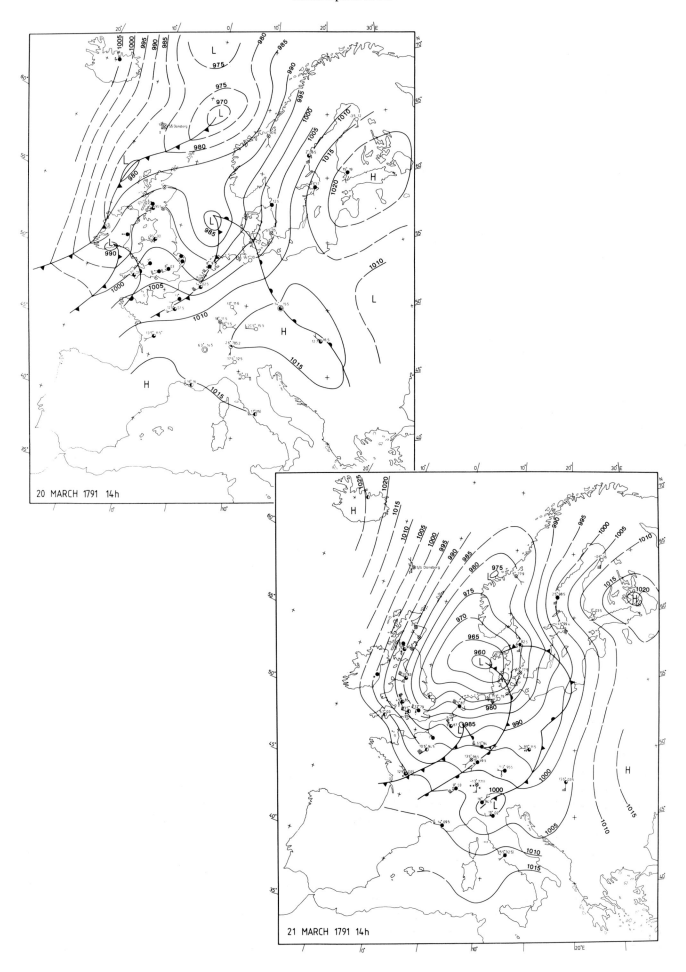

20 MARCH 1791 14h

21 MARCH 1791 14h

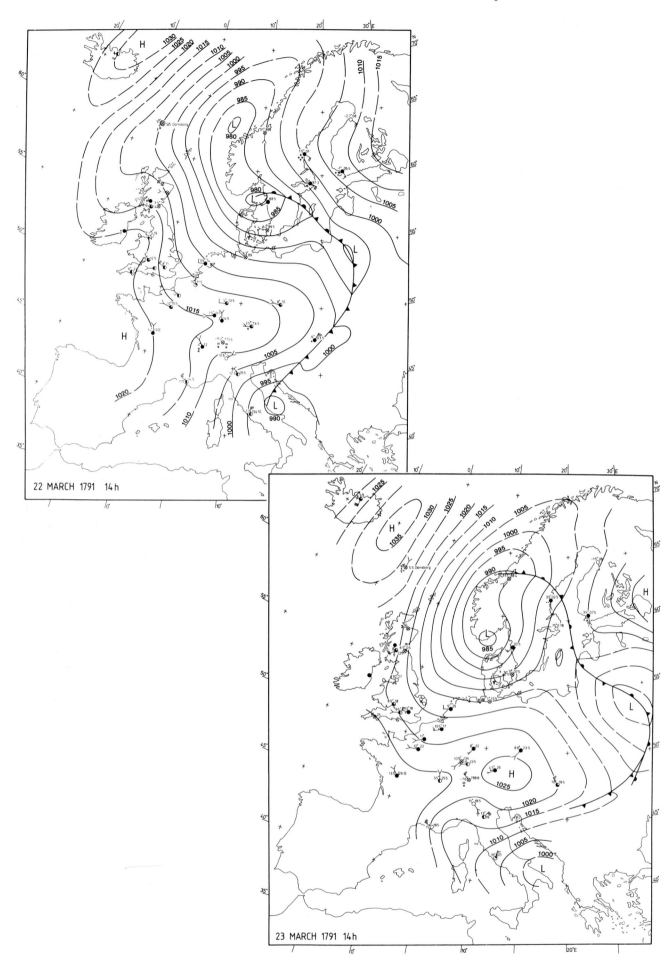

22 MARCH 1791 14h

23 MARCH 1791 14h

The main conclusions from these tests seem to be that the map analysis pattern was basically right throughout the month across the gaps of 500 to 1000 km in the observation network, though the pressure gradients might be in error by an amount which would correspond to a standard error of 5 to 10 knots in the gradient wind.

A deterioration of almost twice this magnitude sets in beyond the limit of the last stations entered on the maps and means that only the general shape of the pattern, for which conventionally broken lines should be used, can be accepted beyond 200 km beyond the last stations. The maps should doubtless be altogether discontinued 400 or 500 km beyond the last stations.

Similarly broken lines, indicating acceptance only of the broad pattern, must in many cases be appropriate in the wide gap between the Iceland and European observations.

## Note on the stormy month of December 1792

The month of December 1792 is remarkable for the frequency with which gales and storm-force winds were reported from many parts of Europe. Deep depressions, in some cases the central pressures probably going below 950 mb, were passing eastwards in sub-Arctic latitudes sometimes at speeds unusual for such deep cyclones and therefore indicating a very strong jetstream. Most of the gales of the month were of correspondingly brief duration as the systems passed mostly quite quickly over each area, but the intervals between the gales were also short, particularly between 5 and 12 and between 18 and 23 December.

Great thermal contrast between the warmest and coldest airmasses is indicated by temperatures up to 11–13 °C in the warm air passing over England and the western fringe of Europe south of 50–53° N and the fact that, despite lack of still, cold periods of clear weather, afternoon temperatures as low as −9 °C occurred in Stockholm in a N'ly wind on 12 December and by the 14th −13 °C was registered in the afternoon at Trondheim and on the 15th −20 °C at Härnösand (62° N) on the Baltic in essentially the same airmass.

Stormy winds were reported on 10 days of the month in northern Holland, 8 days in Paris, 7 at Marseille, 9 in Moscow, and 6 in Hamburg. Storm force winds probably occurred somewhere or other over the North Sea on 14 days of the month. Over the northernmost part of the North Sea and also off the coast of northern Holland, Friesland and north Germany Beaufort force 9 or more probably occured on at least 12 days.

Among the gales of the month most were such as might occur at least once within any 5-year period nowadays, but the storm of 10–11 December 1792 would certainly be considered a rare event. Maximum gust strengths at the surface in the German Bight region in that night were probably in the range 100–120 knots.

## 5 December 1792

*Area*

Southern North Sea and coasts from Holland to Hamburg and the German Bight.

*Observations*

This was the first storm of a remarkable series over the following month, especially during the next week, and already produced an alarmingly high flood tide in the Elbe at Hamburg.

*Meteorology*

A high pressure system (about 1038 mb) centred over Sweden had dominated Europe at the beginning of the month with light winds and a frost situation over northern Germany. The frost at Hamburg had lasted from 24 November to the morning of 4 December, and by 1 December the Elbe was ice-covered. The high pressure had extended as far west as England, where temperatures were also low, and Ireland where the weather on the 2nd was described as fine, mild and open.

The situation was breaking down on the 3 December as the anticyclone moved south into central Europe and westerly winds spread in from the ocean, preceded even in Ireland by sleet and then prolonged rain. A small but deep depression with warm air from Biscay (temperatures reached 12 °C in London on the 5th) advanced quickly across southern England, deepening rapidly and accompanied by pressure falls of 25–30 mb in 24 hours, to be centred (about 980 mb) near 54° N 5° E over the southern North Sea by 14h on the 5th and reached the border of Russia near 55° N 24 hours later. Its movements seem to have been about 750 and 900 nautical miles in the successive 24-hour periods from 14h on the 4th to 14h on the 6th. Geostrophic winds from WSW of about 100–150 knots seem to have occurred in the warm sector as it passed over Holland and northwest Germany and 100 knots from WNW in the air behind the cold front. These wind speeds are mostly similar to what might be expected in extreme cases in today's climate in the month of December. It was in these same regions that the strongest surface winds were reported, though observers as far south as Paris reported 'a great wind' from the SW. The wind at Hamburg was strengthening from the afternoon of the 4th onwards, as a generally cyclonic situation approached, associated with activity on other frontal systems farther north. By the evening of the 5th the WSW wind at Hamburg was reported as a 'hefty storm'. The barometer at Hamburg fell about 40 mb in 24 hours as the storm approached.

*Maximum wind strengths*

The most reliable indications are those of the 100–150 knot geostrophic winds on the afternoon of the 5th, though the pressure gradients cannot be fixed exactly. The duration of winds of this strength was probably nowhere more than 8–12 hours as the small depression passed quickly east.

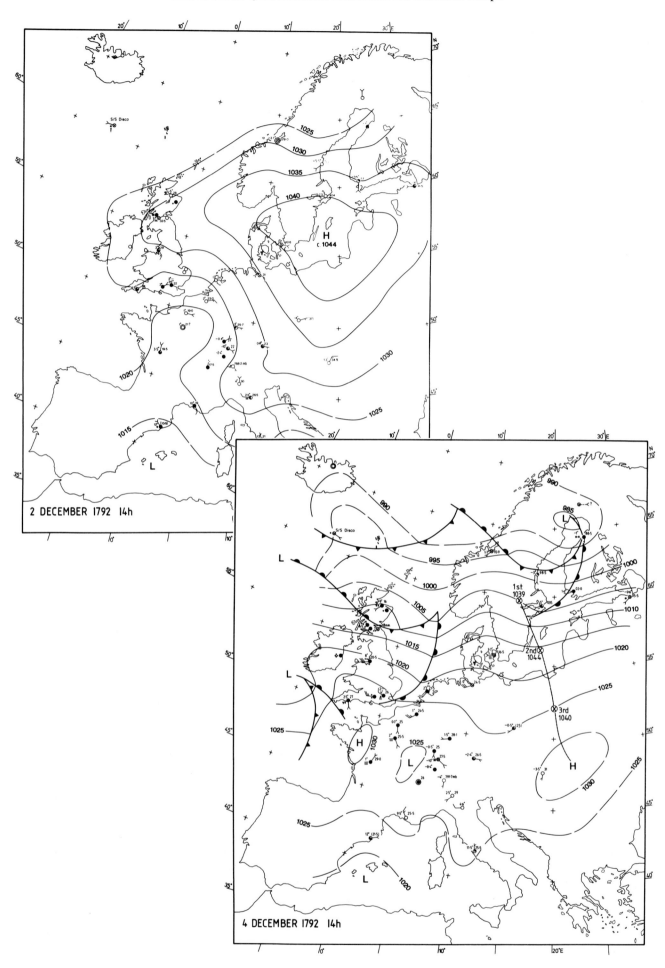

2 DECEMBER 1792 14h

4 DECEMBER 1792 14h

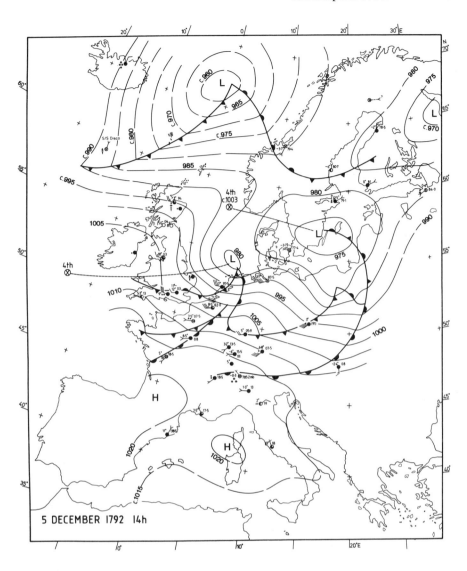

5 DECEMBER 1792 14h

## 7–8 December 1792

### Area

Whole North Sea, also Denmark and the southern Baltic.

### Observations

Stormy winds from NW (or W to NW over the region from northern Germany to the Baltic) and a notably high flood tide in the Elbe at Hamburg (though this was exceeded by about 1 m in the next storm which produced an outstanding extreme flood in the night of 10–11 December).

### Meteorology

Another cyclonic centre, with an associated warm air sector crossing England on the 6th, amalgamated with systems from farther north to form a deep low pressure system, (probably) with two centres about 960 mb, crossing south Sweden on the 7th. This was a fast-moving system, the northern low pressure centre having covered about 800 nautical miles in 24 hours from the 6th to the 7th and covering 700 nautical miles farther to be near St Petersburg by the 8th.

### Maximum wind strengths

The strongest winds developed in this storm were described as 'very high indeed' in Norfolk, where the climax was at about 3h on the 7th, and a 'hard wind' in Holland. At Hamburg the strong W wind on the afternoon of the 7th increased to a storm in the evening and brought a 5 m high tide in the Elbe before abating later in the night. Geostrophic winds of 120 knots or more were indicated over a belt from the sea off the west coast of south Norway and across Denmark to the south Baltic by the barometric pressures read at 14h on the 17th, and about 140 knots over northern and eastern Denmark. Winds of gale force seem, however, not to have lasted for more than 5–8 hours.

99

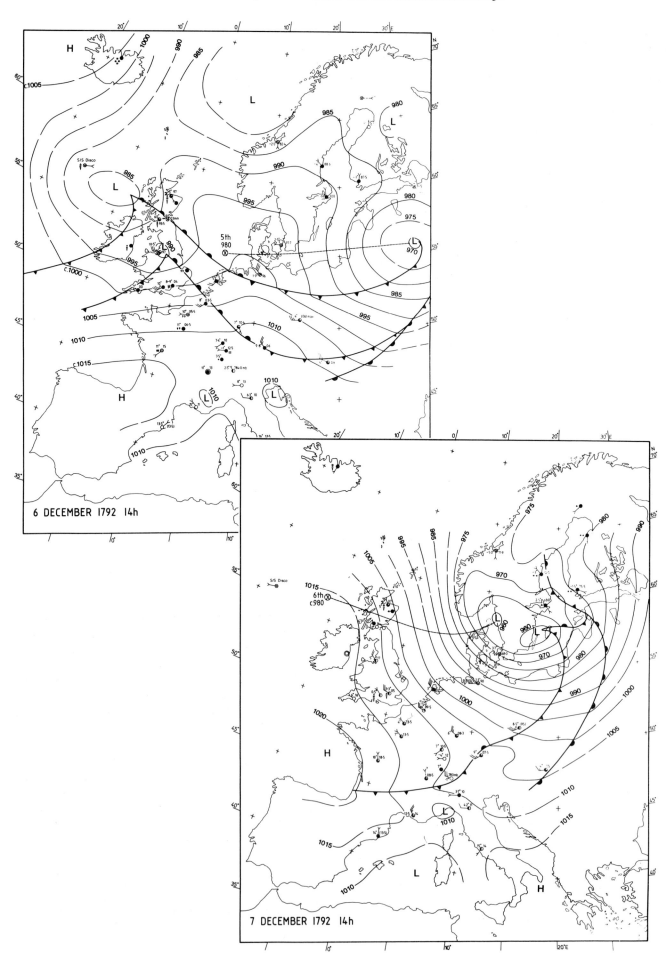

6 DECEMBER 1792 14h

7 DECEMBER 1792 14h

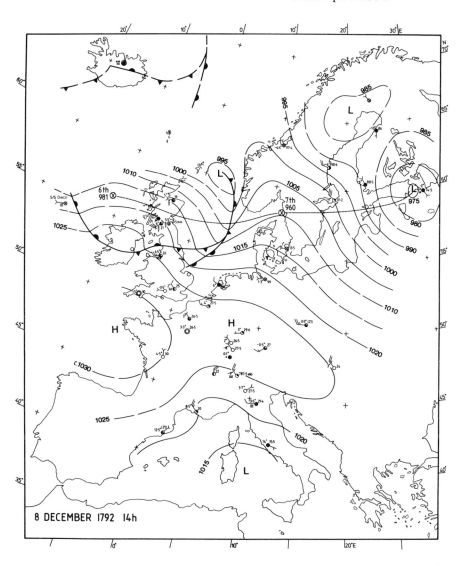

8 DECEMBER 1792  14h

## 10–12 December 1792

*Area*

Whole North Sea and as far north as the waters south of the Faeroe Islands and east to Norway.

*Observations*

Winds of great violence were reported, mainly W'ly over the southern North Sea on the 10th, but soon NW'ly farther north and later over the whole region. On the night of 10–11 December this storm produced the highest flood tide of the century (5.88 m*) in the Elbe at Hamburg with great damage to the dikes protecting the coast.

As in the preceding storm on 7 December, cyclonic systems bringing a warm airstream over England, the Low Countries and northern France with temperatures up to 11–13 °C widely occurring, amalgamated with major cyclonic activity farther north. On the 9th and 10th centres with pressure below 950 mb were indicated travelling east near latitude 64–65° N. During the 10th another centre associated with an open frontal wave beginning to occlude probably reached a depth of 950 mb as it passed Shetland towards northwest Norway. Westerly gradient winds of the order of 150 knots are indicated on the 10th

over the Shetland–Faeroe Islands region, and perhaps reaching this same strength from SW or WSW later that day over northern Scotland, as the frontal wave passed. A report from inland of Aberdeen tells of violent winds all day on the 10th; 'people remember nothing like the winds of these three days'. Strong gales and 'a great wind' from the W were reported on the 10th widely over southern England, Holland and northern France as well as in Dublin and the Glasgow region, though there seem to have been less strong winds for part of the day over northern and eastern England. At Hamburg the wind backed to SW about 17h and became very strong before 'turning somewhat to the North' about 21h. A hefty storm continued all night, probably from about NW or WNW. The flood tide which accompanied this storm exceeded that on 22 March 1791 by about 10 cm.

*Unfortunately, there is no record of the base level above which the tide measurements recorded in the newpaper *Hamburgische Adresse-Comptoir Nachrichten* were measured, nor of the exact location of the measurements. A summary of the eighteenth century high flood tide level records is given by H. Rohde in *Nachrichten über Sturmfluten in frühere Jahrhunderte nach Aufzeichnungen Tönniger Organisten*, *Die Küste* 12, pp. 113–132, 1964 in Heide i.H. (Westholstein Verlag Boyen).

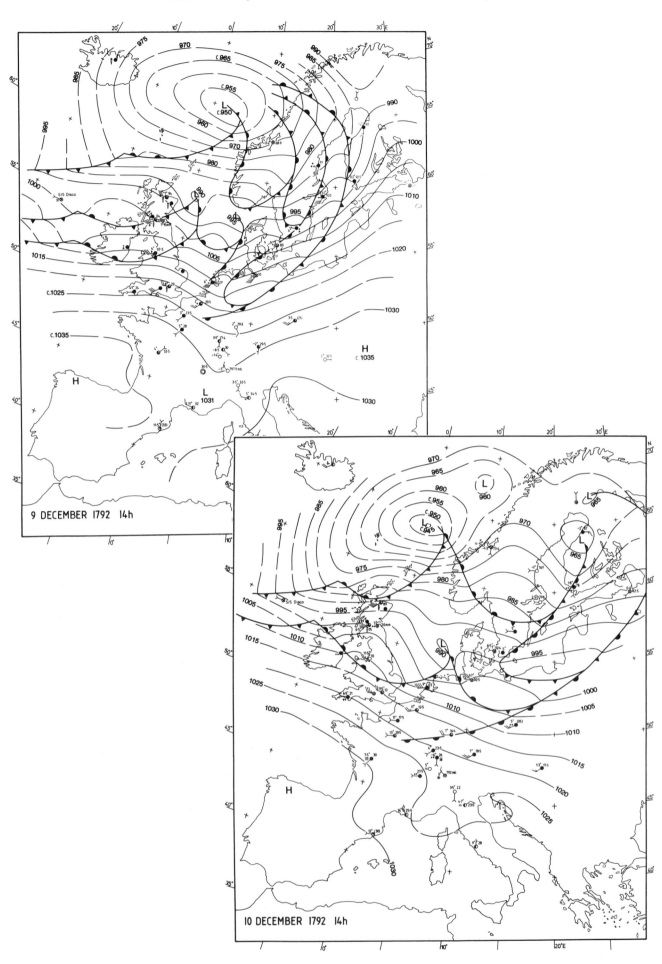

9 DECEMBER 1792 14h

10 DECEMBER 1792 14h

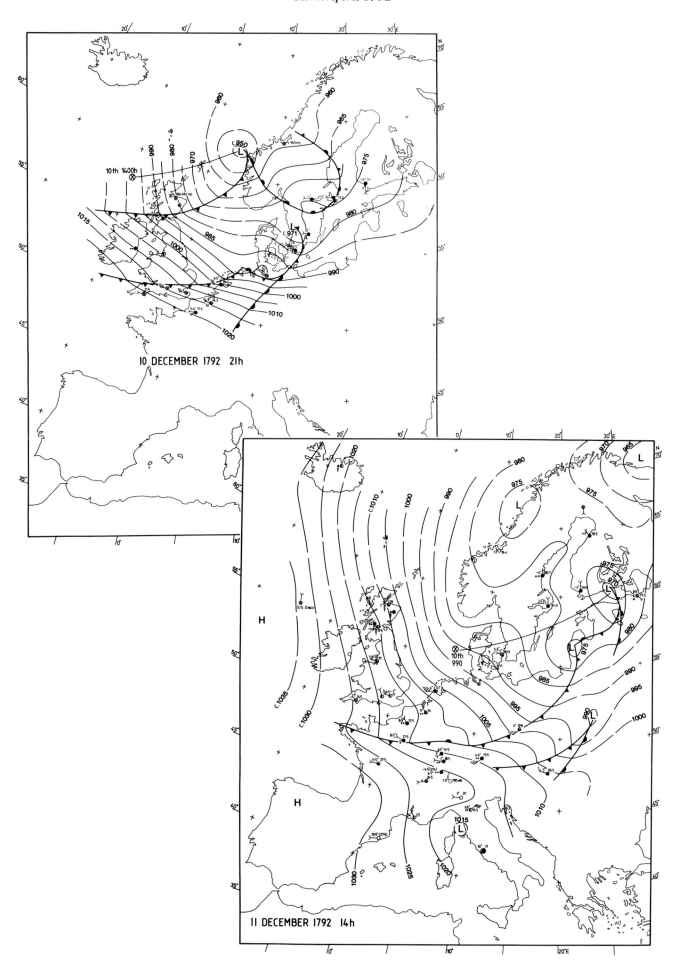

10 DECEMBER 1792  21h

11 DECEMBER 1792  14h

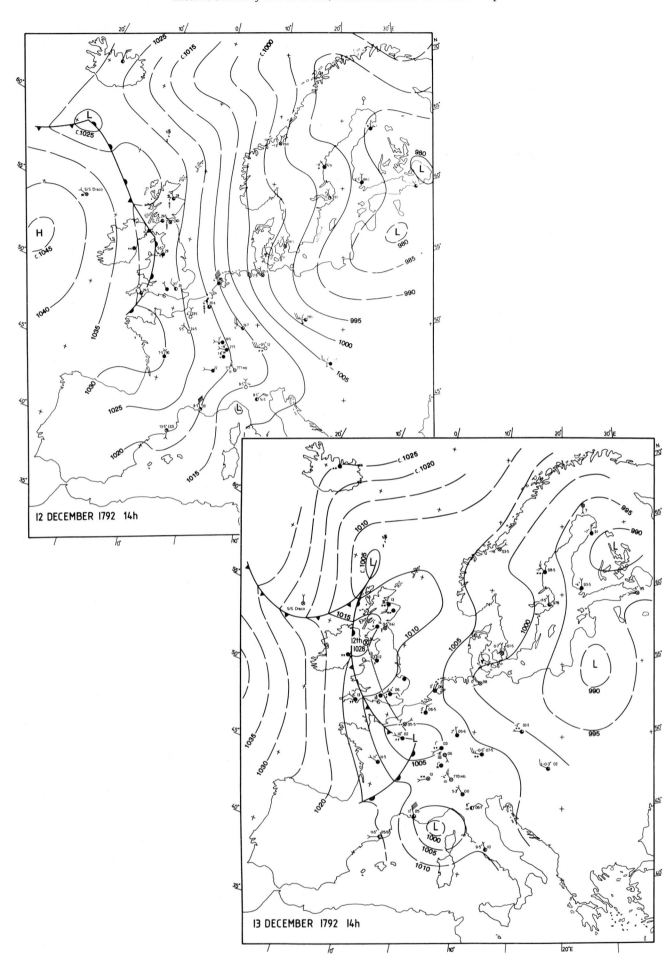

12 DECEMBER 1792  14h

13 DECEMBER 1792  14h

Winds still of gale force over the North Sea finally became NW to N'ly on the 12th before decreasing.

*Maximum wind strength*

No direct measure of the wind strengths is possible, though the strongest gradient winds indicated by the barometric pressures exceeded 100 knots at some time on the 10th at most points over England, Holland and the northern fringe of France northwards to latitude 64° N; a gradient of 140–150 knots is indicated in the afternoon over the region north of about 58° N. In the evening 150 knots is indicated over Holland and northwest Germany. The paths of the wind in the areas of strongest pressure gradient in this depression were nearly straight, so that gradient wind and geostrophic wind strengths were alike. It is probable that the surface winds reached Beaufort force 11 to 12 at times over the waters between Scotland, the Faeroe Islands and Norway in the westerly windstream during much of the 10th and also in the NW to N winds during the later part of that day and the following night. Force 11 to 12 also seems likely to have occurred over the southern North Sea near the Dutch and German coasts. The strongest gusts of the surface wind probably exceeded 100 knots (50 m/sec) over both these regions as well as

at exposed places on land in northern Scotland and the islands and in the Netherlands and northwest Germany.

This storm was also of longer duration than the previous ones in the sequence, though there had been little intermission (only about 24 hours on two occasions) since the storm on the 5th. Gradient winds of 60–75 knots or more were continuously present over the North Sea for 50 hours or more from the middle of the day on the 10th to the later part of the day on the 12th.

Winds were of gale force or more in many parts of the region for 3–4 days and with little intermission over the 7 days from 5–12 December.

This was clearly the severest storm of the month over the North Sea area. The gradient winds measured support the indication of this given by the highest flood tide of the century in the Elbe at Hamburg. Reasonable estimates of the error margins suggest that the true figures for strongest gradient winds on the 10th must have been W'ly $150 \pm 30$ knots over the Faeroes–Shetland region and W to NW $150 \pm 20$ knots in the evening or early part of the following night over the German Bight and Frisian and northwest German coastlands. Strongest surface gusts over 100 knots would be likely and possibly in the range 110–120 knots.

## 19–23 December 1792 (and a note on 30 December)

*Area*

Initially southern North Sea and Dutch and German coasts, then the whole North Sea on 21–22 December, and concentrated over the central, northern and western North Sea on 23 December.

*Observations*

On the 18th, with falling pressures over the whole region, W'ly winds were reported strengthening in East Anglia, the Netherlands, northwest Germany and as far south as the northernmost part of France. (The analysis suggests that another belt of very strong winds, in this case, from W to WNW, developed on the 18th over the northern North Sea and coast of Norway between about 58° and 62° N, but there are no actual observations available on the coast south of Trondheim.) In the night of 18–19th the W to NW winds were described as very strong (or 'great wind') everywhere from Hamburg to Paris. On the 20th the winds in this area backed to SW and were again strong, especially about north Holland. On the 21st these places reported 'very hard wind' from between W and NW, which on the 22nd backed to SW while remaining strong and gusty. At Hamburg the tide level again became notably high. At the same time the NW, and later (on the 23rd) N, winds on and near the eastern coasts of Scotland and England became 'very high indeed', or 'as high as has been known for a long time'.

*Meteorology*

As in the storm sequence between 5 and 12 December, vigorous cyclonic activity and frontal systems in latitudes

about 60–65° N were accompanied by other systems passing east across Britain and the southern North Sea, which were bringing very mild, humid air in the warm sectors over England and parts of the continent south of 53° N. Temperatures up to 11 °C were measured as far north as Liverpool and 9 °C, or over, as far into the continent as Mannheim and Munich. There were prolonged and quite heavy falls of rain with the fronts over England and the Netherlands.

A strong jetstream was indicated by one of the main depressions travelling eastsoutheast 700 nautical miles in 24 hours on a track from near Iceland to the eastern Baltic States on the 18th–19th. Later the cyclonic centres moved on more meridional tracks from the North Sea to north Norway and from east of Iceland to Holland. The deepest centre may have attained a depth of 950–955 mb east of Iceland on the 20th. The sequence ended with Arctic cold air brought south with the final depression, giving temperatures of −1 °C to −2 °C in the strong N wind over Scotland on the afternoon of the 23rd and considerable snowfall as far south as Norfolk. Some days of rather cold weather over Britain and western Europe followed, with rain, sleet and snow at times, lighter variable winds and temperatures generally between −2 °C and +4 °C. On the 30th another, small depression moving quickly (over 600 nautical miles in 24 hours) southeastwards across the North Sea brought another gale, though briefly, over parts of the region.

*Maximum wind strength*

Gradient winds over 70 knots were first indicated off the coast of west Norway and northern Holland and Germany on the 18th. Over the 19th and 20th the strongest gradient

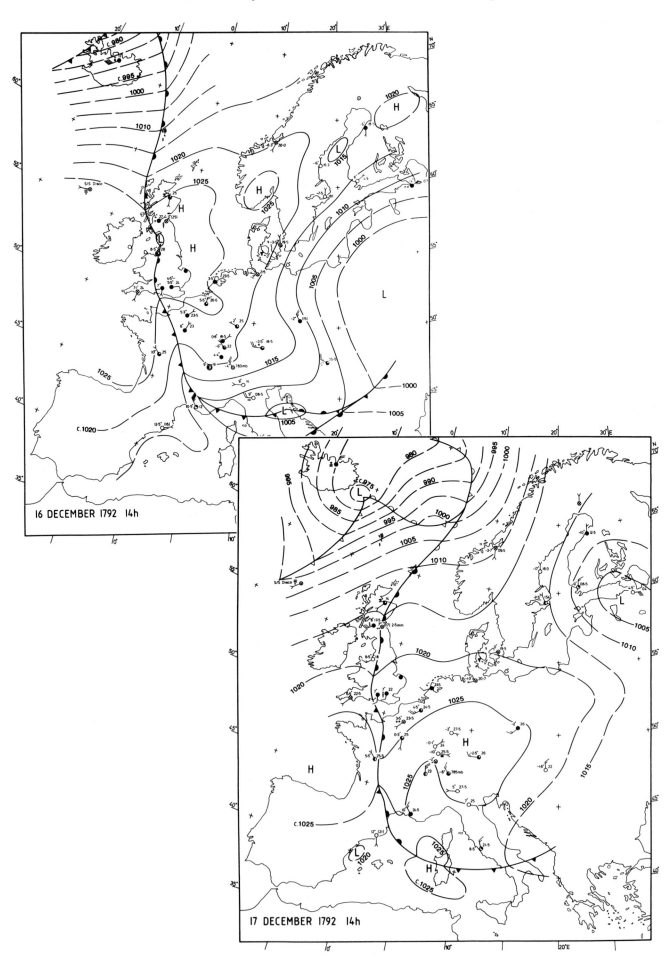

16 DECEMBER 1792 14h

17 DECEMBER 1792 14h

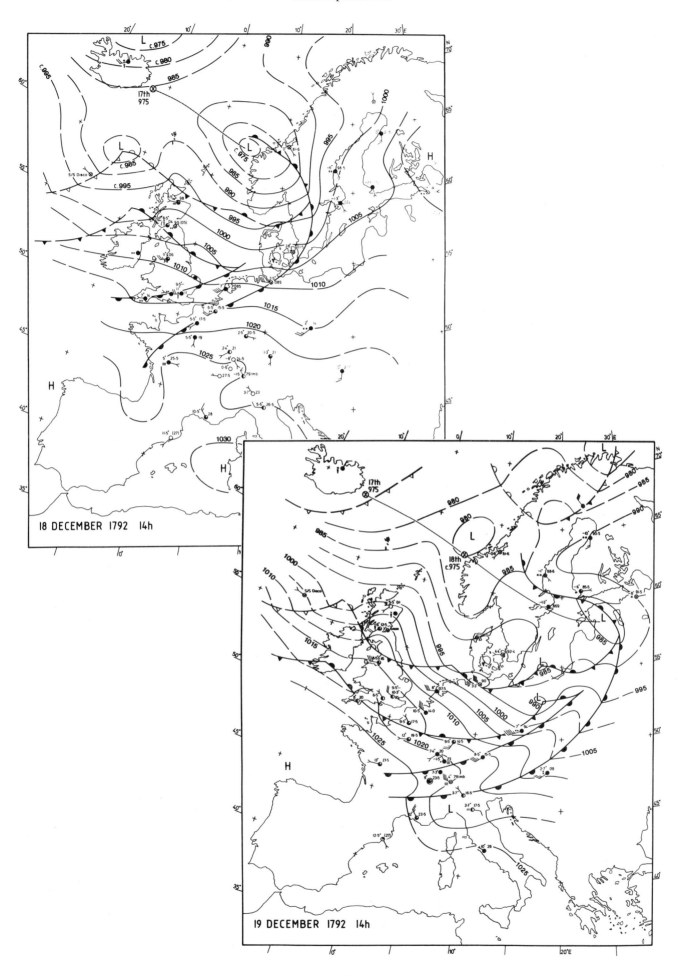

18 DECEMBER 1792  14h

19 DECEMBER 1792  14h

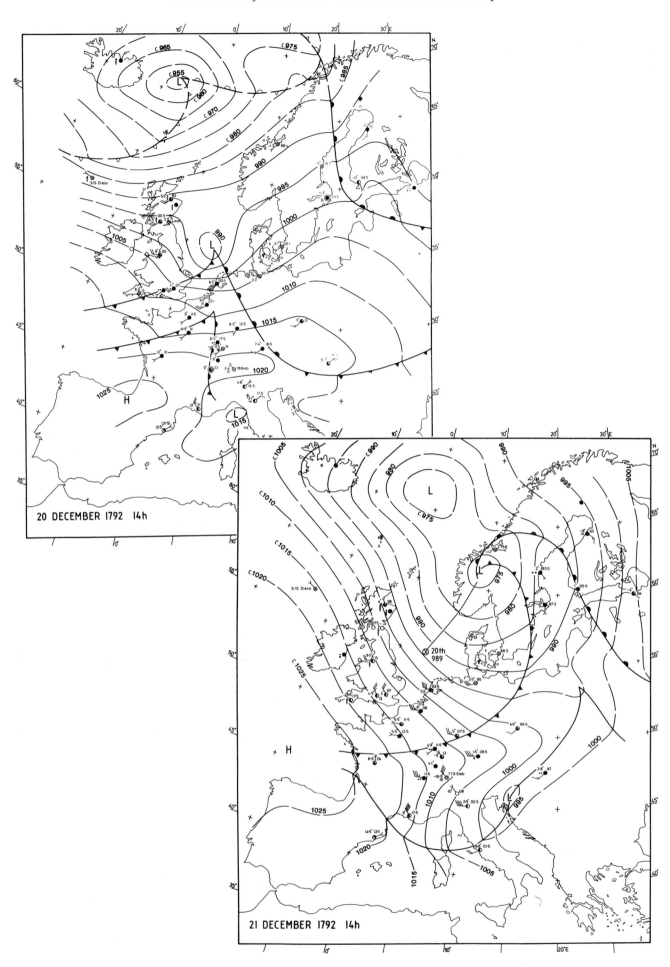

20 DECEMBER 1792  14h

21 DECEMBER 1792  14h

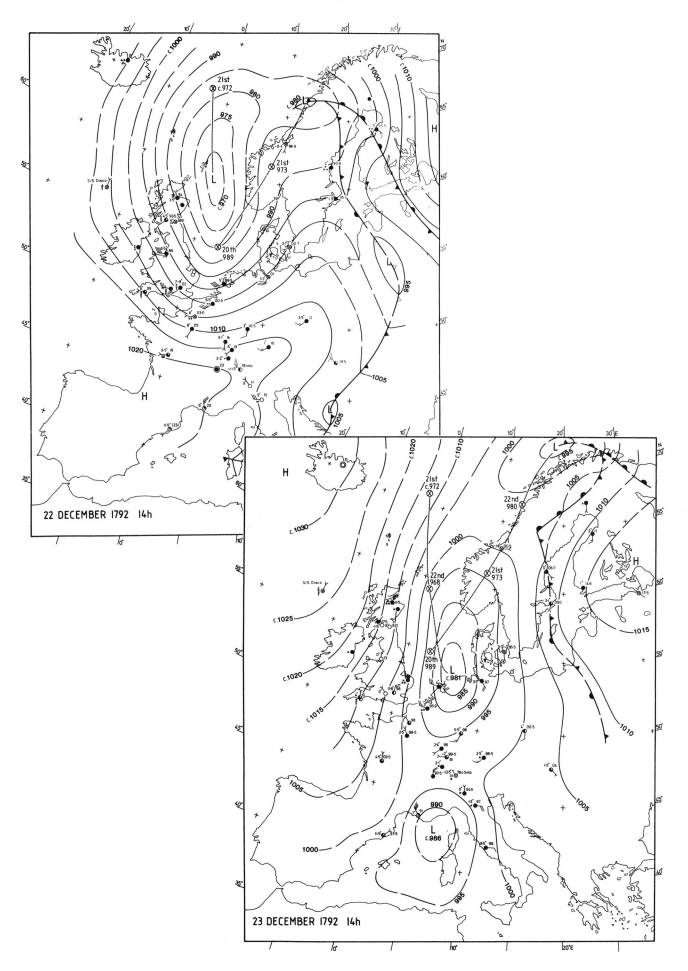

22 DECEMBER 1792  14h

23 DECEMBER 1792  14h

winds seem to have been over the southern North Sea and Dutch and German coast region: on the 19th WNW to NW'ly gradient wind of 80–90 knots was followed by WSW around 90 knots on the 20th. On the 21st a continuous NW'ly windstream from Iceland to the Low Countries was associated with a barometric pressure gradient indicating gradient winds of 60 knots over the central North Sea and 70 knots over Scotland and near the British coast. By the 22nd the gradient wind from NW to NNW over waters off eastern Scotland and northeast England was about 80 knots, while the gradient winds over north

Holland–German Bight–eastern North Sea had become SW to SSW 70–75 knots, as the latest cyclonic centre moved south towards the central North Sea. On the 23rd, with lowest pressure about 981 mb near the Frisian Islands off north Holland, a gradient wind of 100 knots from the north was indicated just west of the centre. And the weather in Norfolk with a N'ly gradient wind of 75–90 knots was described as 'wind very high . . . very cold. Wind rough all day . . . much snow'. The winds abated that evening.

---

## 6–12 May 1795

*Area*

North Sea and Scandinavia, from the east coast and coastal regions of Britain to the east coast of Sweden, also on the 8th probably the whole Norwegian Sea between longitude 2° W and the coast of Norway and on the 8th–9th the whole northern Baltic.

*Observations*

Knowledge of this storm comes mainly from records of the severity of the damage to the forests in Sweden (Sernander 1936). This was the severest of about 40 catastrophic storm events in the Swedish forests reviewed between 1731 and 1935. Particularly north of Uppsala and widespread in the province of Västmanland, west to northwest of Stockholm, whole forests were felled by the wind, many trees broken half-way up their stems and heaps of up to five trees left lying on top of each other. (Some of these descriptions suggest that in some localities tornado activity was also involved.) Huge stores were set up in 1795 and 1796 to keep the timber until it could be dealt with, and sawmills were still being built as late as 1806 for the purpose. It seems that it was not until about 1820, 25 years later, that the devastation from this one storm was cleared.

*Meteorology*

A period of quiet, warm weather over most of Europe, especially western, central and northern Europe, from 25 April or earlier brought unusually warm summer-like weather in southern and central Sweden at its climax in early May, with 26 °C at Stockholm on the 1st and high readings on several of the following days (24 °C to 25 °C at Hamburg and 21 °C to 22 °C at Stockholm on the 4th). On the 2nd and 3rd even in northern Iceland fine weather with snow melting in the sunshine was reported. Winds had been SW'ly or S'ly over western and northern Europe in late April, and on 3 May a belt of high pressure extended from southwest of Ireland across England, the North Sea and southern Sweden to Finland.

Already on the 4th a belt of winds from WSW to W across Scotland and Scandinavia was strengthening, the pressure differences indicating a gradient wind of 70–80 knots south of Trondheim and across central Sweden, and probably 80 knots over the northern North Sea about Shetland. In northern Iceland, the wind had turned NE'ly and by the 5th N'ly, with prolonged snowfall followed by hard

frost. Very strong surface winds from NW were reported in central Sweden on the morning of the 6th, continuing with increasing force and turning N and later NE by the 9th. Winds were lighter and more variable over the North Sea on the 7th, before a small frontal wave depression appearing near northern Scotland moved southeast over the North Sea and later east to the Stockholm area of Sweden, deepening to 990 mb on the 8th and about 984 mb over Finland on the 9th. On the 8th a massive N'ly outbreak of very cold Arctic air covered the whole width of the Norwegian Sea and the North Sea, a continuous N windstream from latitudes north of 70° N to the coasts of Germany and the southern North Sea from Norfolk to Denmark, where 'very high' (or 'hard') winds were reported; but it was farther north that the greatest strengths were attained. This windstream developed in the rear of a depression sequence, which was seen on the 4th as developing wave disturbances passing northeastwards on a very strongly concentrated polar front between Europe and Iceland.

On the 10–11 May another frontal wave disturbance travelled southeast over the North Sea from north of Scotland. On the 11th and 12th this cyclonic centre (lowest pressure 1010 mb, deepening to 995 mb) turned eastwards across the fringe of northern Germany to the Baltic states to reach the Gulf of Finland on the 13th. The previous depression recurved, or extended, back to the west over Lapland, and a separate centre of low pressure was maintained over or near southern Sweden, perhaps developing initially over the Skagerrak in the lee of the Norwegian mountains. These developments led to a renewal of the strong N'ly winds over the North Sea on the 11th, continuing until some time on 12th May.

The sequence ended with the low pressure over central Scandinavia extending for a time southwest as far as the southern North Sea by the 15th, and gradually weakening before moving away quite gently northeast over southern Scandinavia and central Finland to the Arctic. Strong N'ly wind was again reported north of the Dutch coast on the 16th.

Further gales followed later in the month, but none were of comparable severity.

*Maximum wind strengths*

Gradient winds from W or WSW were of up to 80 knots over waters between the Faeroes–Shetland–Orkney region

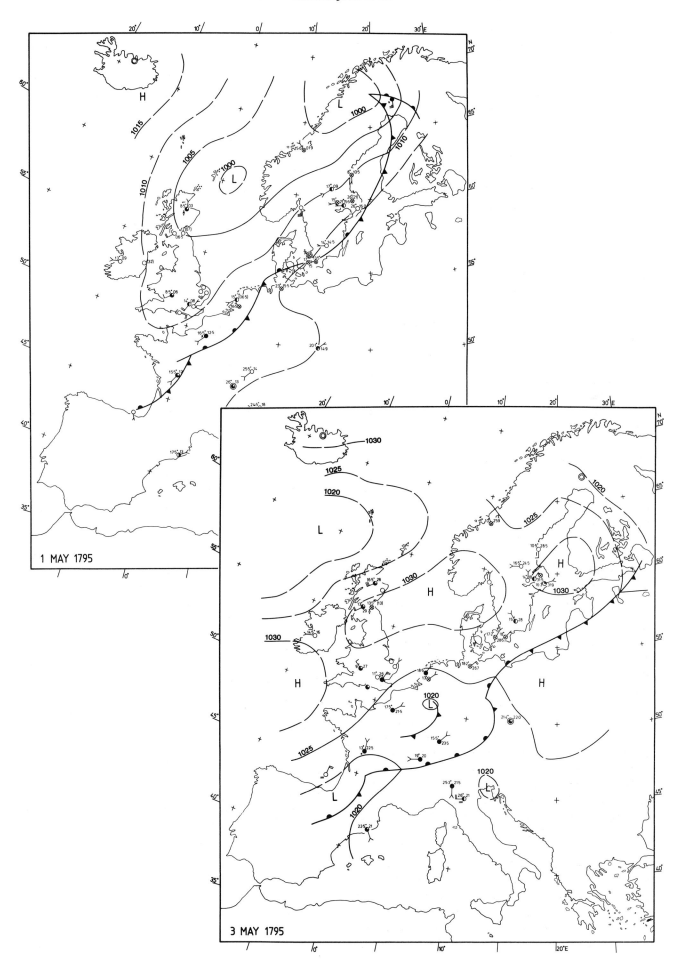

1 MAY 1795

3 MAY 1795

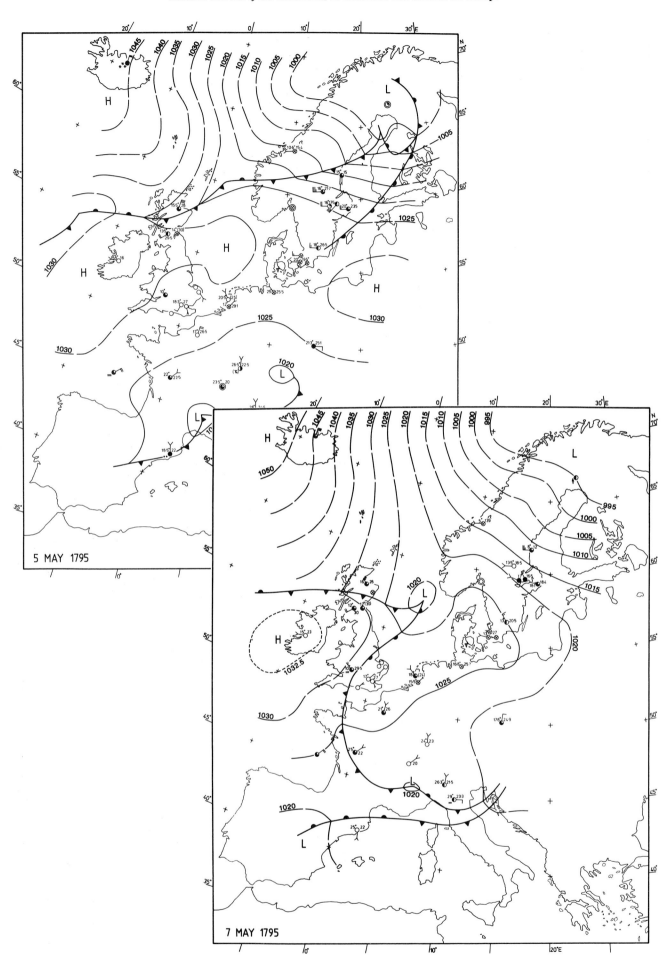

5 MAY 1795

7 MAY 1795

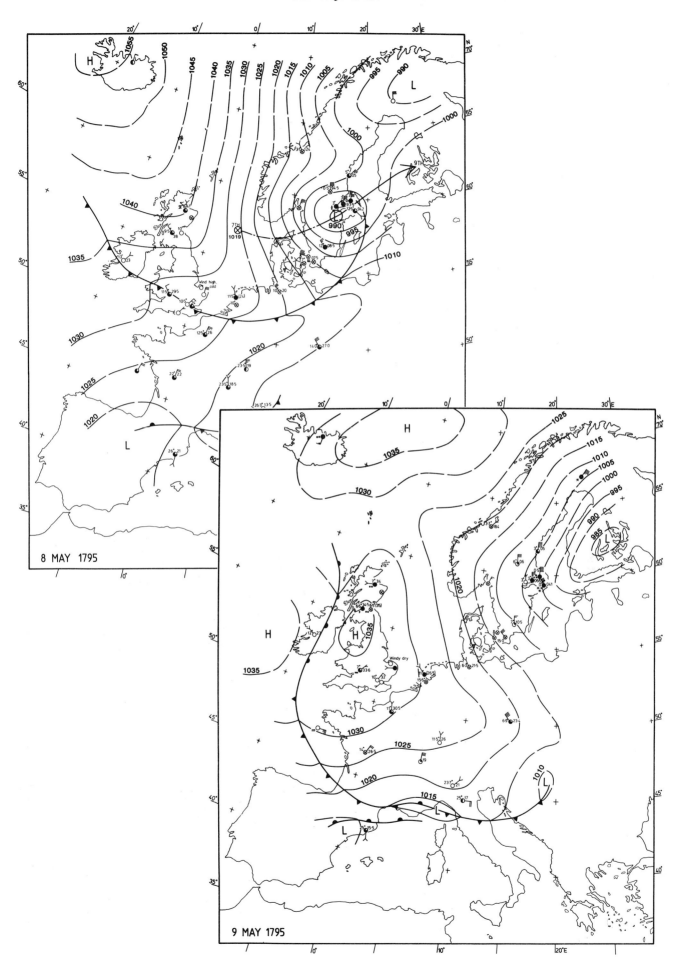

8 MAY 1795

9 MAY 1795

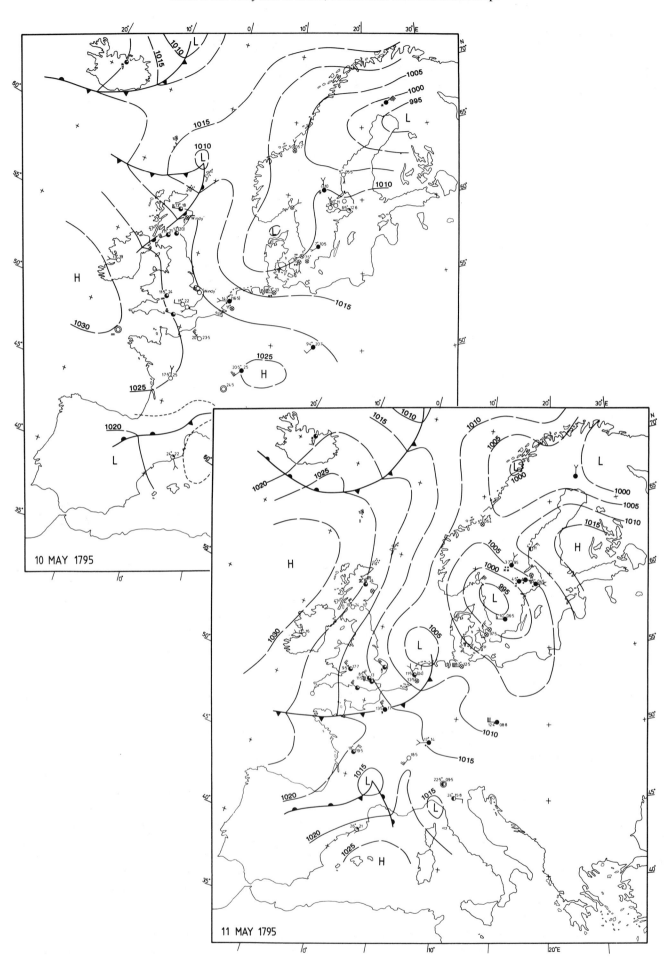

10 MAY 1795

11 MAY 1795

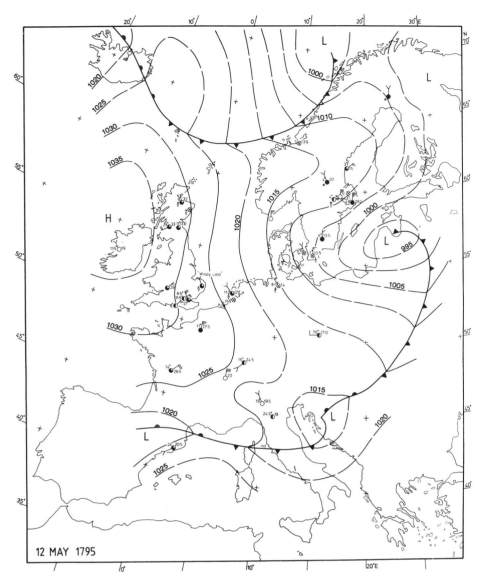

12 MAY 1795

and the coast of Norway on the 4th. Again on the 8th gradient winds, this time from the N, were of 70–80 knots over a huge area of the Norwegian Sea and probably reached 80 knots or more over the northern North Sea between Shetland and Norway.

On the 8th over central Sweden, and also over the Kattegat–southwest Sweden area, geostrophic winds of about 100–120 knots were indicated. The gradient wind

was probably of the order of 70–80 knots here also. Strongest gusts over the somewhat irregular terrain of central Sweden were probably in the range of 75–90 knots, but damage to the forests was probably increased by the weight of the snowfall. Strongest gusts of the surface winds over all the sea areas mentioned in this section were probably of the order of 80 to 90 knots. (Reference: Sernander, 1936.)

## 23–28 October 1805

*Area*

The Atlantic sea areas west of Gibraltar.

*Observations*

After anticyclonic weather between the 10th and 18th with NW'ly winds of Beaufort force 4 to 5, the winds off Cadiz became lighter on the 18th and then backed SSE, veering to SW on the 20th. That night the wind became squally, rain was widely reported and lightning was observed. During the battle of Trafalgar on the 21st the wind was SW'ly moderate during the forenoon, but rising to strong and with an increasing W'ly swell in the night

that followed and all next day. By noon on the 23rd the wind had reached gale force, and the damaged ships of the British navy with the 18 prize ships that they had captured were exposed to a six-day storm as they limped toward Gibraltar. Gales force 8 to 9, always at SW or W, were observed at least once on most days from the 23rd to the 28th, the winds moderating temporarily on the 25th though the 26th was the worst day of the spell. Only four of the captured vessels reached Gibraltar.

*Meteorology*

As a British naval vessel off Brittany all through this time reported SE'ly winds, frequently of gale force, it is clear that

115

there was a more or less stationary centre of cyclonic activity over the ocean just west of the Iberian peninsula. A break in the spell on the 25th was signalled not only by the weakening of the winds affecting the fleet west of Gibraltar but by a short-lived change reported by the ship near the Brittany coast: there the wind veered to SW and fell to force 4 on the 25th, before returning to ESE and freshening on the 26th. At that stage presumably one low pressure centre moved away to the northeast and was soon replaced by another.

Conditions over and to the northwest, north, or northeast of the British Isles were apparently anticyclonic throughout the dates mentioned in this account.

More details of this storm sequence and the weather of the last 20 days of October 1805 have been analysed and published by D.A. Wheeler in his article on weather at the battle of Trafalgar, in *Weather 40*(11), pp. 338–46 (1985).

This storm was outside the areas with which this study is generally concerned, but is included for its historic interest and because it is thought to be an example typical of the storm sequences liable to occur with the cyclonic activity – often associated with cut-off cold-pool depressions – to the south of ('blocking') anticyclones over the region of the British Isles and northern Europe.

## 12–16 January 1818

*Area*

Scotland and all Scottish coasts, Denmark and Danish waters generally, also on the night of the 12th–13th and on the 15th the southwestern coasts of England. By implication there must have been gales or severe gales at various times during these days in nearly every part of the North Sea, the Channel and southern Baltic, as well as neighbouring parts of the northern Atlantic.

*Observations*

Buildings in Edinburgh were reported severely damaged by W'ly gales on the 12th and 14th–15th. The observer at Gordon Castle near the Moray Firth in northern Scotland (57°38′ N 3°4′ W) reported trees blown down in the storm on the 13th. Strong SW and W'ly gales were reported in the Danish Sound (Øresund) on the 13th, 15th and 16th and a SE'ly storm in the afternoon of the 12th.

*Meteorology*

This very unsettled sequence is illustrated by once daily maps from the 12th–16th. The only observations shown are the wind directions at Reykjavik, Iceland which contribute to the analysis although with no pressure observations they cannot be directly incorporated in the isobar pattern.

The barometric pressure level was steady over France throughout, Poitiers reporting mild, cloudy weather in W or WSW winds blowing from the Atlantic, and MSL pressure always between 1026 and 1030 mb. By contrast, over all the northern part of our maps the pressure level was continually changing. Pressures over northern Scotland fell 30–40 mb from the 12th to the 13th, recovering almost fully by the 14th only to fall 25–35 mb again by the 15th and rise sharply on the 16th. The changes of barometric pressure level in Denmark were of a similar order, but with lower pressures on the 12th than on the 13th and the lowest pressures of the sequence about 7h on the 16th.

This was a very mobile spell of weather in the zone about the latitudes of the British Isles and Denmark to southern Sweden with the rapidly occluding low pressure centres travelling 600 to, in two cases on the 14–16 January, not less than 800 nautical miles in 24 hours. A major low pressure system (*Zentraltief*) seems, however, to have been more or less stationary over the Norwegian Sea at least from the 12th to the 15th.

The warm SW'ly air from the subtropical Atlantic brought temperatures up to 11 °C in Edinburgh and London on the 13th, and 5 °C was reported in Copenhagen on the 15th even in a colder air stream, which had also produced 8 °C at Perth in central Scotland on the afternoon of the 14th.

The sequence seems to have been characterized by much colder maritime Arctic air repeatedly intruding southwards in the rear of each depression, bringing temperatures down to the freezing point in Scotland on the 12th and 16th and wintry showers to northern Scotland on the 15th and to Denmark on the 16th.

The winds reported in Iceland register these advances of very cold air from the Norwegian and Greenland Sea; although SW or S at Reykjavik for much of the 12th and more briefly on 14th–15th, a stiff NW or N'ly was blowing on the evening of the 11th, again N'ly all day on the 13th becoming strong before turning NE and lighter on the 14th, and yet again all through the afternoon and into the night on the 16th. On the 15th the breeze was NE'ly and the observer reported the cold was severe.

*Maximum wind strengths*

Gradient winds seem from our analysis to have approached 100 knots in the W'ly over the North Sea on the 12th and may have reached 120 knots over Denmark on the 15th. These were the only measurements in this sequence that may be based on an adequately controlled isobar pattern, but similar values may be conjectured over Scotland at times (a) early on the 12th and (b) in the evening of the 14th and following night.

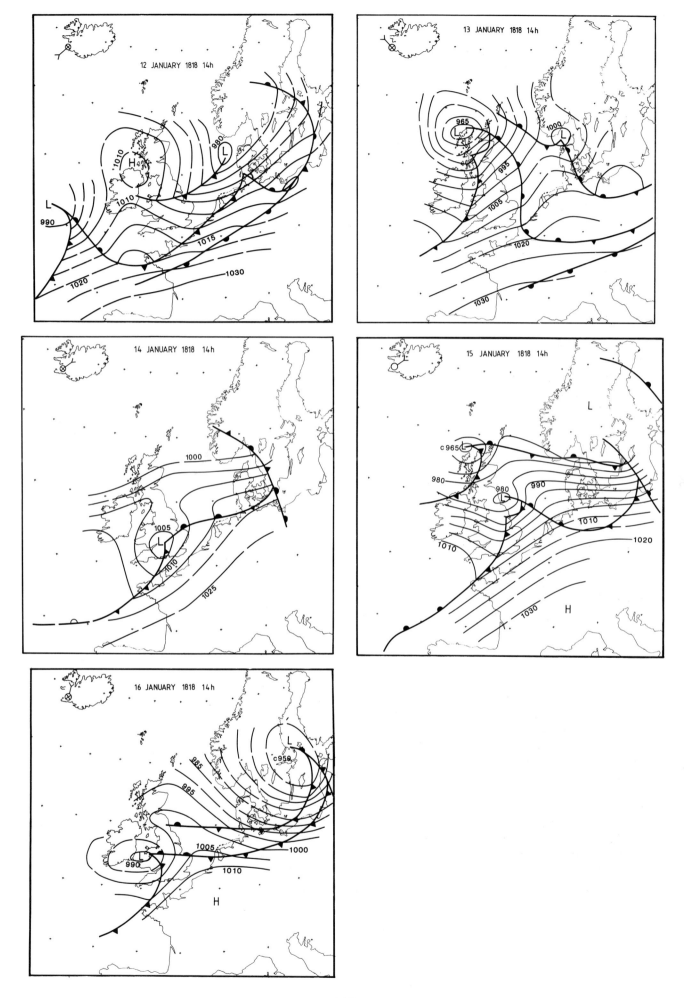

## 11 March 1822

### Area

West Norway coast and all Denmark, especially northern parts and near the west coast of Jutland. Probably at some stage on the 10 and 11 March all parts of the North Sea. Gale or storm force winds were also reported for a time in Sweden. Strong effects on sea level, but all the damage reported seems attributable to the wind. Much of the loss of life in Norway was due to the suddenness of the deterioration and onset of violent storm conditions following beautiful weather earlier in the day.

### Observations

Reports from Denmark (e.g. the lighthouse at the northernmost point, Skagen or The Skaw, and from ships in the Kattegat) show that strong winds from directions varying between SW and NW, mainly of Beaufort force 6 or above, and occasionally up to perhaps force 10, had prevailed there since 6 March. A record from Bogense on the middle island (Fyn) mentions a strong wind from NNW all day on the 7th and a storm from WNW with 'frightful hail and snow showers night and day' on the 8th. Winds were also very strong, with rain and hail showers, on the 10th.

The situation development on the 11th in Norway has been described and analysed by E. Wishman formerly of the Norske Meteorologiske Institutt and now a curator at the Archaeological Museum in Stavanger (Wishman, 1986), and the Norwegian part of the map here given for 11 March largely follows his analysis. His report is taken from eye-witness accounts. One was compiled, evidently in the days immediately after the event, by a farmer, I. Gullikson, who was also a respected local politician interested in all aspects of his home scene and a man so watchful of wind and weather that he had kept, since 1808, regular notes of the weather development, seasons and harvests where he lived at Grude in the Klepp area near the coast just south of Stavanger. The other account, from a very different character, is the personal, proud story of his own deeds on the day, written some time afterwards in prison in Oslo by a 'master thief', Gjest Baardsen, many times convicted. At the date of the storm he was under arrest at the sheriff's residence at a point on the west coast 200 km north of Stavanger, at the entrance to Sognefjord. In the special circumstances of the day, this man was temporarily released from custody by the sheriff's wife, in the absence on duty of the sheriff at the time of the emergency and the lack of other able-bodied men on the site, to get together a crew to row out through the storm to rescue children and others stranded on islands. His story fits so well with Gullikson's that it seems to add trustworthy details about the storm and its timing.

At the points mentioned on the west coast, the day began with a beautiful clear dawn, almost no cloud, and no wind in the more northern area, outside Sognefjord, and only a light drift of air from the SE near Stavanger. Consequently, the local fishing boats went out despite the exceptionally low barometer reading: some of them rowed several miles out to sea and were followed by the womenfolk and boys, who rowed to various small islands about the fjord entrance to cut heather as food for the cattle.

About 8 o'clock in the Stavanger area and 9 o'clock outside Sognefjord dark clouds came up swiftly, the wind rose and soon afterwards the storm was raging, initially from SW, later from NW, and with sleet falling at times. Conditions were worst between about 11 a.m. and 2 p.m. People could hardly walk or see far in the open, houses and barns were blown down, some houses were blown clean away from their foundations, and roofs flew off. Gjest Baardsen and a fellow prisoner were having to do all they could to secure the sheriff's roof before being called upon to organize the rescue boat. All the fishermen who had gone to sea were drowned, as were three men trying to make fast a vessel that had drifted a stone's throw out from the harbour in the sheltered fjord at Sandnes, behind Stavanger. Eight children's lives were saved by Baardsen's boat going from island to island from about 4 p.m. onwards, in the failing light at the end of the day, as the storm moderated.

Reports from Denmark tell that the storm from about WNW was at its most violent in north Jutland between 10h and 14h, although the lowest barometric pressure at Copenhagen (probably around 980 mb) was measured early in the day. By the midday observation it was some 1.5 mb higher. (Willaume-Jantzen (1896) states that the first reliable pressure measurements in Copenhagen were in 1842, but changes during the day and from day to day indicated by the instruments in use earlier should be reasonably trustworthy.) There was damage to buildings in Denmark as far south as Ribe on the west side of Jutland near latitude 55° N, and northern parts of the central island, Fyn, in about the same latitude. In north Jutland the churches were damaged and other buildings collapsed in many places; even on the eastern side as far south as Randers (56.5° N) houses fell and big trees were uprooted. In the Hjørring district (57.5° N) 'some farmhouses were thrown down in almost every parish'. The rain and hail also damaged buildings in various parts of the country. The storm on the 11th was considered the worst for many years past.

The farmer at Grude in west Norway, Gullikson, mentions also (quoted from Wishman, 1986 – my translation) that 'on the day of the hurricane there was such a flood tide that the sea rose beyond its natural bounds' and that on the English coast the ebb tide fell so exceptionally low that people walked out on the sea bed and collected articles, including some things of value, that had been lost a hundred years or more earlier.

### Meteorology

This storm came after a famous mild winter in Britain and western Europe with continual stormy winds. Gram-Jensen (1985, p. 46), recorded a North Sea storm flood in Ribe in October 1821 and thereafter with every full moon all through the winter. (The moon was full on 8 March 1822.)

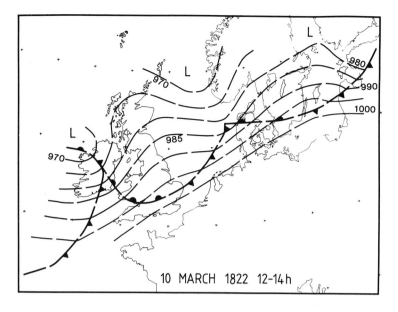

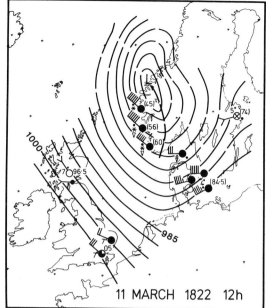

It is clear that from 6 to 11 March Denmark was under the influence of the wind systems of deep depressions, the fronts of which were passing from west to east across the country, with winds veering between SW and NNW. The situation seems to have become successively more stormy over those days. On the 10th pressures became low again in Scotland, and it seems likely that a further cyclonic system, developed from waves on the front that had been active over a zone from southwest of Britain to Scandinavia and the Baltic, travelled northeast over or near northern Scotland to join the very deep centre farther north over the Scandinavian region. This new disturbance gave strong winds in southern Britain and gales in Scotland, presumably also over the whole North Sea, as it passed. Over west Norway pressures were already becoming exceptionally low. The map here presented for the 10th has not been fully analysed but indicates the general nature of the situation. On the 11th barometric pressures in the Norwegian coast region mapped were described by Gullikson as the lowest anyone could remember: Wishman

(1986) considers this must mean a value about 940 mb. (Since that time 939.5 mb, on 27 January 1884, has been the lowest pressure measured in Bergen.) The fronts shown on our map for the 10th had probably been swept away east and southeast of the region by the 11th. Wishman's trough line here shown over west Norway may possibly have been a bent-back occlusion, but seems not to have extended as a sharp feature south of about 58° N. Barometric pressures over the area rose rapidly after this, and there was a single day of hard frost in Denmark on the 12th before the wind turned southerly again.

The vigour of the development clearly has something to do with the strong airmass contrast between the N'ly cold air from the Norwegian Sea that brought frost as far south as Denmark on the 12th and the warm air which produced temperatures as high as 12 °C over England and the Channel on the 10th. The vigour may also to some important extent have to be explained by inherited features of the situation (the general pressure level and vorticity distribution) built up over the previous days and months.

## 2–5 February 1825

### Area

Whole North Sea. Disastrous storm flooding of the coast from Holland to Denmark (Jutland), notably around Hamburg and the lower Elbe–German North Sea coast in the night of 3–4 February.

### Observations

A strong storm raged all day on 3 February at Hamburg and continued until midday on the 4th. The wind had been WSW with much rain in the forenoon of the 3rd, and veered through W to WNW in the afternoon and evening: all day it was a great storm with heavy squalls. After midday on the 4th the wind veered N and dropped lighter.

From the afternoon of the 3rd onwards the air was mostly very clear but with some black clouds and hail and

snow showers. Late in the evening on the 3rd, or in the night, thunder was reported and in some places continued for a considerable time, while in others it was only a couple of sudden fearful lightning flashes and heavy thunder claps. It seems to have thundered everywhere from south Germany to Norway and Sweden as the storm passed. The contemporary report remarks on the unusualness of this, since thunderstorms are usually well separated from each other. In some places thunder occurred again on the following days.

High tide in the River Elbe was about the normal height at midday on the 3rd, or registered at most a moderate excess, but the water level went down very little with the ebb. The returning tide that evening reached the top of the dykes three to four hours before the time of high water.

From that time on, floods over the coastal lowlands

developed rapidly, damaging the low-lying villages whether they were protected by dykes and sluices or not. Thousands had to flee their homes in haste in their night clothes, and some had to spend three days on the flooded low-lying islands before the storm had abated enough for them to undertake the journey to the mainland in safety. The disaster in these areas was fully comparable with the Christmas 1717 storm and according to one report* the height reached by the flood was comparable with those of 8 October 1756, 22 March 1791 and 11 December 1792 also.

Petersen and Rohde (1977) have it that the flood reached a higher level than in the eighteenth century cases, although the storm was 'by a long way not so severe as on 11.12.1792'. The height of the flood on 3–4 February, 1825, clearly owed a great deal to the fact that the storm surge closely coincided with the spring tide.

This storm was also remarkable for its duration: three days near Hamburg and in parts of the North Sea four days.

The whole coast from Holland to Jutland was affected by the flooding. The death toll was not as great as with the great floods of the previous century, but even so more than 800 people died and many thousands of cattle. Thousands of buildings were destroyed. The dykes had been strengthened since those earlier times and 'raised 3 or 4 feet' (1 m or rather more) since 1792.

The entire 1824–5 winter was stormy, the first coastal flooding occurring on 3 November and some unprotected coastal islands being reported twenty times under water during that winter. Continued saturation and wave action may have weakened the dykes.

*Meteorology*

In late January barometric pressure was generally high over Ireland and the whole southwestern sector of the British Isles. By 28 January 1825 an anticyclone with central pressure over 1040 mb was centred over southern Ireland and stretched across Wales to influence the weather over the whole of England and Scotland also. Farther north a sequence of intense depressions was passing from west to east in the Arctic, from north of Iceland to the Barents Sea north of Norway and beyond. The broad stream of westerly winds between Scotland and Iceland was of gale, or storm, strength with gradient winds approaching 100 knots: mean sea level barometric pressure at 14h on 28 January was 1030 mb north of Inverness and 982 mb in southwest Iceland.

By 30 and 31 January the whole pattern had shifted somewhat farther south, with vigorous depression centres moving east over the ocean from south of Iceland to pass over Lapland and the northern Baltic, and the highest pressure (still about 1040 mb) was centred over Biscay and the south-west approaches to the British Isles. On 1st February, for the first time in this sequence, a vigorous NW'ly windstream broke through from around Iceland

into the North Sea with NW'ly gradient wind strengths up to 140 knots over and near central and northeast Scotland and about 90 knots from W to WNW over the central North Sea and Denmark. The depression off the Norwegian coast near Trondheim at that time had a central pressure of about 960 mb. On the 2nd this centre had moved to northern Sweden and deepened to about 950 mb, with geostrophic winds of about 150 knots around the western and southern sides of the depression over Scandinavia and the northern Baltic.

On the 2nd the winds over the North Sea had slackened temporarily, but a new NW'ly outbreak was approaching Scotland behind a cold front, with gradient winds near Iceland already 120 knots from the NW and temperatures as low as $-20\,°C$ with the strong winds near Reykjavik. The new depression centre associated with this further outbreak of Arctic air moved from near southeast Iceland to lie over central Sweden with mean sea level pressure near the centre 945 mb or thereabouts at 14h on 3 February. Gradient winds at that time over the western and southern North Sea were NW'ly in the range 80–90 knots, increasing to about 100 knots over the coast of the Netherlands. On 4 February the depression centre had divided, leaving a complex pattern over Scandinavia, the lowest pressure being about 950–955 mb over the Gulf of Finland while subsidiary small centres were left behind at the Norwegian coast. Over the western and central North Sea the gradient wind was now from NNW up to 120 knots.

On the 5th the pattern was still complex in the central area of low pressure over Scandinavia and north Russia, while gradient winds continued very strong over the western and southern North Sea and over part of northern Britain probably still 100–120 knots. A feature of this storm sequence, as in 1791, was that the strongest winds blew on nearly straight (great circle) paths.

There seem to be indirect indications that the surface waters of the Bay of Biscay and the North Sea, possibly also the Baltic, were rather warm for the season when this sequence began. This may be judged from the air temperatures observed at the coasts when the winds were blowing onshore and the tendency for cyclonic curvature to develop in the isobar pattern over these seas whenever the pressure gradients were light. A very significant feature of this sequence must be the exceptionally cold air which arrived over Iceland and the northern part of the map and the greatly enhanced thermal gradient then created between latitudes about 50–55° and 65° N. In this respect the situation resembles that deduced in the case of the storms in 1588.

*Maximum wind strengths*

The strongest gradient winds indicated in this North Sea storm were mainly around 120 knots, only locally up to 140 knots near northeast Scotland on 1 February; to this extent the report that the storm did not reach the severity of the cases in March 1791 and December 1792 seems to be substantiated. But the duration of the storm over the North Sea was very long – most of 5 days, i.e. 5 days apart from the temporary lull on 2 February. On the 5th gradient

*Report mentioned by W. Müller in his *Beschreibung der Sturmfluthen an den Ufern der Nordsee und der sich darin ergiessenden Ströme und Flüsse am 3 und 4 Februar 1825*, Hanover, 1825.

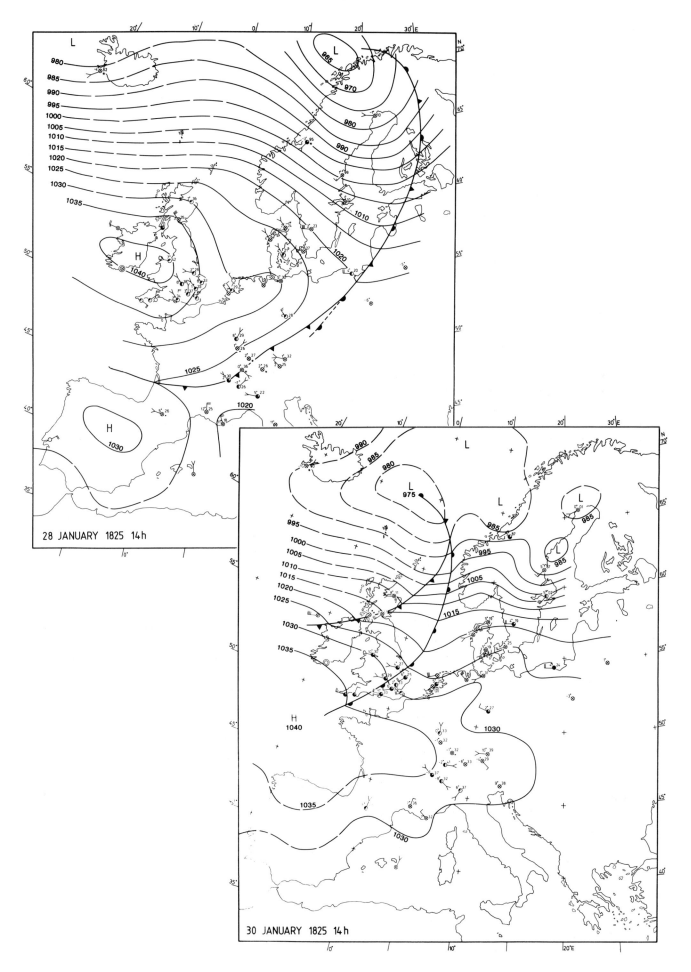

28 JANUARY 1825 14h

30 JANUARY 1825 14h

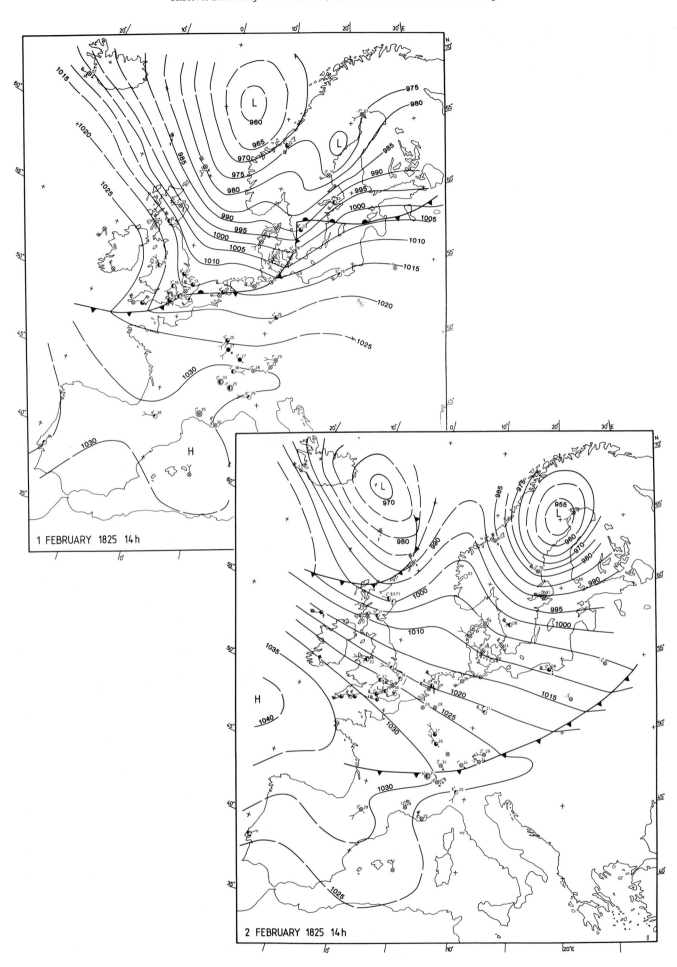

1 FEBRUARY 1825 14h

2 FEBRUARY 1825 14h

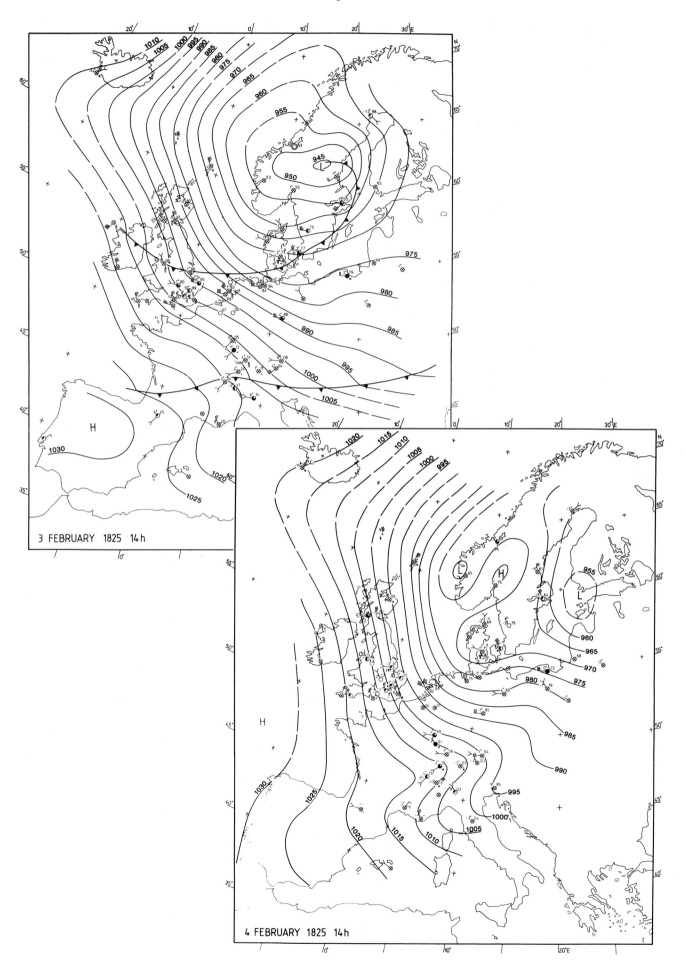

3 FEBRUARY 1825 14 h

4 FEBRUARY 1825 14 h

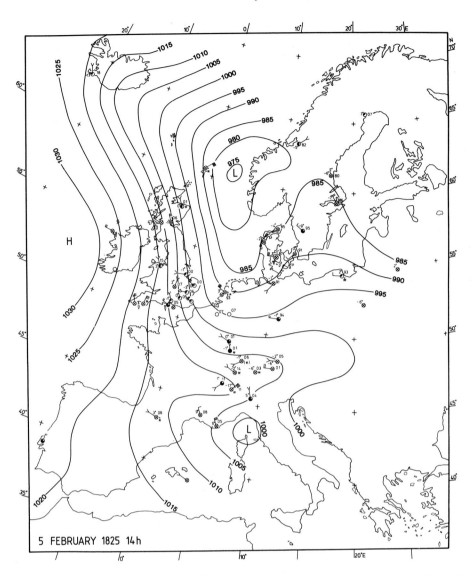

5 FEBRUARY 1825 14h

winds from NNW were still 100–120 knots over northern England and Scotland and strong gales were still blowing at the surface onto the Dutch coast.

The strongest gusts at the surface over the North Sea during this sequence were probably of the same order as the gradient winds, i.e. up to 100 to about 120 knots or perhaps locally a bit more near the coasts of northeast Scotland and the Orkney Isles on the 1st.

The still stronger winds blowing on the 2nd over northern Norway and Sweden, parts of the northern Norwegian

Sea and northern Baltic, may have had some influence on the situation by introducing very cold air in depth over the northern part of the region discussed.

*Data sources*

Basic details of this storm were obtained from an article 'Zur grossen Sturmflut vom Februar 1825' by W. Schröder in *Acta Hydrophysica*, Band XVII (Heft 1), Berlin: Institut für Physikalische Hydrographie der Deutschen Akademie der Wissenschaften zu Berlin, Akademie Verlag, 1972.

## 3–4 August 1829

*Area*

Northeast and central Scotland, including Orkney. Gales also in eastern England and all waters off Scotland's and England's north and east coasts.

*Observations*

This storm has been reported (Lauder, 1830, 1873) and remembered chiefly as an outstanding rainfall event, but the 'furious NE wind' raised high seas, which caused shipwrecks and broke open one Scottish harbour (Garmouth, near Elgin) so that part of the village was washed away

in the early hours of 4 August. There seems to have been little wind damage other than through the intermediary of the stormy sea. There was an interval of some hours to probably most of one day of westerly wind over northern Scotland before the wind direction returned to N and NE and the gales set in.

Measurements of the rainfall are only available from places in relatively sheltered situations, the greatest being $3\frac{3}{4}$ in or 95 mm in 24 hours from 5 a.m. on the 3rd to 5 a.m. on the 4th at Huntly (57°28′ N 2°47′ W), near the junction of the Deveron and Bogie Valleys, south of the main mountain areas and watersheds where by far the

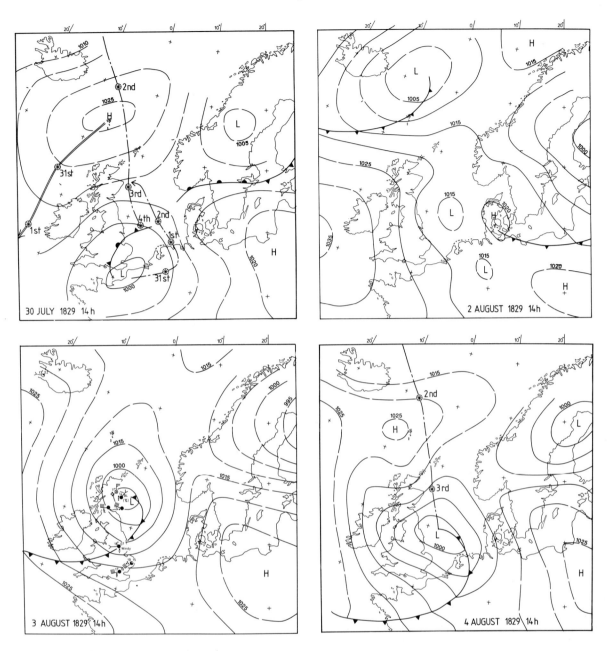

30 JULY 1829 14h

2 AUGUST 1829 14h

3 AUGUST 1829 14h

4 AUGUST 1829 14h

heaviest rainfall fell and the greatest river floods were caused. Those areas where the exceptional rainfall gave rise to flooding of the lochs and rivers to levels never previously reported, and, it is believed, unequalled since, were the Monadhliath mountains (about 57° N to 57°10′ N near 4° W), especially in the heads of the valleys running down from their eastern face, and more generally in Strathspey eastward from the point closest to those mountains.*

Flooding of a similar order seems to have been caused by torrents coming down to Speyside from the north face of the Cairngorms (56°50′ to 57° N 3°30′ to 4° W). The

flooding south of these mountains in Deeside and Donside, and farther south in Glen Esk, seems to have been less extreme, though also enough to wash away bridges. Old people in the areas worst affected were positive that the flood levels exceeded those of the greatest remembered previous floods, in 1768.

At least nine bridges, some of them rather recently built main road bridges, and many houses and other buildings, were destroyed by the flood at various places in the eastern Highlands of Scotland and many lives were lost. Damage was also done by the wind and rain at Kirkwall in the Orkney Islands and at various places in mainland Scotland between Inverness and Wick.

It had been an earlier characteristic of the summer of 1829 in the northern Highlands of Scotland that the fine warm weather of May and June, that had been accompanied by drought, began to be interrupted from early July onwards by sudden, intense falls of rain which eroded the water courses and broke bridges. There was a repetition

*Water entering through a (2900 cm²) hole blown open in the ridge of a house-roof at Tomintoul, over 500 m above sea level near 57°15′ N 3°23′ W, accumulated to the extent of 180 litres in 12 hours during the day on 3 August – equivalent to a depth of 870 mm! The circumstances of the collection do not, however, permit acceptance of this except as a very rough indication of the order of the local rainfall intensity.

on 27 August of that year which bore noticeable resemblances to the events of 3–4 August, although somewhat less extensive and less severe. Once again the storm was preceded by an interval of W'ly winds which then changed to N'ly. There was again damage, mostly due to river floods, widespread over Scotland north and northwest of a line from Killin to Nairn: and some bridges were knocked down and mills and cottages damaged. Again the worst of the rainfall seems to have been on the Monadhliath group of mountains, but this time affecting the rivers on the northern side, and also farther north in Ross-shire, in Strathglass. Nairn itself suffered worse than in the previous flood, but the rivers south of there were less affected. There is some report of the coldness of the wind and exposure of those on the mountains who fled their homes near the streams but no reports of damage due to strength of the wind on the 27th are known to the writer.

*Meteorology*

The daily sequence of weather maps analysed is here illustrated by the situations on 30 July and 2–4 August. Cyclonic centres had been moving from Biscay to the Low Countries, northwest Germany and the southern North Sea at least since 25 July, occasionally bringing very high temperatures (up to 33 °C) to low-lying continental situa-

tions and many thunderstorms over England on the 25, 29 and 30 July. As high pressure systems centred between Scotland and Iceland on the 25th and around the end of the month withdrew south and southwestwards, Arctic air spread from the Norwegian Sea and the northernmost Atlantic, first over Scandinavia and the Baltic by 1–2 August, and swept south over Scotland, England and the western and southern North Sea from the 3 August onwards. A vigorous depression moved south from the Arctic north of Iceland on the 1st to northeast Scotland on the 3rd, reaching the southern North Sea on the 4th and Holland by the 5th. Its track, and that of the departing anticyclone, is seen on our map for 30 July, which also shows that the last of the series of cyclonic centres from Biscay recurved northwards over the North Sea and probably coalesced with the Arctic depression near northeast Scotland on the 3 August.

This merging of the southern and northern cyclonic activity presumably sharpened the thermal gradients near that point. The highly localized extreme rainfall intensity over parts of the Scottish mountains also points to a probability of great thermal instability. The strongest gradient winds indicated by our analysis are no more than 70–80 knots from the NE on 4 August near the Scottish coast.

---

## 14 October 1829

*Area*

Much of the North Sea, especially northern and central parts, northeast Scotland and also Denmark and Danish waters.

*Observations*

Many ships were lost in a NE'ly and N'ly gale at and near the coast of northeast Scotland. Severe gales from between NE and E were also reported by ships in Danish waters south of about latitude 56° N. Over southern Denmark the winds shifted abruptly during the day from WSW'ly to ENE and rose to force 9 in the evening, force 10 in Bornholm in the southern Baltic.

*Meteorology*

Cyclonic activity over the North Sea region had been 'fed' by frontal wave disturbances coming from the southwest, from Biscay and the Atlantic. The system deepened while still complex over the North Sea on the 14th. At the same time it was being forced somewhat farther south than before by changes in the high pressure system to the north of it. As the depressions moved east, there was a substantial rise of pressure over Scotland and Ireland and a sharp recovery of the high pressure level over Scandinavia later. N'ly winds set in during the 14th and 15th over the whole region here mapped but soon became light.

After the cool wet summer of 1829 in Britain, the remaining months of the year were all colder than usual

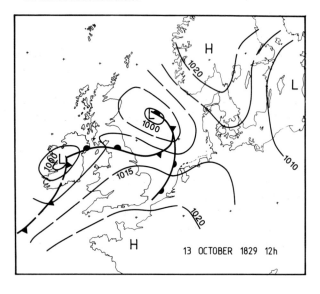

13 OCTOBER 1829 12h

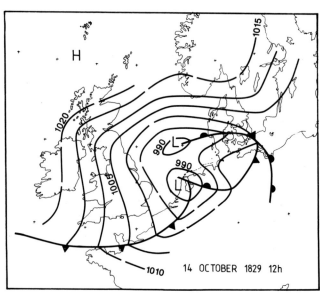

14 OCTOBER 1829 12h

and the winter that followed was severe – one of the historic occasions when the Bodensee (Lake Constance) in central Europe froze over completely. It seems likely that at the time of this storm an enlarged Arctic high pressure system was already resisting the encroachments of depressions from farther south and keeping them to tracks in lower latitudes than usual.

## 25 November 1829

*Area*

Coast of Scotland and probably all the North Sea coast of Britain on the 24th and 25th.

*Observations*

Reported that an ENE'ly gale on this date wrecked many ships.

*Meteorology*

The situation, here illustrated by the map for the 24th, is of a very broad belt of E'ly winds between high pressure over Scandinavia, and to the north of Scotland, and low pressure over France and central Europe.

The severity of the wind reported from Scotland seems a little surprising, but at a number of points on the maps on the 23, 24 and 25 November the E'ly winds reported were stronger than expected from the pressure gradients here determined from our very limited network of pressure observations. Strong gales up to force 10 were reported on the 24th and 25th from southern and eastern Denmark and Bornholm.

It is likely that there was an uneven distribution of wind velocities across the width of the belt of E'ly winds. Some of the concentrations of strong wind may have been close to the lines of various old fronts that had come up over southern Europe, and some may have been topographically engendered by the coasts of the Baltic. It is also possible that one or more such narrow zones, or 'filaments', of strong wind was continuous from Denmark or the coast of south Norway to Scotland.

The northern anticyclone was intensifying slowly from the 23rd to the 25th. It is possibly that the same background remarks given for the storm in October 1829 apply also in this case, save that the cyclonic activity was by now confined farther south over Europe.

It is recorded (De Boer 1968) that a storm in November 1829, presumably this one (since the weather map encourages this suggestion), also quickly undermined the tower of the Spurn Head (near Hull) lighthouse of that day. (The tower was, in the event, saved from falling by the hasty shovelling of sand and shingle underneath it, but little more than a month later it was found that the building had to be abandoned and a temporary light set up.)

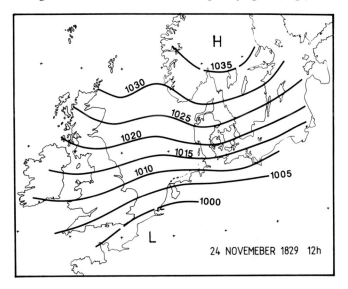

24 NOVEMEBER 1829 12h

## 18 November 1835

*Area*

North Sea, especially the German Bight.

*Observations*

Petersen and Rohde (1977) report 'a very severe storm flood' on this date, reaching 3.1 m above the mean high water level at Cuxhaven. (The lower limit of their 'very severe' class is 2.9 m above the mean, and of the 'severe' class 2.16 m above the mean.)

A NW'ly gale was reported at Aberdeen earlier on the 18th and in London a strong W'ly wind, which became stormy in the evening. Gales from between W and NW in Denmark on the 17th were followed by a lull, with winds backing to between W and S until past midday on the 18th, but at 8h on the 19th gales from between NW and NE were reported all over Denmark, up to force 10 in the Kattegat and at Bornholm, in the Baltic.

The storm flood in the German Bight presumably occurred through coincidence of this N'ly gale with the evening tide.

*Meteorology*

The situation is illustrated by our maps for midday on the 18th and 8h on the 19th, by which time the Low centre had moved from the northern North Sea to the southern Baltic and probably deepened by about 15 mb. By the afternoon of the 19th much colder air was reaching Denmark from Russia and the northern Baltic, temperatures near Copenhagen falling to −1 °C by 16h. No exceptionally strong pressure gradients are indicated, however. A sustained pressure rise followed, bringing mean sea level pressures in Denmark to 1030 mb by the 24th.

On the 17th Denmark had been experiencing W to NW'ly gales in the rear side of a previous depression, which had moved from west to east at a rather higher latitude.

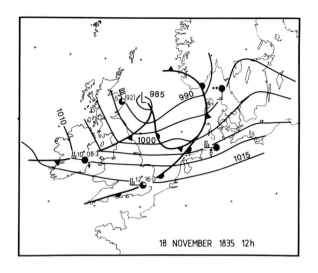

18 NOVEMBER 1835 12h

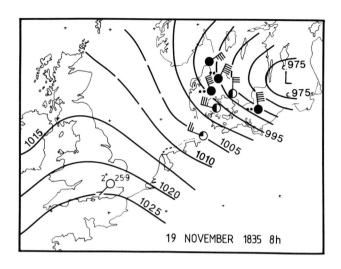

19 NOVEMBER 1835 8h

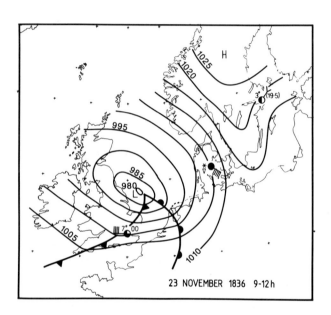

23 NOVEMBER 1836 9-12h

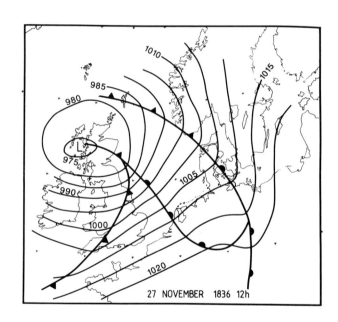

27 NOVEMBER 1836 12h

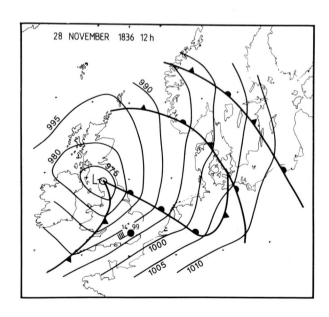

28 NOVEMBER 1836 12h

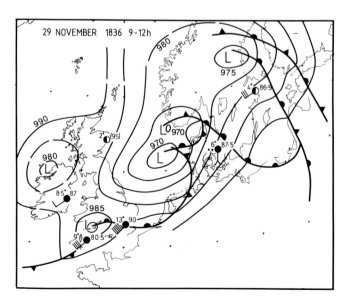

29 NOVEMBER 1836 9-12h

## 23 and 27–29 November 1836

*Area*

The Channel, southern England and southern North Sea.

*Observations*

On the 23rd 'a heavy westerly gale' was reported in the Channel and 'a dreadful storm at Calais'. Also on the 23rd severe ESE to SE gales, with occasional snow showers, were reported in many parts of Denmark.

From the 27th to 29th there were 'continued heavy gales from the west' in the Channel, gradually turning SW'ly. Among a collection of 'early Greenwich observations 1807–1840' in the library of the Royal Greenwich Observatory (and cited in Brazell's *London weather* (1968)), the gale on the 29th unroofed houses and blew down trees. The Greenwich observer reported that the wind on the 29th 'rose to a hurricane for about three hours, doing great damage . . . a dreadful tempest' from 11–14h. It had dropped to force 4 by 22h. The barometer rose from the time of strongest wind, and mean sea level pressure seems to have reached about 1017 mb by the evening of the 30th.

*Meteorology*

Our map for the 23rd shows an intense depression centre, about 980 mb, having crossed England at latitude 53–54° N and heading towards Denmark. The W'ly gradient winds over southern England south of the centre may have approached 100 knots.

Further frontal cyclones crossed Britain in the following days with strong winds and various places reporting heavy rain. On the 26th, as a centre crossed the country in the latitude of Wales, temperatures in London being at first low (4 °C) rose to 10 °C by late evening. A warm, misty night followed. By midday on the 27th a bigger depression was centred in the region of the Hebrides and mild air was spreading across England. On the 28th what appears to have been another centre, with pressure now down to about 970 mb, was crossing the country near the Scottish border and temperatures in the warm, moist air over England were around 14 °C. Colder air had reached Dublin from the west, with heavy showers. There were 'heavy gales' over southern England, and an observer near Plymouth reported 'storm all night and all day'.

The broader scale situation was changing, as pressures over central Scandinavia fell 30–50 mb from the 23rd to the 29th. The cyclonic activity progressed from the British Isles and North Sea towards the northern Baltic, and the map for the 29th shows Arctic cold air from the north having reached Aberdeen (temperature 2 °C at midday) and spreading south to all parts of the British Isles. A polar Low is suspected near northwest Ireland and an intense frontal wave depression was moving rapidly east over southernmost parts of England, with gradient winds in the warm sector again approaching 100 knots.

## Winter 1837–8

*Area*

Orkney and the northern North Sea.

*Observations*

A prolonged NE'ly gale in Orkney, at Otterswick on Sanday (59°15′ N 2°30′ W) caused the sea to scour the beach, laying bare an ancient forest floor with black moss and big fallen trees (trunks up to 60 cm thick – bigger than any trees that now grow in Orkney), which had evidently been submerged by the sea and buried by sand for some thousands of years. (Reported by W. Traill (1868) in a note 'On submarine forests and other remains of indigenous wood in Orkney'.)

The exact date of this storm seems not to be recorded, but indirect evidence suggests it was most likely in February 1838. In that winter a great anticyclone dominated Scandinavia with monthly mean pressure in January over 1040 mb in Finland. Ice, presumably driven or spreading from the Baltic, covered the Skagerrak between Denmark and Norway and proceeded to drift in a broad belt northwards up the west coast of Norway: at the end of February near Stavanger it stretched out to sea as far as the eye could see. In March the drift changed and the ice was carried back southwards along the west Norwegian coast.

## 7 September 1838

*Area*

Northern North Sea, Northumberland and Scottish coasts.

*Observations*

Storm made famous by the wreck on the Farne Islands, off Northumberland, of the passenger steamship S/S *Forfarshire* and the rescue of nine of the passengers by the Farnes lighthouse keeper and his 23-year-old daughter, Grace Darling. (This led to the founding of the Royal National Lifeboat Institution.)

The ship, of about 400 tons, built in 1834, was an

example of the early steamers with auxiliary sail – not especially large for her day. The boilers and engine could deliver 190 horse power to the two paddles, one on either side, which drove the vessel. The two steeply raked masts were not typical, nor could she carry as much sail as the sailing ships of that time. She was regarded as something of a luxury vessel.

She had left Hull on schedule at 6.30 p.m. on the 5th for Dundee. Northbound along the coast, with a SE'ly breeze and showery weather, she was off Flamborough Head after ten hours when one of the boilers was found to be leaking despite repairs which had been done before

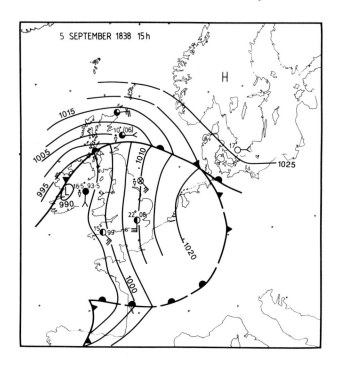

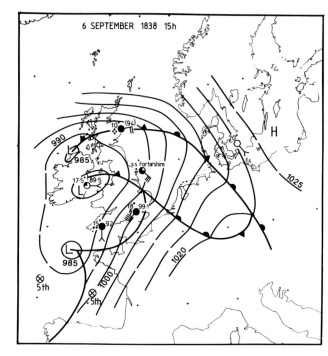

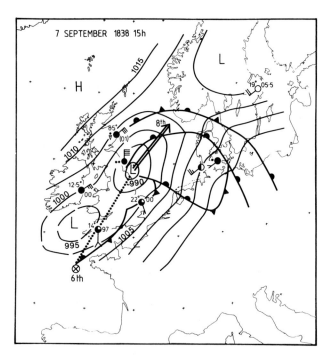

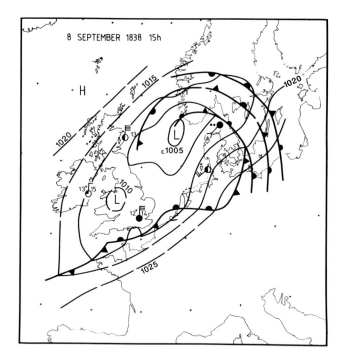

departure, and there was some loss of power. The ship nevertheless reached the Berwick area by dusk on the 6th. The weather had deteriorated with a stiff breeze and the sea was getting up, but wind and tide still favoured the ship's progress and there seems to have been no anxiety. But soon the wind changed sharply to NNE, of gale force, and the leak from the boiler worsened. At 1 a.m. on the 7th, off St Abb's Head, the engine stopped altogether. Sails were raised, but when the tide turned both wind and current carried the awkward hulk swiftly south till she struck the rocks of the outer Farnes at Big Harcar, the landward end of a reef marked by the lighthouse farther out to sea.

Some survivors were soon taken off by a passing sloop

from Montrose and landed safely at Shields, near Tynemouth. Another vessel, the Diana of Newcastle, was also driven onto the rocks by the same storm, but is believed to have got off later.

No one survived in the after section of the ship when it broke adrift behind the paddles, but those left on board the bow section clambered onto the rock, which was about 100 yards wide, and were mostly saved when daylight came.

Clearly, the disaster was largely associated with the vulnerability of the vessel and the quick, unexpected change of wind and weather. To judge the severity of the storm we need an analysis of the meteorological situation.

*Meteorology*

Unfortunately, our observational coverage in the 1830s is less satisfactory than in some earlier epochs, particularly as regards details of the observation sites and the instruments. Barometer corrections to sea level therefore cannot be precise and seem rougher in the case of this storm than the others analysed in the same decade. Nevertheless the broad outlines of the situation are clear.

Our analysis is illustrated by once daily maps for 5–8 September 1838. Cyclonic activity centred over the British Isles, with generally S'ly winds over England on the 5th, produced the lowest pressures (below 990 mb) during the 6th, and its focus shifted stage by stage to the North Sea as an intensifying outbreak of N'ly to NE'ly winds was drawn into its rear side on the 7th. Warm air from the SW producing temperatures up to 22 °C in southern England was associated with a series of frontal waves and warm sectors arriving during these days from the Bay of Biscay: one of these became the main centre of the cyclonic activity as it crossed England northeastwards on the fateful night of 6–7 September.

On the 8th, a N'ly gale was still blowing on parts of the Scottish east coast and the cold air brought afternoon temperatures as low as 12–14 °C to the south of England, while further small frontal waves seem to have maintained strong winds near the south coast.

The pressure difference indicated between the low pressure centre near northeast England on the 7th and the roughly known pressure level over Denmark suggests a S'ly gradient wind of over 60, perhaps 80, knots. But judging by the resistence of the cold NE'ly windstream over the northern North Sea to the advancing fronts from the south, the NE'ly to N'ly gradient winds at some point near the depression centre, in the coastal area where the ship came to grief, may have reached about the same strength.

The rescue took place some hours later in the localized and limited shelter provided by the islands and rocks.

*Acknowledgements*

I am much indebted to D.A. Wheeler of Sunderland Polytechnic for much detailed information about the last voyage of the S/S *Forfarshire* and to Mr Calderwood, Curator of the RNLI's Grace Darling Museum, Bamburgh for other details, published in the *Berwick Advertiser* on 15 September 1838.

*References*

*Encyclopaedia Britannica*, Bamburgh and Grace Darling entries.

Original newspaper reports in the *Berwick Advertiser* of 15 September 1838, the *Morning Chronicle* of the 11th, and *The Times* of the 13th of that month.

Russell Goddard, T. 1937, *Guide to the Farne Islands*, 4th edn 1956 (reprinted several times subsequently), Newcastle-upon-Tyne, Hindson and Andrew Reid, 31 pp.

# 6–7 January 1839

*Area*

The Atlantic fringe of the British Isles, especially western and northern Ireland, much of Scotland (some reports suggest especially southern Scotland), northwest England and North Wales; a little later much of England, then the whole width of the North Sea and on to Denmark and the Baltic.

*Observations*

A great storm: in Ireland considered the severest of the whole record (far less effect was felt there from the great 1703 and 1987 storms which are much more famous in England). Barometric pressure values are available from at least 25 points in the British Isles besides others on the continent and in Scandinavia. This was clearly one of the deepest depressions ever recorded so near the British Isles.

Observations assembled by the Irish Meteorological Service for a study of the storm to mark its 150th anniversary (Shields and Fitzgerald, 1989) revealed lowest pressures about 922.8 mb at Sumburgh Head, Shetland about 14h on the afternoon of the 7th and 925.2 mb at Cape Wrath around 0h on that day. At Aberdeen 936 mb was observed. The storm cyclone is thought to have been at its deepest around midnight 6–7 January, centred near $58\frac{1}{2}°$ N 11° W, off the Hebrides, with a central pressure at that time about 918 mb (and possibly rivalling the Atlantic storm of 5 December 1986).

The storm was credibly reported as exceptionally gusty and there was evidence of whirlwind and tornado activity besides. Many ships were wrecked at sea, on the coasts, and in harbour. Buildings were blown in or blown down and very many were unroofed.

Estimates of the death toll cannot be certain or complete. The Irish meteorological study cited indicates a probable figure upwards of 90 in Ireland on land and sea and perhaps 400 in all in the British Isles.

The greatest localized losses seem to have been in and about Liverpool, where there were 115 deaths and damage to shipping in the area estimated at nearly £500 000 in the currency of that time. The newly built Menai Bridge linking the island of Anglesey to North Wales was damaged. Damage to shipping and port structures in Limerick was estimated at £30 000 (Shields and Fitzgerald, 1989), and all the ships in Portaferry, Co. Down in northeast Ireland were driven from their moorings on to the shore. Many ships were wrecked in Lough Swilly, Co. Donegal, in the extreme north of Ireland. Losses on the west coast of the island were severe, those at ports and in towns in the south and east being much less.

No numbers or totals have been seen for the losses on farms, but numerous ricks of corn and hay were scattered, besides, no doubt, much damage to buildings and livestock.

About 100 000 trees were destroyed on one estate alone in Co. Fermanagh in northwest Ireland, and numbers reported for sundry other places suggest an overall total for Ireland of at least a quarter to half a million trees.

It seems that the equivalent total insurance losses from this storm just in the British Isles and around the coasts – our judgement being based on just the limited sampling

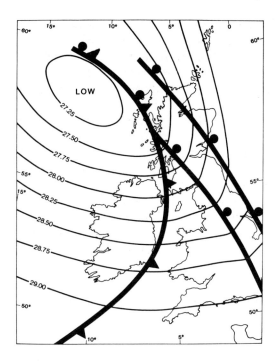

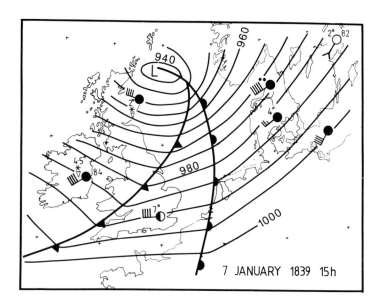

Pressure distribution and weather map analysis for 0h 7 January 1839, proposed by L. Shields and D. Fitzgerald of the Irish Meteorological Service (1989). (Barometric pressures at sea level in inches of mercury.)

known to us – were probably from one to five million pounds in the money of that time (and possibly more). The figure should be multiplied by perhaps 50 to bring it to the equivalent in today's money. The *Dublin Evening Post* of the 12 January 1839 reported that 'every part of Ireland – every field, every town, every village has felt its dire effects'. And clearly the losses in Britain, on the North Sea, and beyond, are likely to have been considerable.

*Meteorology*

A useful test of the possibilities of analysing this storm (and, by implication, others also) in this compilation is afforded by comparing the map for midnight (0h) on the 7 January in the study by the Irish authors (cited above) with our map for later the same day (15h), when the low was centred in the North Sea southeast of Shetland. The present author and the Irish analysts were working without knowledge of each other. (Both maps are reproduced here.)

The Irish map prints the pressure values of the isobars in the units in use in 1839: 27.25 in of mercury is equivalent to 922.8 mb, 29.00 in to 990.5 mb. Our map made use of data from fewer observation points (barely half as many) but spread over a wider region, from Ireland to Sweden.

The very deep depression arrived from the Atlantic on a path from about WSW or W by S, quite similar to the 1703 storm though farther north. The 1839 storm was probably travelling fast as it approached. The weather in Ireland had been quiet and cold as late as the morning of the 6th with snow or sleet in places – in Dublin the weather in the forenoon was described as 'very calm and gloomy' – but in mid-afternoon the wind began to freshen

and the weather to feel warm: by evening it was blowing a W'ly gale and between 2h and 4h on the 7th – in some other places from about midnight to 5h – hurricane force was reported. (The death toll may have been reduced somewhat because the worst impact of the storm came in the night.)

The advance of the system slowed (from probably 40–50 to about 20 knots as it passed the British Isles. This seems to have been the point also where the low-pressure centre began to fill a little. It approached the north Denmark–south Norway and Sweden region near 58–60° N later on the 7th and 8th as a 940 mb centre. It may, however, from then on have been rejuvenated somewhat, to become an almost stationary intense system dominating the eastern Baltic for some days.

In Ireland the strongest winds seem to have been generally W'ly, but in Scotland and Scandinavia the SE to S winds reached strong gale or storm force as the system approached, and a new gale from about NW up to force 10 or more set in after the centre had passed. After the time of our map SE force 10, with rain falling, was reported at Skagen at the north tip of Denmark all through the evening of the 7th, followed by W force 12 from 4h to 8h on the 8th and force 10 through the night of 8th–9th: the storm there continued as N force 10 through the 9th and until 4h on the 10th. At Denmark's easternmost island, Bornholm (near 15° E), the rain did not clear until early on the 9th and was followed by NW force 10, gradually declining as the skies cleared on the 10th (but followed by further rain and a renewal of the storm on the 11 and 12 January).

*Strongest winds*

The strongest pressure gradients measured in this storm over the northern half of the British Isles indicate gradient winds in the region of 90–100 knots. This agrees well with Shields and Fitzgerald (1989) who suggest that gusts ranging from 75–90 knots would probably have occurred widely over Ireland, with some exceeding 100 knots off the island. Similar values would apply to northern Europe, and gusts could have reached 110 to 120 knots at some exposed places in western and northeastern Scotland.

The duration of really damaging winds in Ireland was put at just 'two to four hours of winds of singular ferocity' at some places and eight to ten hours of 'very hard gales and squalls' on the Irish Sea between Dublin and Holyhead (Shields and Fitzgerald, 1989 from Espy, 1841). Farther east the duration seems to have been longer. The winds in London were already reported as 'a violent hurricane' before the warm front on the 7th, and in Denmark later many hours of forces 10 to 12 were reported.

Our map of the situation at 15h on the 7th shows the winds freshening over Denmark, and already Beaufort force 10 in the north, far ahead of the cyclone centre. This map shows winds simultaneously of gale strength over a huge area, from Ireland in the west to Bornholm in the Baltic, and from northern Scotland to southern England.

Damaging strength also was attained over a probably exceptionally large area at times in this storm.

The energy of this storm seems to have been related, as in other cases, to notably strong temperature contrasts between the airmasses involved. Temperatures in Dublin were $+3\,°C$ or below in the early hours of the 6th, rose to nearly 12 °C in the late evening of that day and briefly reached 9 °C that night at Pentland Skerries north of the Scottish mainland, before falling back sharply to about 4 °C in the morning hours of the 7th. The excessive gusts and squalls widely reported, however, seem to indicate also a notably big lapse rate of temperature with height, presumably most of all in the cold air circulating around the low pressure centre behind the cold front: this, too, may have played a part in the exceptional deepening of the centre.

## 20–21 October 1846

### Area

Ireland, probably mostly west and south coasts, also coasts of the Channel and southwest England on the 21st.

### Observations

Violent storm reported in Ireland, thought to have originated as a tropical hurricane, although there seems to have been little effect in Dublin apart from low pressure readings (mean sea level pressure may have fallen to 960–965 mb) in the night of 20th–21st. The storm is not mentioned in the register of the observatory in Phoenix Park, Dublin: a fresh W breeze reported there in the morning of the 20th later decreased and backed SW and then fell light from SE during the night.

The winds reported in Ireland and England were generally W'ly on the 20th, later backing SW or S. By 9h on the 21st the wind in London was S'ly force 6 and gusty, with rain, and it later veered W force 6, the weather continuing cloudy or overcast. An observer near Plymouth, on the coast of southwest England, reported strong W'ly winds all day on the 21st, and gusty, sometimes with heavy showers and in the evening with continuous rain,

and remarked that it became 'very stormy by night'. Temperatures were up to 12 °C there and in London in the wet, overcast weather, but no notably high or low temperature reports have been received.

### Interpretation

The date of this reported storm is consistent with a possible origin as a tropical hurricane. It seems in any case that a deep Atlantic depression certainly approached Ireland, and a frontal system crossing the southern part of the British Isles from west to east between about midnight and midday on the 21st ushered in a period of many hours of strong or stormy winds, mainly from about SW to W at places near the south (and especially southwest) coasts. These winds may have been kept going by wave disturbances on a trailing cold front.

The main cyclonic centre seems to have continued northeast, reaching the North Sea coast of Scotland with central pressure probably still below 970 mb, but no reports of storm winds from northern Britain are known to the writer. A secondary centre may have formed over the southern North Sea later on the 21st with pressure values similar to the main centre.

## 10 January 1849

### Area

Northeast coasts of Scotland.

### Observations

A great E'ly gale and heavy seas washed away heavy harbour defences at Peterhead in northeast Aberdeenshire.

This is from a report by T. Stevenson in his *Design and construction of harbours*, quoted by Sir Archibald Geikie in *The scenery of Scotland* (1901). A length of more than 100 m of harbour wall built 3 m above the high water mark of spring tides was carried away, one section weighing 13 tons was moved 15 m. No other reports of damage by this storm are known to the writer.

This E'ly storm seems to have affected northern Scotland while milder air and SW to W winds, which also developed to gale strength, were pushing up from the south. E'ly winds had prevailed generally over the British Isles from 1–7 January 1849. As this wind current was forced to retreat northwards, the winds strengthened until the 10–11th, when a depression with central pressure 970 mb or below passed eastwards across Scotland. Dublin had its strongest wind of the 1848–9 winter with the W'ly gale at 21h on the 10th. On the 11th the winds at Dublin and Edinburgh veered NW and became much lighter, and temperatures in Edinburgh were near or below the freezing point that day.

## 28 December 1849

*Area*

The North Sea coast of Britain and evidently much of the North Sea, especially western side.

*Observations*

A great storm tore a 400 m wide breach across the neck of the peninsula connecting Spurn Head at the mouth of the Humber (53.6° N 0.2° E) to the mainland. Spurn peninsula was reduced to a row of islands at high tide. A rampart defending the lighthouse was washed away.* (This was by no means the first time in history that a storm had broken the connection between Spurn Head and the mainland (De Boer, 1968): the link had been re-formed by drift of sediments and wind-blown material, each time farther west, and latterly protected by engineering works. In the first decades of the nineteenth century Spurn peninsula had been in a more broken condition.)

No storm at this date seems to be mentioned in the records from the continental side of the North Sea.

*Meteorology*

This storm came after nine days of quiet anticyclonic weather with about normal temperatures in Britain, with some sunshine and variable wind directions. It marked a sharp polar outbreak with NW'ly and N'ly winds spreading south over all the British Isles and the near continent, as a rather deep, and probably deepening, low pressure system moved southeast or southsoutheast into the North Sea and thereafter declined in intensity. Pressures in the north and west of the British Isles were generally lowest in the night of 27–28 December and during the 28th farther south, when the depression was probably deepest and closest to the southern and southwestern North Sea.

The Spurn Head lightkeeper reported that the gale was from about NNW, with rain and sleet at times. The tide level was exceptionally high. Afternoon temperatures on the 28th, after the passage of two cold fronts south over the country, were as low as −4 °C to −7 °C over all Scotland, −4 °C in London and −1.5 °C in Dublin. It was the coldest day of the month, but ushered in a period of four weeks of almost unbroken cold to very cold weather in England. The meteorological situations and sequence seem to have resembled the spell of about the same length in February 1956.

Winds were lighter over the following days but continued generally NW'ly.

This seems not to have been one of the severest storms, although severe in its effects at Spurn, since storm damage seems not to have been widely reported. There may have been local conditions in the configuration of the near-shore sea-bed at that time which made the swell especially effective in causing damage at Spurn. Nevertheless, with the very sharp fall of prevailing temperatures in Britain from +5 °C to +6 °C on the 26th, a very rapid transport of the colder air from the Arctic is implied, at speeds which were only maintained on the 28th. And, as the frequent snow showers reported in Scotland indicate, the windstream was probably very rough and gusty.

I am indebted to Dr G. De Boer of Hull University and Mr P.W. Spink for local details of this storm at Spurn.

*Similarly great masses of stone had been shifted by the sea during the building of the Bell Rock lighthouse on a reef in the North Sea, 19 km east of Dundee, in an otherwise unremembered storm in May 1807 (Geikie, 1901, p. 61, from R. Stevenson's *Account of the erection of Bell Rock lighthouse*).

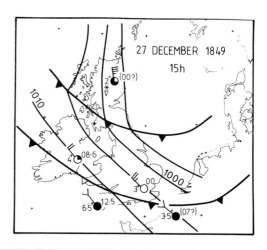

## 1 January 1855

*Area*

North Sea, notably southern half, and adjacent lands, as well as most or all of central Europe.

*Observations*

This great storm on New Year's Day followed a month or more of mild weather in Europe and prevailingly low barometric pressures. It was immediately preceded by such a great fall of pressure on the 1st as H.W. Dove said could 'seldom be met with'. The stormy winds seem to have accompanied a cold front which progressed across central Europe on the 1st and 2nd, accompanied by 'violent showers of rain and heavy falls of snow: at Berlin these rains had completely the character of heavy thundershowers: from Silesia to Hamburg thunder and lightning were reported to have been observed at several places'. Then a 'powerful frigid aerial current prevailed to such an extent that the cold weather of the latter half of January and February' – in Britain as in central Europe – became 'unusually intense'.

There was an unusually high tide level along the entire German North Sea coast, only 10–20 cm below that which caused the sea flood in February 1825, and part of the island of Wangeroge – at the eastern end of the Frisian islands, not far from the mouth of the river Weser – 'was washed away by the furious beating of the waves, and the most substantial embankments on the north-west coast of

Germany could scarcely resist their violence'. But repairs to the sea defences had just been completed and the situation nowhere came to severe damage or losses (Petersen and Rohde, 1977).

Winds – presumably from a S'ly or SW'ly point – raged as far east as Austria and the southern slopes of the Riesengebirge in present-day Czechoslovakia in the early hours of 1 January. Over 30 000 trees were uprooted within a limited area of about 400 ha near Kremsmünster.

*Meteorology*

H.W. Dove (1858) has given a collection of barometric pressure measurements on this date at more than a hundred places widely distributed over Europe in the form of departures from the overall average level for January. Although it is not made really clear these seem likely from the context to be departures from the averages of the previous four to seven years. (A similar technique for revealing barometric pressure patterns on weather maps seems to have been first used by H.W. Brandes in 1820 in a series of daily synoptic maps for the year 1783 using the European data published in the yearbook of the Societas Meteorologica Palatina of Mannheim.) As a carefully derived map of the average January pressures, reduced to sea level and standard gravity, for the decade 1850–9, from the world network of observing stations then available, has been given by Lamb and Johnson (1959, 1966), Dove's figures can be used to give an approximate map of the pressure situation on 1 January 1855 which can be regarded as fully corrected values. The time of day to

which the observations apply is unfortunately not known.

The map shows an unmistakable NW'ly storm situation over the North Sea. Gradient winds measured over England seem to be, in some areas, about 100 knots and over East Anglia stronger than that. Over northern Scotland and northern, western and southern parts of the North Sea similar strengths probably prevailed at some time during the day.

A strong mistral situation also seems indicated in the classic area for that wind.

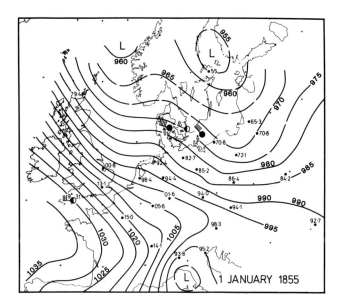

# 25–27 October 1859

*Area*

Biscay, British Isles, western and northern North Sea. The 'Royal Charter storm'.

*Observations*

First indications of the storm depression were noted in the Bay of Biscay near Cape Finisterre, northwest Spain, on the 24th–25th. The centre progressed nearly northnortheastwards over Britain on the 25th from Cornwall to the Yorkshire coast near Flamborough Head on the 26th, and thence onward, near the coast of Scotland, to between the Shetland Isles and Norway on the 27th.

The strongest winds in the system developed as a rather narrow stream from the N or NNE over the Irish Sea on the 25th and 26th and then over the coasts and islands of Scotland on the 26th. Admiral Fitzroy (1860, 1861), the first Director of the Meteorological Office, estimated the strongest winds as 60–100 m.p.h. (about 52–87 knots). On the 25th, there were also strong to gale force winds from SW and SE over a huge area from Portugal, Spain, and France to southern England and Wales. The windshifts from SE to between SW and NW over southern Britain and from SE to NE or N over northern England, Wales and Scotland on the 25th and 26th were sharp.

The iron-clad ship S/S *Royal Charter*, which had sailed round the world and anchored (with lowered sails and

steam shut off) off the north coast of Anglesey, was wrecked about 7h on the 26th with the loss of 500 lives, while other vessels under sail nearby were able to stand off to westward and largely escaped damage. The N'ly gale hardly extended beyond the east coast of Ireland, 150 km to the west, and did not reach Liverpool until 12 hours after the wreck of the *Royal Charter*, only 100 km away to the east. Thus, the windstream was quite narrow and only making very slow progress eastwards. The main advance of the system was towards about NE by N. As the cyclone centre moved across England, there was a calm area about 15–25 km wide in its centre.

The pier at Brighton was destroyed by the gales on the other side of this same depression. (These incidents led to the introduction of gale warnings by the Meteorological Office.)

*Meteorology*

Fitzroy (1860, 1861) has provided maps of the weather situation, with numerous entries of wind direction and force, as well as pressure and temperature values, and weather descriptions (rain, hail, snow, fog, etc.), from ships at sea and places on land, several times a day, from 21 October to 2 November 1859. Some of the maps cover the width of the Atlantic between the British Isles and Newfoundland, north to Greenland and Iceland and south to Spain and Portugal – probably the earliest attempt at weather mapping the whole ocean area. (The sequence includes the passage of another depression, from west to east across northern England to Denmark, on 1–2 November.)

Unfortunately, although pressure values were already adjusted to sea level, the techniques of weather mapping had not yet been worked out, and Fitzroy's papers seem to have been partly aimed at securing discussion to improve this. It seems impossible to derive with assurance a precise pressure distribution from the way the measured values are represented on his maps.

Temperature values representative of the measurements in different parts of the map are less problematical and have been used here to indicate what is certainly a reliable portrayal of the wind pattern, with a suggested frontal analysis, and approximate pressure distribution, on the mornings of 25 and 26 October.

After a warm, dry summer that year, pressures and temperatures over and about the British Isles had fallen from the middle of October onwards, and the eastern North Atlantic region generally seems to have been covered on 25–27 October by a N'ly wind or 'polar current' (Fitzroy's term). Weather near Ireland's east coast turned sharply colder on the 20th and there were snow and hail squalls. By the 25 October, much of northwest Ireland was under snow, and there was an unusually sharp frost under clear skies near Galway. This was followed by heavy rain and

light winds between S and W, here interpreted as associated with a small polar Low moving south near the west coast of Ireland.

The sharpest wind development, and the change of wind, farther east, which sank the *Royal Charter*, is tentatively shown here as linked to a slow-moving cold front which may have marked the eastern limit of the coldest, Arctic air. Its passage brought snow in various places in northern England and Ireland, in Orkney, and probably all over the high ground in Scotland. But this front cannot be traced farther south than about 51° N on the available evidence.

This storm must be written down as one of relatively small span, the entire width of the area with strong winds at any one time being no more than 300–500 km (less than 300 nautical miles). The width of the belts with winds of gale force seems to have been much less than this. The reported disaster and other incidents of damage seem to have been quite closely associated with squalls accompanying the sudden wind-shifts.

This seems to be a case of a storm arising near the axis of a meridionally orientated upper cold trough extending south to near latitude 45° N. There is little indication of any well-formed jetstream near the path of the storm, though some concentration of a SSW'ly upper wind at jetstream level cannot be ruled out.

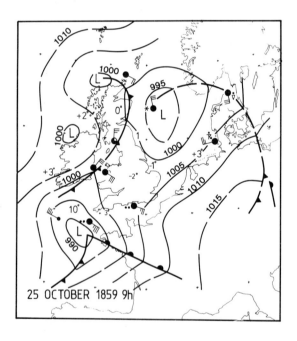

25 OCTOBER 1859 9h

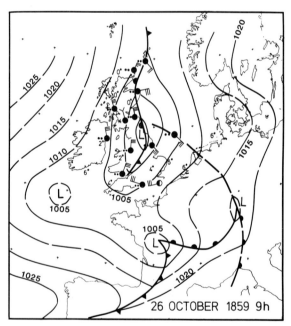

26 OCTOBER 1859 9h

## 21 February 1861

### Area

The British Isles, especially southern parts, and by deduction much of the North Sea as well as neighbouring parts of the Atlantic between latitudes about 48° and 56° N.

### Observations

Gales, mostly from points between about S and WSW, were reported from places as far apart as Galway, Plymouth and the Channel area, Dover, London and Great Yarmouth.

There were also strong S'ly winds in the Clyde and about northern Ireland.

The central tower of Chichester cathedral was blown down, and one wing of Crystal Palace in London was demolished by this storm.

### Meteorology

Weather maps reconstructed from the weather observations listed in the manuscript daily weather reports, which were then already being prepared in the British Meteoro-

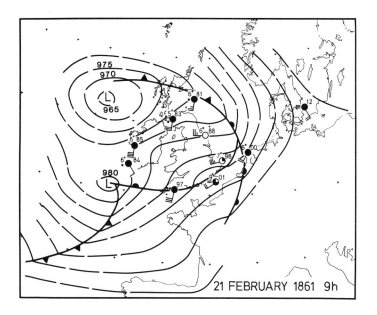

21 FEBRUARY 1861 9h

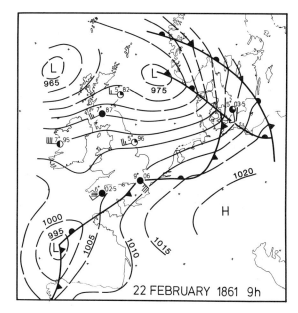

22 FEBRUARY 1861 9h

logical Office, indicate a cyclonic southerly wind situation over the British Isles on the 20th and low pressure continuing over the Atlantic west and north of Ireland till the 22nd. Other quite deep depressions were approaching from the southwest, apparently originating as waves on a cold front, although the surface temperatures on the 9h maps were no lower than 5–7 °C in the colder air as compared to 8–10 °C in the warmer and more humid airmass coming from the SW that brought mist and fog in various places on the south coasts.

The existence of a colder air supply – presumably associated with an upper cold trough extending south over the Atlantic west of these islands – can be deduced from the continuing cyclonicity there and the fact that pressure at Lisbon, Portugal, fell 12–13 mb between the 21st and 22nd, ending the anticyclonic conditions that had

prevailed there. By the 24th, N'ly winds were blowing over the whole width of the British Isles.

The damaging squalls of wind over the southernmost part of England were probably associated with the trailing front seen close to the south coast on the maps for both the 21st and 22nd, with active cyclonic development and one smaller frontal wave which seems also to have been associated with a more localized strengthening of the wind.

Gradient winds probably in the order of 80 knots seem to be suggested near the point of occlusion of the warm sector over the southern North Sea, and in the newly developing warm sector depression south of Ireland, by the pressure pattern derived from the observed values on the map for the 21st and again in the post-frontal westerly wind stream over the southern North Sea on the 22nd.

## 26–27 December 1862

### Area

North Sea, and English, Scottish and continental coasts.

### Observations

NW'ly gales were reported in the Clyde and on the east coasts of Scotland and England, severe gale in some parts (e.g. Tynemouth/Shields) and gales from W or SW in the Channel, southern North Sea, Dutch coast and German Bight.

On the north-east coast of Norfolk, near Happisburgh (52°49′ N), the desolate remains (cottages, roads and wheel-tracks, ploughed fields and ditches) of the older village of Eccles, lost to the sea and buried in sand in the seventeenth and eighteenth centuries, were laid bare by the exceptional scouring action of this storm. (The ruined church had been overwhelmed by the sand in a storm in or about 1794, but the belfry tower was still standing.)

### Meteorology

The maps reconstructed from the observations listed in the

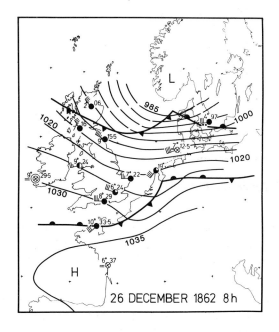

26 DECEMBER 1862 8h

daily weather reports of the Meteorological Office reveal a W'ly storm situation, gradually turning to NW, over the North Sea. Pressure was low or very low over Scandinavia and an anticyclone with its axis over Biscay, France and the Alps, showed sea level pressures approaching 1040 mb.

Petersen and Rohde (1977) record another storm flood tide in the German Bight–River Elbe just seven days earlier. The situation was one of generally NW'ly to N'ly winds over the British Isles from 19 to 30 December.

The storm situation of 26 December seems to have been a notably severe one with gradient wind strengths up to 150 knots over parts of the North Sea and Denmark and up to 100 knots near the Scottish, English and Dutch–German North Sea coasts.

Airmass contrasts were quite strong and two main frontal systems seem to have been involved in the cyclonic developments. Maritime tropical air was giving temperatures of 10–11 °C with mist and fog on the southern coasts of England and Ireland and the Atlantic coasts of France. Elsewhere, over England and Ireland, temperatures ranged from 6–9 or 10 °C, but north of a cold front seen over mid-Scotland on the 26th only 2–3 °C, and on the morning of the 27th a returning front brought snow over that area.

## 24 January 1868

### Area

Scotland and northern North Sea, also windward coasts of Ireland, Wales and west Cornwall.

### Observations

S'ly and SW'ly gales were reported from early on the 24th from the southwest and south coasts of Ireland, western Cornwall, exposed parts of the coasts of Wales and Scotland. Severe gale from SSW was reported at Aberdeen at 8h. Later in the day severe gale from WSW was reported at Valentia Observatory near the southwest tip of Ireland and from S at Nairn in northeast Scotland.

Parts of stone buildings were blown down in Edinburgh, where the wind force was presumably also that of a strong gale or storm.

### Meteorology

The reconstructed weather map shows a wintry high pressure situation that morning over nearly all southern and eastern England in a ridge extending from an anticyclone over Scandinavia. The situation over the British Isles had been generally cyclonic for some days previously. A rapidly sharpening pressure gradient for S'ly and SW'ly winds existed over the northwestern half, or more, of the British Isles as warm air pressed in from the Atlantic. Temperatures on the Atlantic fringe were from 7–11 °C, from 9–11 °C in the full warm air with coastal fog, but freezing over most of England and the continent with snow in places.

By afternoon, a cold front had passed the Atlantic coasts of Ireland with the winds turning westerly, but still strong: WSW force 9 at Valentia. Pressures had fallen very rapidly, up to 20 mb (or more) in six hours. Next day a cyclonic centre had come in over the British Isles from the Atlantic, and on the 26th a NW'ly situation prevailed briefly over these islands, followed by a long run of westerly weather sometimes combined with some anticyclonic influence.

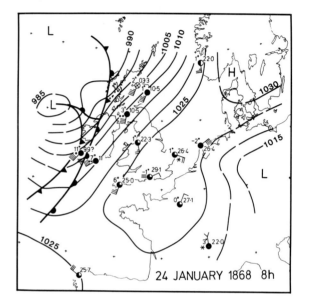

24 JANUARY 1868 8h

The storm on the 24th was associated with an unusually intense development, presumably with a strong jetstream from about SW in the upper atmosphere, accompanying the long narrow warm sector detected over Ireland that day. Surface pressure gradient winds over northern and western districts of the British Isles increased from about 70–80 knots to perhaps 140 knots.

There is some possibility that Edinburgh was affected by localized and periodic concentrations of excess strength of the surface wind as a result of a mountain-wave phenomenon downwind from the Pentland and other hills in the southern uplands of Scotland, as in the case of the Sheffield disaster of February 1962 (attributed to lee waves from the Pennines).

The predominantly westerly situations which dominated over the British Isles after this storm, from 27 January and through February were often marked by anticyclonic influence.

*Plate 1.* Bell Rock Lighthouse during a storm from the NE. The lighthouse is in the North Sea about 56°26′ N 2°23′ W, 17.5 km off the Scottish coast, southwest of Arbroath. (Drawing by J.M.W. Turner, engraved by J. Horsburgh. Early nineteenth century: depiction of a storm witnessed by Turner, probably before 1820.)

*Plate 2.* The wreck of the S/S *Forfarshire* in 1838 on the Farne Islands, off the northeast coast of England, made famous by the rescue of some of the passengers by Grace Darling and her father, the lightkeeper. (From a painting, probably by J.W. Carmichael, in the possession of the Royal National Lifeboat Institution, at the Grace Darling Museum in Bamburgh.)

*Plate 3.* Shipwreck in a storm – thought to be off the south coast of England – in the 1880s. (From a painting by T.R. Miles, reproduced by the Royal National Lifeboat Institution by kind permission of David Oliver May Esq., FRINA of Lymington Marina.)

*Plate 4a.* Strong swell running off the small island of Heligoland in the German Bight. (Photograph by the late F.A. Schensky, with acknowledgement to Dr H. Rohde of Hamburg.)

*Plate 4b.* Storm waves washing over part of Heligoland in 1926 during a N'ly storm. The island has been largely destroyed by storm seas in the last 1000 years. (Photograph by F.A. Schensky, with acknowledgement to Dr H. Rohde.)

*Plate 5.* Waters driven by a North Sea storm on 3 January 1976 breaching the dyke protecting the Haseldorf Marsch beside the River Elbe near Hamburg. A few minutes later the dyke gave way. (Photograph reproduced by kind permission of Dr H. Rohde.)

*Plate 6a. Above*: The Channel ferry *Hengist* driven aground on the Kent Coast at The Warren, 1 to 2 km east of Folkestone, by the storm on 16 October 1987. (Photograph by courtesy of South Kent Newspapers Ltd.)

*Plate 6b. Right*: Houses and cars in Addlestone, Surrey, badly damaged by a falling tree in the great storm of 16 October 1987. (Photograph by Tony Andrews, *Surrey Herald*, kindly supplied for this book.)

*Plate 7.* Huge waves pounding the coast and harbour at Porthleven, near Helston, Cornwall in the S'ly storm on 17 December 1989. Gusts to 104 knots were measured on the Cornish coast. (Photograph kindly supplied by Cornish Photonews, Falmouth. Copyright: David Brenchley.)

*Plate 8.* Five sea floods in 100 years recorded by flood-level marks at the door of the great church in King's Lynn, Norfolk. (Photograph by H.H. Lamb.)

*Plate 9.* The Sands of Forvie: a 30-metre-high dune covering a once important medieval township on the east coast of Scotland north of Aberdeen, buried in a great S'ly storm in August 1413. (Photograph by H.H. Lamb.)

*Plate 10.* The old town at Skagen, north Denmark, seen buried in blown sand. (Painting by Vilhelm Melbye, dated 1848.)

*Plate 11a,b.* Two views of a step, or shingle and sand platform, created along the north-facing coast of the Skagen peninsula by the storm of 16 October 1987. In plate 11b the dimensions of the platform are indicated by a metre-rule held by Dr Hauerbach. (Photographs by H.H. Lamb.)

*Plate 12.* Forest damage by the storm of 31 January 1953 in eastern Scotland with many trees felled in a section of pines in planted woodland on the Sidlaw Hills near Dundee. (Photograph kindly supplied by the *Scots Magazine* (D.C. Thompson & Co. Ltd), Dundee.)

*Plate 13a,b.* Two views of trees felled in a normally sheltered city-centre site by the storm of 2 April 1973: canal scene in Amsterdam (Herengracht) next morning. (Photographs by H.H. Lamb.)

## 15–16 June 1869

*Area*

Coasts of the British Isles and France: strongest on the east coast of Scotland and northeast coast of England in the night of 15th–16th.

*Observations*

A gale set in suddenly, blowing from the E, NE, or N on various parts of the Scottish North Sea coast and northeast England. There were many shipwrecks and great loss of life.

Severe gales or storm force winds were also reported at some stage during the 15th or 16th at Liverpool and on the French coast of the Channel, as well as at Brussels and London and at Toulon on the Mediterranean coast of France (mistral).

*Meteorology*

After predominantly anticyclonic situations over the British Isles, occasionally combined with a N'ly wind drift from the last days of May onwards, pressure was falling generally from the 12th to the 14th. On the 14th the situation had become cyclonic and mostly N'ly over Britain and Ireland, with several centres of low pressure around 995 mb over the Channel, East Anglia and the central North Sea, and another, probably small, centre off the Hebrides. The N'ly and NW'ly winds became strong later on the 14th in Ireland, as the centre of gravity of the complex system moved east. The N'ly wind stream gradually dominated more of Britain over the 15th and 16th and spread farther south to affect part of the Bay of Biscay and northern France, where gale force winds from the W increasingly turned NW'ly.

Some of the wind-shifts were abrupt changes of direction associated with frontal passages. Gradient wind strengths seem not to have exceeded about 60–70 knots anywhere in the system, though gust strengths near fronts and coasts may have exceeded this.

Northerly winds, often combined with anticyclonic conditions over the British Isles, dominated the situation over this part of Europe for the rest of the month.

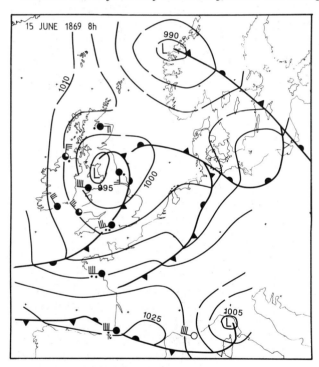

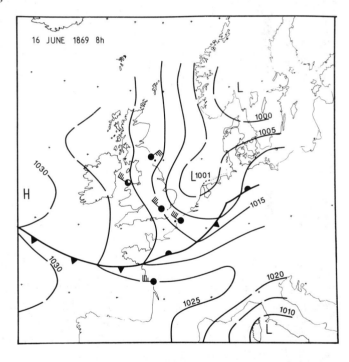

## 24 March 1878

*Area*

The British Isles, especially coasts of Scotland, England, Irish Sea and the Channel.

*Observations*

Reports of gale, and even hard or severe gale, almost confined to sudden squalls, one of which caused a famous disaster, the sinking of the naval training ship HMS *Eurydice* off the Isle of Wight with the loss of everyone on board.

*Meteorology*

A NW'ly and N'ly wind situation prevailed over the British Isles and parts of the Norwegian Sea and North Sea from 22 March to the first days of April. As seen on the maps for the 24th and 25th, this was a very cold Arctic airstream, which ultimately extended south over Biscay to Spain and Portugal. Similar air covered Scandinavia on those days, with temperatures as low as − 14 °C in northern Sweden and snow, or snow showers, falling as far south as Paris. There were hail showers on the coasts of Ireland, Wales and southwest England. The squalliness is evidence of thermal instability and presumably very cold air aloft, since the pressure gradients were not exceptionally strong (hardly anywhere more than about 60 knots).

In the weekly supplement section of the Meteorological Office's *Daily weather report* it is stated that on the 24th 'Weather was fine over the greater part of the country, but in the north of Scotland and Ireland showers had commenced at 8 a.m. In the course of the day, a small depression passed south from the north of Scotland along our east coast, causing an extremely sudden change in the weather, bright clear skies being followed very quickly by heavy snow, while the wind rose to a gale or hard gale in the west of Scotland and northwest of England. In addi-

tion to this, a very sudden NW'ly squall passed with great rapidity (southwards) over England, the wind for a few minutes rising to the force of a fresh gale. This change occurred at 10.20 a.m. at Bedale in (north) Yorkshire, at 3.50 p.m. in London and at about the same hour in the Channel.'

(This situation, possibly involving a polar Low, seems to provide a parallel for the one reported to have caused a disaster to the fishing fleet near the southwest coast of Norway on 11 March 1822, q.v.)

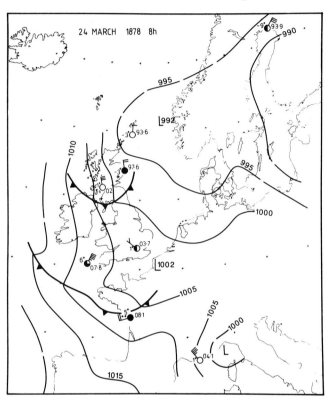

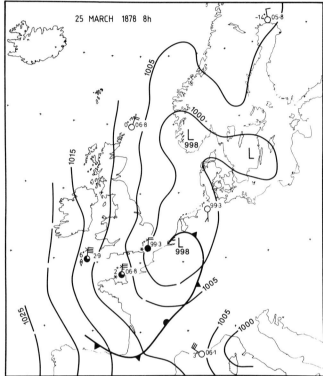

## 28 December 1879

### Area

Great gales widely reported over the British Isles, the North Sea, and southern Scandinavia. The Tay Bridge disaster storm.

### Observations

Gales from S or SW were already reported at most points on the Atlantic coasts of the British Isles and on the west coast of south Norway as early as the morning of the 27th. Severe S'ly gale force 9 was reported that evening at the Scilly Isles and on the west coasts of Ireland and the Hebrides, and SSE force 9 near Stavanger, Norway. On the 28th the storm eased during the first half of the day over most of Britain, especially in Scotland and northern England, where the winds became quite light, and generally W'ly, for a time. Later in the day the winds backed and freshened rapidly. After dark the wind was squally and at Aberdeen was reported to have backed to S, at Wick and Stornoway to SE, before working its way round to W or WSW again.

In the Firth of Tay, where part of the nearly new railway

bridge was blown down with a train on it, the veer to WSW was probably sudden and squalls were occurring about every 10 minutes with the wind at WSW force 10, according to the inquiry which was set up after the disaster. The signalmen at Newport on the south side of the Firth had watched anxiously as the train headed north on to the bridge at about 7.30 p.m.; despite the darkness of the night, with heavy clouds and rain and the furious wind, sparks were seen to come from the wheels when the train was about half-way across and then disappeared. One witness, looking from the west, near the north end of the bridge, saw 'a mass of fire fall from the bridge'. It was about 8 p.m. when conditions improved enough to see that part of the bridge was gone. Owing to the fury of the storm nobody heard any noise of the accident.

At the height of the storm people in the Scottish towns were afraid to go out of doors because of the falling trees and masonry. But the severe stages of this storm in Scotland seem mostly to have lasted no more than 3–4 hours.

Although this account is based on the weather observation reports from 50 stations in the British Isles and neighbouring countries, analysis is still made difficult at

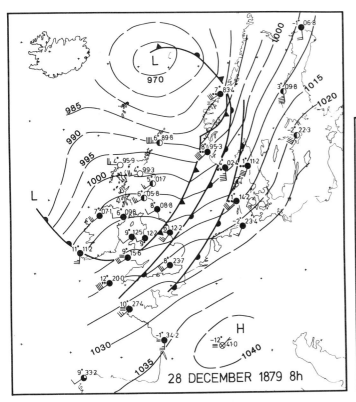

28 DECEMBER 1879 8h

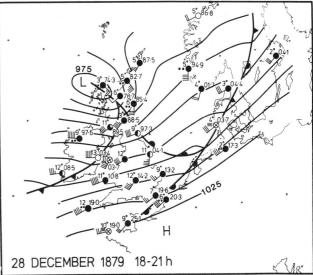

28 DECEMBER 1879 18-21h

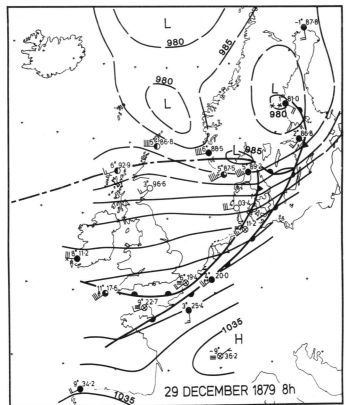

29 DECEMBER 1879 8h

this stage in the development of the meteorological observation network by the roughness of the wind force estimates at many places – an understandable difficulty at a time when very few of the observers can have had experience of instrument measurements of wind and the instruments themselves were inferior to those that have used later techniques. As in other, earlier situations analysed in these studies, full use had to be made of the requirements of logical consistency of pattern development and movement of the weather systems with the previous and later maps.

R.H. Scott, later Director of the Meteorological Office and at that time Secretary to the Meteorological Council, gave evidence at the Court of Inquiry. Gusts, reported from the instruments available at the time as 120 m.p.h. at Glasgow in the same gale were corrected to 100 m.p.h. or 87 knots. The same figure was presumed to apply to the Firth of Tay about the time of the accident, but as our analysis shows the latter site to have been nearer to the tip of an occluding warm sector the squalls in the Firth of Tay may have been somewhat fiercer.

Synoptic weather maps drawn from data published in the *Daily weather reports* show a very rapidly developing situation. Barometric pressure fell at Stockholm from over 1040 mb on the morning of the 27th to about 986 mb 48 hours later. Very severe frost had developed over the continent in the high pressure regime, with $-16\,°C$ at Lyons on the morning of the 27th, $-10\,°C$ at Paris and Brussels and $-4\,°C$ in parts of England. But very mild air was approaching from the ocean, and temperature rose as high as $+13\,°C$ in Edinburgh and in Dublin on the evening when the storm was reaching its height. Sea fog was quite widespread on the coasts of the Channel, Biscay and South Wales.

The analysis indicates two warm fronts advancing, as the mild air entered the European scene on the morning of the 28th, followed closely by a cold front which led to a veer and easing of the winds but rather little fall of temperature. Curvature of the wind-flow pattern hinted at another storm system approaching from the west. The evening observations on the 28th, made at 18h in the British Isles and 20h in Scandinavia, cover the time of the accident and height of the storm in the Firth of Tay. The map for that time shows perhaps the most intense phase of the storm, with a 974 mb low pressure centre close to the Outer Hebrides.

The pressure gradients in areas with nearly straight isobars suggest gradient winds of up to 130 knots in the warm sector over southern Scotland, also over the southern Hebrides and near the north coast of Ireland, and about 80 knots over southern England.

Next morning the depression had advanced from the Hebrides to near the mouth of Oslo Fjord, an average speed of nearly 50 knots (which seems to have been registered in the previous 12 hours also, when this low pressure centre arrived from the Atlantic). Gradient wind speeds of 100–110 knots are indicated over parts of Denmark and south Sweden on the morning of the 29th.

The speed of advance suggests a strong jetstream in the upper atmosphere.

By any reckoning, this must be counted as one of the great storms, for the overall extent of the regions experiencing gale force winds, for the strength of the extreme winds, and for the suddenness of onset of several of its phases.

(An account of the Tay Bridge disaster and subsequent inquiry is given by Carr Laughton and Heddon, 1927.)

## 14–15 October 1881

*Area*

The North Sea, Denmark and south Baltic, also much of the British Isles – especially the North Sea coast (Eyemouth fishing fleet disaster) and Channel, Dutch and north German coastlines.

*Observations*

Great loss of life at sea and destruction of shipping, attributed particularly to the suddenness with which the storm came on and the violence of the squalls.

A special meeting of the Royal Meteorological Society in 1881 noted that 'the area over which injury was produced was very large, and, although not without precedent, it is happily rare'. In the greater part of Ireland and southwest England structural damage was not unusual in extent or character, 'but along the east coast and in the East Midlands the damage was excessive' and, 'on the northeast coast unprecedented. In Scotland the destruction of trees was enormous.' About 35 trees were blown down in Greenwich Park (London).

At Eyemouth in Berwickshire where the morning of the 14th had dawned fine, with clear sky and calm sea, the fleet of 41 fishing boats, mostly big deep-sea boats, sailed out under a steady breeze. It was a fleet famous for the size and finish of the boats. In the middle of the day the wind fell light, and then the storm struck suddenly. Nineteen of the boats were lost and 129 of the little town's manhood. (The barometer had been very low, and warnings had been issued. At Arbroath and Berwick the fleets stayed in port.) Inland this gale was considered a 'great storm' in many districts, perhaps 'the severest since 1859'. A factory was blown down near the railway station at Banbury (Oxfordshire) and damage to buildings was reported in Hampshire and near the Welsh border. Hundreds of trees were felled by the wind in many parts of England and ricks unthatched (*Meteorological Magazine* 16: 179).

French and German shipping in the Channel and the North Sea was either halted or much delayed by the gales there.

On the North Sea coasts of Denmark and Germany there was a considerable sea flood in the night of 14–15 October, in one place 'one foot higher than in 1825, the highest for over 40 years. Large numbers of sheep, cattle and horses were drowned. Buildings were deeply flooded on both sides of Jutland. The storm wind was SW on the afternoon of the 14th and turned to WNW in the morning of the 15th.

*Meteorology*

The cyclone centre which produced this storm was first identified about 150 miles south of Nova Scotia on the 10th. By noon on the 13th it had only deepened to 997 mb and was in longitude $22°$ W. It deepened greatly to about 960 mb as it crossed the British Isles and reached the North Sea, drawing Arctic cold air into its rear side. The temperature at Aberdeen was only $3\,°C$ on the evening of the 14th with a N'ly gale: that morning there had been E'ly gales in northern Scotland with $6–7\,°C$. The central pressure was still about 960 mb – a notably low value for the area concerned – as the centre approached Jutland from the west on the evening of the 14th, but no lower than about 975–980 mb over southeast Sweden the next evening. On the 16th it was centred (985 mb) over the northern Baltic.

The system advanced at a fairly steady speed of about 35 knots as it crossed the Atlantic, but its progress across the North Sea and southern Scandinavia was slower than this (as is quite usual for a mature cyclone). There were no longer any winds of near hurricane strength in the system by the evening of the 15th. The breadth of the belt of destruction as the storm crossed Britain was estimated by the *Illustrated London News* at 200 miles (about 320 km).

*Maximum wind strengths*

At noon on the 14th strong SW'ly gales force 9 were reported by ships on the central and southern North Sea and inland in Kent. Beaufort force 9 from NW to NNW was also reported at various places in Ireland in the rear of the depression, while ENE force 11 was reported near Aberdeen as the system approached. Force 10 was attained in the Channel Islands on the morning of the 14th. Force 10 to 11 was also reported in the NW'ly winds on the

northern North Sea and near the coast of Jutland when the depression centre was in the Skagerrak on the morning of the 15th. At that time force 10 from W or WSW was also reported at the North Sea and western Baltic coasts of Germany and later on Bornholm and the south coast of Sweden.

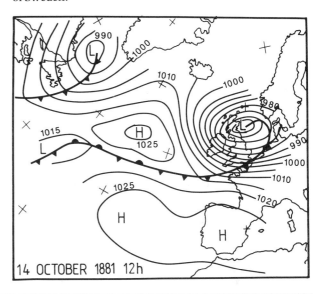

14 OCTOBER 1881 12h

# 6 March 1883

*Area*

Whole North Sea, especially northern and western parts, the Hebridean Sea and northern part of the Irish Sea.

*Observations*

Northerly gales were observed across the whole area from the Hebrides (Lewis) to the coast of Norway and south as far as Holyhead and Hull.

The newspaper *Hull News* reported later in the month (24 March 1883) unparalleled destruction of Hull fishing smacks and loss of life in this storm: 23 fishing smacks from Hull were lost with 135 crew members drowned in the central North Sea somewhere near the Dogger Bank, and over 70 other Hull smacks were damaged. Besides these, 18 Dutch boats belonging to fishing villages on the north coast of Friesland were lost on the night of 5–6 March with 83 lives lost.

*Meteorology*

On 5 March a broad airstream from between about W and NW covered the area between the Hebrides and the northern Baltic, on the flank of an anticyclone (central pressure 1043 mb) centred over the western part of Ireland. Barometric pressure on the morning of the 5th was falling quickly over the whole region of the NW'ly winds north of Scotland and across Scandinavia, and later that day over the North Sea, apparently as a depression moved southeast from the Norwegian Sea. Pressure at Hamburg fell 34 mb in 24 hours, and by the morning of the 6th the centre of lowest pressure seems to have been about 984 mb in the southeastern Baltic region with a great N'ly gale over the North Sea. Temperatures over the British Isles had been about normal for the first 5 days of the month, but

it then became very cold with frequent severe frosts, snow and hail falling on many days and often strong winds, and so continued until the end of the month. Winds from between N and E prevailed during that time. With a monthly mean temperature of 1.9 °C in central England this ranks as one of the four coldest Marches in over 300 years of record (only 1785 with 1.2 °C, and probably 1674

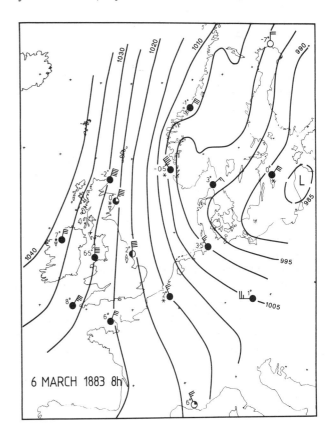

6 MARCH 1883 8h

and 1667, having been colder). This storm therefore marked the introduction of an unusually cold spell and long dominance of cold air from the Arctic over this part of Europe. On the 6th there were hail and snow showers in most parts of the British Isles and temperatures in northern Scotland and in Shetland remained below freezing point all day. The anticyclone had acquired a meridional orientation over the eastern Atlantic with pressures at mean sea level still 1038 mb or more on the west coast of Ireland. The gale was also strong (up to force 9) on the west coast of Norway and affected all parts of Denmark as well as the coasts of the German Bight and northern parts of the Netherlands.

*Maximum wind strengths*

At the morning observation Beaufort force 10 was reported at Wick in northeast Scotland and force 9 at Aberdeen. At Hamburg, NW force 7 was observed. The gradient wind from the north was 80–90 knots over the northern North Sea and in the region from Shetland south to Aberdeenshire, and east of there a belt of gradient wind up to 80–90 knots extended south to about the region between Hull and the Dogger Bank. The wind was reported to be squally at many of the land stations and surface gusts up to 90 knots or a little over may have occurred over the regions named in this section.

*Note.* At least one further gale this month was of some note. The *Meteorological Magazine*, vol. 18 (No. 207), p. 46, reports an unusually high tide in Boston (Lincolnshire) and in the Humber (up to 1.3 m above normal spring tide level) on the 10th and 11th after several days of very strong gales from between NE and NW: this was felt as a noteworthy N'ly gale in the Danish islands too on the 11th.

---

## 26 January 1884

*Area*

Atlantic Ocean between latitudes 45° and 60° N, the British Isles, all the North Sea, southern Scandinavia, and much of western and central Europe.

*Observations*

Very widespread strong gales. The barometric depression associated crossed Scotland, giving the lowest mean sea level pressure value ever measured in the British Isles, 925.6 mb, at Ochtertyre, Perthshire (near Crieff), in the late evening (21:45h) of the 26th.

SW and W winds force 10 were reported by many ships on the Atlantic on the 25th and 26th and on the Flemish coast near the Strait of Dover. On the morning of the 27th there were W to NW winds over the British Isles force 8 to 9 on most coasts and force 10 locally in the Channel, while SW to S winds were reported at some places on the coasts of the Netherlands, Denmark and south Norway and even on the coasts of the southern Baltic. At some stage on the 27th strong gales from W or SW were reported from southern France (force 12 at one place on the west coast) and in Switzerland and Germany.

At the top of Ben Nevis (1343 m above sea level), when the depression was passing near, the observer emerging from the old observatory on a rope to ensure his return could make no headway against the gale. (Six hours later when they succeeded in reaching the screen they could not read the instruments because of the blinding snowdrift lashing their faces.) The implied strength of the wind on the summit at the times when the observer could not move out of doors must have been above Beaufort force 10.

There was great damage to trees and woodlands in many parts of the British Isles. More than one million trees were reported blown down on just one estate at Cargen, in Galloway (southwest Scotland). Damage to buildings included the newly completed dome of the Grand Pavilion at Llandudno.

*Meteorology*

The whole period from 19 January to the 27th was very stormy in the British Isles. A record fall of barometric pressure by 32.5 mb in just 4 hours, from 1021.5 to 989 mb, was noted at Stornoway in the Hebrides at the beginning of the sequence of storms, on the evening of the 19th.

Another storm cyclone with central pressure down to 960 mb crossed the region on the 25th, the centre then being off west Norway near 64° N in the morning and moving away northeast. Next morning it was near Lofoten.

The morning map for the 26th shows a new depression with a wide warm sector of mild SW'ly winds centred near 55–56° N 15–16° W with a pressure already about 960 mb or slightly below. This had developed from a frontal wave near Newfoundland, lowest pressure about 1010 mb, 24 hours earlier and had travelled some 1500 to 1600 nautical miles in that time – a mean speed of 65 knots, implying the presence of a strong jetstream. A ship in mid-Atlantic experienced an exceptionally strong gale, near 50° N 36° W later on the 25th, evidence that the system was deepening rapidly.

During the 26th the cyclone became largely occluded and its progress slowed to a mere 600 nautical miles in the 24 hours to the morning of the 27th, when the centre was about $59\frac{1}{2}$° N between 1° W and 1° E, near Shetland, the lowest pressure then being 930–932 mb.

Sharply contrasting air masses were engaged in this cyclone, and this doubtless played a part in its explosive deepening from around 1010 mb to 925 mb in 36–40 hours. Temperatures over the European plain ahead of the warm front of the depression were generally in the range 0 °C to +5 °C. In the strong NE winds over and near Iceland −10 °C to −14 °C was general, and in this air reaching mid-Atlantic in the rear of the depression values as low as +4 °C and 5 °C were observed as far south as 47° to 50° N in longitudes 20° to 30° W on the morning of the 26th. At the same time, the temperatures experienced by ships in the same latitudes but in the warm sector air between 5° and 20° W were 11 °C to 13 °C.

It seems from the slowing up of the cyclone's progress and the northward curve of the path of the centre over Britain and the North Sea on the 26th–27th that there must have been an abrupt fanning out of the jetstream flow (rather as demonstrated in this compilation in the case of the October 1987 storm).

This gale was the subject of a special discussion meeting of the Royal Meteorological Society a few weeks later. Besides the extremely low pressures observed near the depression centre, at Aberdeen and Glen Tana as well as at Ochtertyre, features of this storm which attracted comment were

(a) the very great extent of the region swept by gale force winds, and

(b) the occurrence of a region of light winds not far from (actually behind) the centre: 13 knots was reported at Aberdeen just three hours after the passage of the centre and 8 knots at Sandwick in Orkney an hour later.

The strongest winds measured, in terms of mean speed, by the anemometers of the time were also noteworthy, though the highest figure mentioned was observed in

another storm in the same week: 80.2 knots averaged over a period of 65 minutes on the morning of 20 January at Sandwick.

This storm occurred in the midst of a long spell of prevailingly westerly weather in the British Isles that lasted from 5 January to 8 February 1884. Until the 18 January it had been combined with an anticyclonic tendency but then became subject to cyclonic intrusions until 1 February.

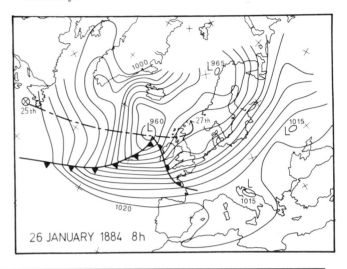

26 JANUARY 1884 8h

---

# 14–16 October 1886

## Area

British Isles and surrounding seas, including eastern Atlantic, Bay of Biscay, and mainly southern and central parts of the North Sea, also Denmark, France, central Europe and the southwestern part of the Baltic.

## Observations

The earlier part of the month was described as very warm in England and notably mild and genial in the British Isles generally. Temperatures in the midlands reached 23–25 °C on most days in the first week. Thunder and lightning had been reported on a number of days from the beginning of the month, and on the 14th–15th the genial weather was broken by high winds which freshened to a heavy gale that lasted for many hours, especially in southern and western parts of these islands, and did much damage to trees and crops. The storm was accompanied by heavy rain and flooding of the rivers in England, Wales, central and southern Ireland, damaging the crops in some areas and delaying the harvest seriously as far north as southern and eastern Scotland. Some bridges were swept away.

Many trees were blown down by the gale in parts of Ireland and several in parts of the English midlands and near the south coast. At Bodmin, Cornwall the heavy gales from WSW and NW on the 15th and 16th were reported to be 'of the longest duration recorded here'.

Gale force winds were also reported in various parts of France, up to force 10 on the 16th on the west coast and the Channel coast, and locally in the east of the country, as well as in the whole of Denmark.

## Meteorology

Strong SW'ly gales were reported by ships in mid-Atlantic on the morning of the 14th in the warm sector of a rather small, but intense depression, lowest pressure about 969 mb centred close to 51° N 24° W. By the morning of the 15th the centre, having deepened only slightly but become somewhat larger, was over central Ireland. Gales up to force 10 were then reported by most ships and some coastal stations between latitudes 48° and 51° N on the Atlantic and in the Channel and an ENE'ly gale of similar strength in southwest Scotland. The system then proceeded to move only slowly to central southern England on the morning of the 16th. This gave N'ly gales force 8 to 10 in the Hebrides, NW'ly gales over much of the west

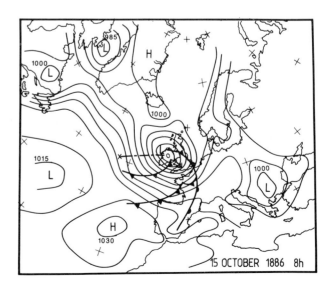

15 OCTOBER 1886 8h

of the British Isles, also in the Bay of Biscay and as far as 20° W, W'ly to SW'ly gales in the Channel and in France and SE'ly gales over Denmark.

No very noteworthy thermal contrasts were detected in the surface temperatures reported in this system.

It seems that this storm, which was followed by continuous high winds for several days in the British Isles is to be seen as marking a change of type in the large-scale atmospheric circulation, such that the rest of October was very wet in the British Isles. Only in the northwest of Scotland the month was reported to be very fine. At Loch Broom (near Ullapool) it was described as 'beautiful' and 'the farmers were (able) to secure their crops in a perfect state'. Since on the eastern side of the country the weather of the month was described as 'exceedingly dark, damp, dismal' at Braemar, and 'damp and unseasonable throughout the greater part' at Aberdeen, we may deduce that the winds were mainly E'ly or NE'ly over Scotland.

## 8–9 December 1886

### Area

British Isles and surrounding seas, including the eastern Atlantic as far west as 20–25° W and as far south as northwest Spain, also the Channel and the North Sea, Denmark and southern Scandinavia south of latitude 60° N, as well as much of France, the Low Countries and western Germany.

### Observations

Severe gales were reported in many of the above areas: NW'ly force 11 on the Atlantic between 49° and 57° N to as far west as 20° W, force 12 from SW at some points on the west coasts of Ireland and France and in the Channel on the 8th, and the E to SE gales near northern Scotland and in the northern North Sea were up to force 10 on that day. Force 11 to 12 from the SW was reported on the coasts of Belgium, Holland and the German Bight on the 9th.

At Haverfordwest in southwest Wales it was reported that the damage to forest trees 'perhaps exceeds anything in living memory'.

Most commentators regarded the extreme of low pressure attained near the depression centre, rivalling (and in some places surpassing) the values measured in the storm of 26 January 1884, as the most remarkable feature of this storm. At Belfast a mean sea level pressure value of 927 mb was recorded as this storm centre passed over.

### Meteorology

The situation was by no means a very close repetition of that which had produced the cyclone of essentially the same, record-making low pressure over the British Isles just under 3 years before. The most apparent common feature was just that both these exceptional storms occurred in, or marked the beginning of, cyclonic spells of some duration near the British Isles.

On this occasion a rather narrow warm sector in mid-Atlantic (35° W) on the morning of the 7th, fed by warm air circulating around an Azores–Madeira anticyclone (which seems to have produced a succession of smaller frontal waves before this, which were still on the eastern Atlantic west of Iberia), was associated with a sharp V-shaped trough of low pressure. By the next morning this system had merged near northwest Ireland with a low that had been close to Iceland (975–980 mb) on the 7th. The combined system had a central pressure about 930 mb near 55–56° N 10° W on the morning of the 8th. By the 9th the centre (now about 945 mb) had crossed the British Isles near the Scottish border and lay over the North Sea about 56–57° N 2° E, and then slowed up even more to be near the Norwegian coast in the Skagerrak on the 10th (lowest pressure 965 mb) with force 8 gales in that area.

Air mass contrasts were considerable until the 8th, when there was still a fairly wide warm sector over France and Biscay with surface temperatures 11 °C to 13 °C, but little to correspond with that was to be seen on the 9th–10th and the largely occluded system was beginning to fill up. Temperatures in the cold air supply to the rear of the depression on the 7th and 8th were mostly − 3 °C to − 7 °C but rose to + 4 °C to + 7 °C as the air stream reached Scotland and Ireland.

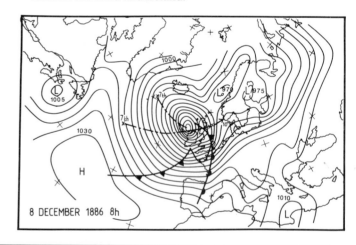

8 DECEMBER 1886 8h

## 9–10 March 1891

### Area

English Channel, southernmost part of the North Sea, and southernmost parts of England and Wales.

### Observations

This storm is often considered as just an outstandingly heavy snowstorm, exceptional for the areas affected, but there was also a severe gale which caused many shipwrecks, uprooted very many trees (probably some thousands), and drifted the snow severely.

There had been a comparable snowstorm affecting much the same areas of Britain on 17–19 January 1881, but this one seems to have exceeded the 1881 case in most areas both as regards the violence of the gale and the quantities of snow.

The snowfall began in Cornwall around midday on the 9th and continued more or less for 36 hours. In south Wales the timing was similar, and by evening 'a furious gale' was blowing. Force 9 gales were reported by ships in the Bay of Biscay and at Brest and near the Scilly Isles. The NE'ly wind reached gale force as far north as Oxfordshire and the north coast of Norfolk. Meanwhile, SW'ly gales and mild weather (temperatures 7 °C to 10 °C) affected Brittany and elsewhere in northern France inland and northwest Spain. In England the gale was accompanied by severe frost (temperatures mostly around − 2 °C) on the 10th, and at least 4 people died from exposure to the gale, the frost and the snow, including one or two on the streets of London and one in Plymouth.

The area which experienced the storm and the snow extended from the south of Ireland to Holland and from about Cheltenham and the extreme south of Wales to the Channel. The level depth of snow in some places was estimated at 50–60 cm, and there were many drifts up to 6 m or roof top level.

Trains were buried in Devon and Cornwall, and in Sussex, and the lines were blocked, in some cases for days. Most roads were blocked too and mail vans buried.

Occasional further snowfall with strong to gale force winds continued till 12h over the snowbound landscape of southern England.

*Meteorology*

Depressions were being generated between latitudes 30° and 40° N in the western Atlantic over a period of a week from 6 or 7 March between a vigorous supply of very cold air from Denmark Strait and Iceland, where temperatures were − 10 °C to − 16 °C with NE'ly gale force winds, and maritime tropical air with temperatures as high as 18 °C near latitude 40° N. One of these cyclones was steered northeast to reach the mouth of the Channel by the night of the 9–10 March, as seen on our maps. There its northward progress was blocked by another supply of extremely cold air which was firmly established all over northern Europe as far south as a nearly stationary front along the Channel and over northern France and Germany into Russia near 55° N, where temperatures as high as 3 °C were general south of the line of the front on the 10th. Gradient winds of about 80 knots are indicated near Cornwall and the Scilly Islands on the 10th.

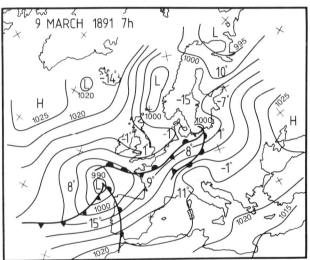

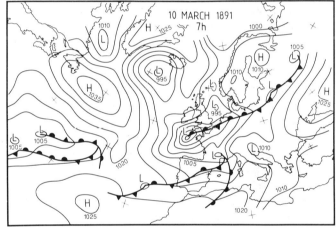

# 16–20 November 1893

*Area*

British Isles, North Sea and neighbouring sea and coastal areas, including on the 17th a wide area of the Atlantic between Ireland and Iceland, with extensions to the Bay of Biscay and French coast on the 17th–19th and along the continental seaboard into the southern Baltic on the 20th.

*Observations*

This storm passed steadily but rather slowly eastwards and gave a very prolonged gale or strong gale in some areas. The severest part was the NW'ly to N'ly gale (see map of the gale limits day by day) in the rear of the accompanying barometric depression, the worst effects being reported in northern Ireland, Scotland, near the coasts of Wales, and off the Scottish North Sea coast. This storm was felt as a

SE'ly to S'ly strong wind or gale on the west coast of Norway on the 17th and the coasts of Denmark on the 18th, but turned N'ly on the 19th, and NE'ly on the 20th.

On the 20th there was a sizeable sea flood on the Baltic coasts between Schleswig-Holstein and Lübeck.

In England there was a great fall of temperature as the wind swung to the NW and N. Highest temperatures on the 18th were about 10 °C lower than on the 17th. At Addington, Surrey on the 18th snow driven by the high wind formed drifts up to 1.5 m deep in the lee of every hedge. Very bitter conditions with drifting snow were also reported from Shropshire and elsewhere near the border of Wales.

Many trees were blown down in exposed districts, in northern Ireland, southwest Wales, and eastern Scotland. Less damage occurred in northwest England and southwest Scotland where there was some shelter from the

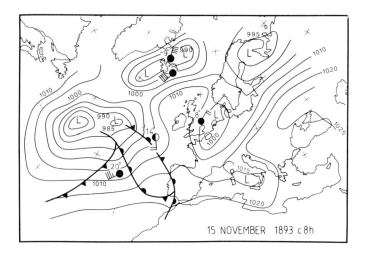

15 NOVEMBER 1893 c 8h

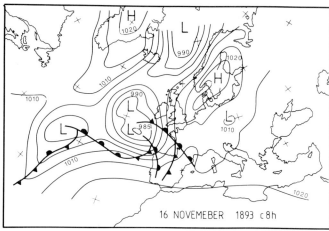

16 NOVEMEBER 1893 c 8h

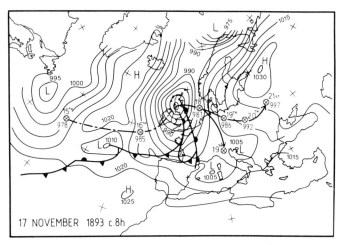

17 NOVEMBER 1893 c 8h

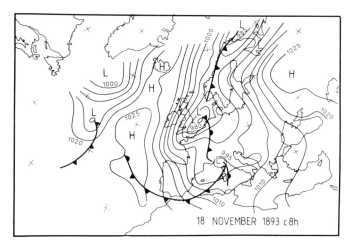

18 NOVEMBER 1893 c 8h

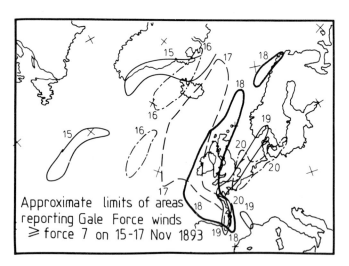

Approximate limits of areas reporting Gale Force winds ≥ force 7 on 15-17 Nov 1893

strongest N'ly gale, but trees were blown down even in the heart of the Southern Uplands of Scotland at Jedburgh. Near Aberdeen there was great destruction in the planted woodlands and great loss of life at sea. The roofs of houses were stripped in some areas. There was less damage near the south coast of England, though the storm was reported as a 'great gale' in Dorset and notably prolonged.

*Meteorology*

This storm was associated with a deep depression, the centre crossing Ireland at its deepest stage on the 16th–

17th with heavy rain and sea level pressure down to about 965 mb. It passed across Scotland into the North Sea on the 17th.

The slow advance east of this storm with its developing belts of S'ly and N'ly winds – the latter well seen over the days from the 15th–18th in the maps – was controlled by a slowly onward marching meridional circulation pattern.

The strongest N'ly gradient winds over the Atlantic and later over Britain and the North Sea on the 17th–19th seem to have been in the region of 90–100 knots, and the S'ly gradient wind near the coast of Norway on the 17th may have approached the same strength for some hours.

At Aberdeen the highest wind speed measured on the 18th was an average of 78 knots for one hour. At Holyhead, Anglesey the anemometer registered a mean wind speed near the surface of 74 knots for one hour and 65 knots over 24 hours. The gale at Haverfordwest in southwest Wales began as a SE'ly, but the worst violence came later when the wind was from between NNW and NNE: the gale there was described as 'terrible' on the 17th, 18th and 19th. Winds up to Beaufort force 10 to 11 from between W and N were also reported in the Bay of Biscay on the 18th and on the Spanish and French coasts near Biarritz on the 18th and 19th. The NE'ly gale in the North Sea and on the coasts of the southern Baltic, Denmark,

German Bight, and eastern part of the Channel on the 20th was also widely reported as force 10.

By the 22nd, with a new depression travelling southeast from northern Greenland across the sea to Norway (lowest pressure 978 mb near Lofoten) and later to the northern Baltic, another N'ly storm had formed over the Norwegian Sea giving force 10 gales as far south as the Faeroe Islands and next day on the Dogger Bank in the central North Sea.

## 10–13 February 1894

### Area

The North Sea and surrounding lands from the British Isles to Denmark and north Germany, and on the 12th–13th the southern Baltic.

### Observations

Strong SW'ly and W'ly gales, ultimately veering to NW, caused sea floods on the coasts of the German Bight and the west coast of Jutland, but not rivalling the water levels of 1825 and other historic floods. The water rose 2 ft (approximately 60 cm) above the quay in Ribe (Jutland). There was also damage inland.

In England, Scotland and Ireland this February was noted as a very stormy month, deep depressions passing in rapid succession. In Shropshire the gale on the 11th was at its most destructive, toppling a large part of a fine church spire, wrecking chimneys, roofs, and farmers' ricks. House roofs were also damaged in the east Midlands of England, and in Norfolk the storm on that day was particularly violent.

In Denmark there was damage to woodlands in all parts of the country, including the eastern island of Bornholm in the Baltic: altogether nearly 400 000 trees (91% of them pines) were felled by this storm; 151 000 of them on Fyn (the middle island of Denmark) and 113 000 on the more easterly island of Sjaelland about Copenhagen.

In Hamburg the hourly mean wind about midday on the 12th was measured as 69 knots and gusts of 82 knots were recorded (Oppermann, 1894).

### Meteorology

The W'ly wind sequence was maintained by cyclone centres mainly in latitudes 60° to 70° N over these days. A secondary depression, giving SW'ly gales up to force 10 in mid-Atlantic near latitude 50° N on the 10th, travelled to a point northwest of Ireland on the morning of the 11th with a central pressure 975–980 mb. On the morning of the 12th this cyclone was centred near the mouth of Oslofjord and had deepened to the unusual figure for that area of about 945 mb. From there it continued to the Gulf of Finland on the 13th, with pressures still rather below 960 mb.

The westerly gradient winds in the warm sector over the southern North Sea, the Low Countries and Germany shown on the map for the 12th seem to have been about 80 knots, and perhaps 90 knots near the front. Surface air temperatures in this warm air were as high as 7 °C to 10 °C far into the centre of the continent.

In the Arctic outbreak from the Norwegian Sea, which developed behind the centre, snow showers spread south over Scotland and as far as the English Midlands with air temperatures falling to 0 °C to 4 °C during the day on the 11th.

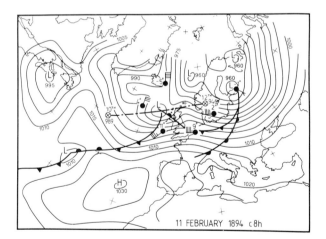

11 FEBRUARY 1894 c 8h

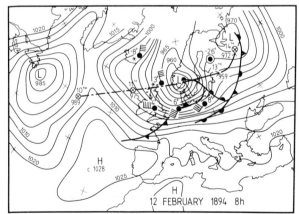

12 FEBRUARY 1894 8h

## 24 March 1895

### Area

England, especially the Midlands and East Anglia, also parts of Wales and Ireland, and later the central and southern North Sea and its coasts, the southern Baltic and its coasts.

### Observations

One account described this SW'ly gale as 'the worst storm of the nineteenth century in the English Midlands'. Houses were unroofed in Northamptonshire and Norfolk – in some cases those same houses whose roofs had been damaged by the gale of 11th February 1894, chimney pots, and stacks, ridge tiles, slates, and thatch were carried away. The lead roof of Blofield church in Norfolk was 'rolled up like parchment'. There was similar damage to buildings and walls in Shropshire and in southwest Wales. Norwich looked in some parts as if the houses had been bombarded.

Large numbers of trees were blown down (elms, poplars, larches, pines, and ash) in Gloucestershire as well as in the counties mentioned above. In Norfolk 2000 trees were reported down on each of three estates, several hundreds on individual farms. The damage there was considered the worst in any storm since 1703. In Suffolk corn and straw stacks were destroyed and 'sheaves of corn were carried great distances'. A yellow-red cloud of soil, 15 or more kilometres in length, was reported blowing across the Breckland north of Bury St Edmunds. There was little damage farther south. Thunder and lightning were reported in various places, including Manchester and Leicester and many places in Ireland.

In the places named the 'hurricane' lasted only a short time: in Northampton three quarters of an hour, the worst of it in just 10 minutes between 13.50 and 14h, associated with passage of the cold front. East of Norwich the worst time was about two hours later, but the duration similar.

*Meteorology*

This storm should, perhaps, be regarded as part of the sequel to the severe winter weather of January and February 1895, with its snows and frozen rivers. The thermal contrasts between air from the Arctic and northern parts of the continent and the warmer seas must have been especially great.

The temperatures over England were 10–12 °C in the warm sector of the depression, and in places on the lee side of the country before the fronts arrived. After the cold front they were generally 6–7 °C.

A big depression on the well-marked polar front moved east from the Atlantic near 60° N 25° W (about 960 mb) on the 23rd towards the coast of Norway, the centre (depth little changed) near the Faeroe Islands by the 24th and then extending as a long trough of low pressure from west of Scotland to Bear Island on the 25th. A frontal wave appeared near 48° N 26° W on the 23rd and soon began showing gale force winds: on the morning of the 24th this was centred near southeast Ireland (lowest pressure about 978 mb) and proceeded northeastward across Wales and England to reach southern Sweden by the 25th, deepening to about 960 mb and accompanied by gales, especially at the northeastern, eastern and southern sides of the intense centre.

The area affected by gale-force winds was perhaps never more than about 300 nautical miles across, but it was estimated that the strongest gusts probably reached 90 knots or more in the English Midlands.

Gradient winds of the order of 75–80 knots are indicated by the isobars in the warm air over the British Isles on the 24th and also by the pressure gradient in the cold air over northern Germany and the Baltic, south of the cyclone centre, on the map for the 25th (not reproduced here).

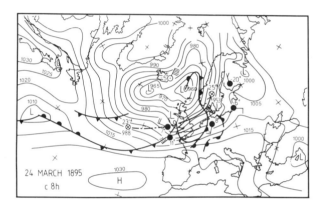

24 MARCH 1895 c 8h

# 28–29 November 1897

*Area*

The whole of the British Isles, later especially coasts exposed to the North, the North Sea and, on the 29th, Denmark also.

*Observations*

This gale was first felt in western Ireland (Co. Galway) as a stormy W wind on the 27th. The storm increased as it advanced east, and all parts of the British Isles experienced a strong NW'ly gale on the 28th, continuing into the 29th in England, Scotland and Wales, and becoming NNW or N'ly except in the extreme south, where the wind continued nearly W'ly. The wind was at gale force for 24 hours or rather more in many districts.

On the north Norfolk coast it was described as a 'furious gale', but remembered most of all for the exceptionally high tide that accompanied it, said to be the highest in living memory. Buildings were flooded to a depth of several feet. Its highest level at Blakeney was just 1.3 m below that of the disastrous 1953 flood. There were shipwrecks around the coast, including one in Fishguard Bay on the coast of southwest Wales.

Inland in East Anglia, and probably in other parts of the country, this was not such a destructive gale as the much briefer one in March 1895, but again farmers' stacks were badly damaged.

When the gales reached Denmark on the 29th, the wind turned NE'ly in northern districts and in south Norway and was accompanied by hail and snow showers: in parts of northern Jutland a fierce snowstorm was reported.

*Meteorology*

November had been a fine, mild, dry month in the British Isles, especially the southern half, with notable blooming of plants in England and Ireland. The breakdown of the anticyclonic W'ly wind situation over Britain into a gale which led to the N'ly storm over these islands and the North Sea on the 29th can be followed from the maps. The map for 26 November shows the frontal positions and isobars at 20-millibar intervals on the morning of that day and, as dotted lines, the fronts on the 27th. Full pressure maps for the 28th and 29th complete the sequence.

Temperatures between 8 °C and 12 °C prevailed in the surface air over southern parts of the British Isles in the warm sectors coming in from the Atlantic, but very much colder air lay over the continent in France and Germany south of the axis of the elongated high pressure system. Temperatures as low as $-7\,°C$ to $-9\,°C$ were reported over a very wide area as far south as central France on the morning of the 26th. The milder air beginning to invade from the ocean seems at first to have advanced east over a (presumably shallow) layer of this air without disturbing the frost regime, and then breakdowns occurred in such a way that preliminary warm fronts appeared ahead of the warmest air in the warm sectors of the oncoming disturbances from the Atlantic. This may be detected in the frontal pattern for the morning of the 27th here shown (in dots on the map for the 26th). The light variable surface winds veered to SW from points mainly between S and E, and the frost gave way to temperatures between $+4\,°C$ and $+6\,°C$ in north Germany, Denmark and the Low Countries. With passage later of the first fronts from the Atlantic the winds veered further to about W and temperatures rose to $+7\,°C$ to $8\,°C$ on the coasts of Holland and Denmark and $+5\,°C$ in the Baltic islands and the coasts of the states in the eastern Baltic.

On the morning of the 26th already the broadly W'ly wind stream crossing the British Isles included a narrow warm sector of maritime tropical air, with surface temperatures about $+12\,°C$, in the trough between the anticyclones over Europe and the Atlantic. Barometric pressures were over 1025 mb south of 55° N over England and Ireland and reached this level also in longitudes 45° to 50° W on the western Atlantic, which was also dominated by an anticyclone, centred 300 nautical miles southeast of Nova Scotia. By the 27th the pressure level over the British Isles had fallen as another trough, and another warm sector within it, advanced quickly from mid-Atlantic (see outline map). A 985 mb cyclone centre just west of Iceland then began to deepen, as pressure rose over Greenland and a cold NE'ly windstream from the Arctic advanced across Iceland by the 28th, when the deepening cyclone centre had moved quickly southeast and was approaching northernmost Scotland. This centre then proceeded farther to cross the North Sea, reaching Denmark with lowest pressure around 960 mb on the 29th. This development led to air from the Canadian sub-Arctic and later from northeast Greenland crossing the Atlantic and air from the Norwegian Sea moving south over England: temperatures fell to 3–5 °C in eastern districts on the 29th, when NW force 9 to 10 was observed on the coasts of East Anglia and Kent and on the French coast in the eastern part of the Channel.

Gradient wind strength is indicated as about 75–80 knots from the N over the western North Sea in the region of the English coast on the morning of 29th November.

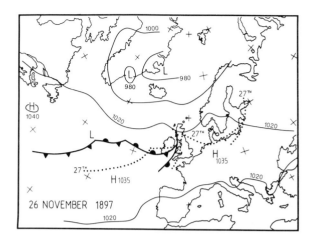

26 NOVEMBER 1897

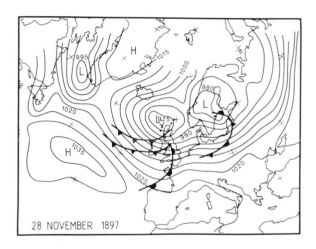

28 NOVEMBER 1897

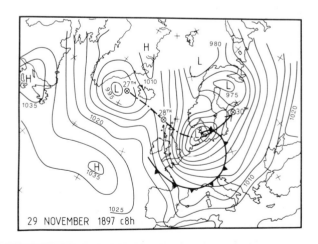

29 NOVEMBER 1897 c8h

## 25–26 December 1902

*Area*

Northern and eastern North Sea, south Norway and Sweden, Denmark and the southern Baltic and its coasts.

*Observations*

A severe storm with unusually high winds for the regions concerned, doing great damage to the woodlands in Denmark, felling rather greater numbers of trees than the February 1894 storm, especially in northern and eastern districts. Of the trees which fell 92% were pines (cf. in 1894 91% pines). It also lifted 300 roofs, destroyed 70 mills, and killed 15 people in Denmark. The winds during this storm were broadly W'ly but SW at first and finally NW'ly.

The storm in Copenhagen was at its height between 0h and 6h on the 26th: between midnight and 1h the mean wind speed there was 56 knots (force 11), gusting to 78 knots. At Hald near Viborg on relatively high ground in central northern Jutland an hourly mean wind speed of 68 knots (force 12) was reported. And on the eastern island Bornholm, in the Baltic, gusts up to 90 knots were measured (Weismann, 1904).

*Meteorology*

The first 13 days of December had produced a rather severe cold spell in Denmark with E'ly winds, followed by S'ly winds in the middle of the month, but W'ly winds dominated the last part of the month there and in the British Isles.

On the morning of 24 December barometric pressure was very low over the Barents Sea north of Norway with a cyclone centre about 960 mb, the end point of a succession of vigorous depressions steered in that direction by a long SW'ly airstream from the western Atlantic near Bermuda and the extensive anticyclonic system (pressures in the ridge over 1030 mb) from near 30° N 50° W to southern England and the European plain as far east 50–53° N 40° E. On the 25th the Barents Sea Low centre was transferred to the northernmost Baltic, and pressures in southwest Norway fell 20 mb in 6 hours as an Arctic cold front moved south from the Norwegian Sea into the North Sea. Later, pressure fell 25 mb in Copenhagen between

midday and midnight, and the Low centre shifted farther south and deepened further to lie over the Gulf of Riga, with lowest pressure 953 mb (a notably low value for that position) on the morning of the 26th.

The wind speeds reported in gusts in this storm were the highest that had ever been measured in Denmark since the introduction of anemometers capable of measuring gusts.

The temperatures had not fallen below normal over the British Isles with the approach of the cold air mentioned, and by the 27th–28th another cyclonic system had arrived from the Atlantic.

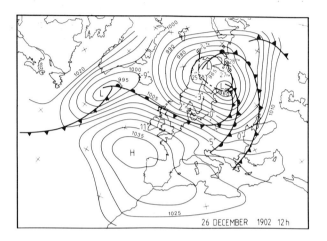

26 DECEMBER 1902 12h

# 26–27 February 1903

*Area*

A belt across Ireland, Wales and most of England, especially the north, and Scotland, to the North Sea and Denmark.

*Observations*

The greatest hourly mean wind speeds recorded in this gale were very high (force 11), up to 57 knots at Kingstown (now Dun Laoghaire) near Dublin and 59 knots at Southport (53.7° N 3.0° W), where gusts up to 80 knots were measured.

There was great damage to buildings and trees, farmers' ricks were destroyed and some cattle killed. Nearly 3000 trees were uprooted in Phoenix Park, Dublin, 4000 on an

estate in Kilkenny farther south. Many trees were uprooted and houses unroofed also in England, Wales and Scotland, though the wind force was much less on the east coast. In Orkney this gale 'was by no means severe'.

The most disastrous incident was in northwest England where a train with 10 passenger coaches and vans, albeit lightly loaded, was overturned and blown off a viaduct near Ulverston, Lancashire. There were also disasters at sea.

This gale, which struck western Ireland on the 26th, was variously described at Derrynane, Co. Kerry in southwest Ireland as 'the worst ever remembered' and near Athlone, in central Ireland, as the greatest storm 'since the night of the big wind' on 6 January 1839. At Douglas, Isle of Man, the storm in the night of 26–27 February was

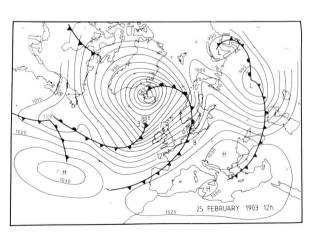

25 FEBRUARY 1903 12h

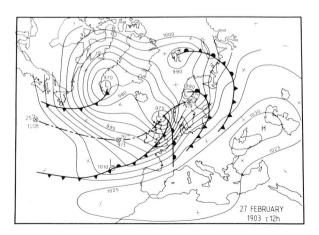

27 FEBRUARY 1903 c12h

considered 'probably of almost unprecedented violence'. Even in the east of England, as also in the Wye valley, the 'terrific gusts' were remarked as exceptional.

It reached Denmark also on the 27th as a SW and later W'ly gale which affected the whole country.

*Meteorology*

A belt of high pressure from a warm anticyclone centred over the eastern USA near Washington DC (over 1035 mb) to another southwest of the Azores (1042 mb) about 30° N 40° W to another over north Africa centred (1031 mb) near 31° N 4° E on the 24th had been keeping a flow of generally mild air going near 40–45° N over eastern North America, 40–55° N over the eastern Atlantic, and 40° to near 60° N right across Europe. A succession of cyclone centres was steered eastwards near 50° N on the western Atlantic and reaching Europe between latitudes 50° and 65° N. On the 25th coalescence of cyclonic centres simplified the situation by producing one single deep cyclone centre (about 940 mb) just north of Iceland, dominating the whole North Atlantic and bringing cold air from the central Arctic towards the Atlantic, where the next cyclonic features were intensified by the 26th–27th. This storm, which passed northeastwards across central Scotland in the early part of the 27th giving pressures down to 960 mb near the centre, was one of the products, but afterwards the general, large-scale meteorological situation continued much as before.

The gradient wind probably reached over 80 knots in the warm sector SW'ly wind over the British Isles and a similar speed in the S'ly gale, before the fronts, over the North Sea. The gradient may have approached 100 knots at the deepest phase of the storm centre on the 27th.

## 12–13 March 1906

*Area*

Whole North Sea and its coasts, especially eastern side, but most of British Isles also affected, later all Denmark, Sweden, and all parts of the Baltic.

*Observations*

On the 12th N'ly winds of force 8, locally 9, were reported through the whole width of the Norwegian Sea south of 64° to 66° N and on the eastern and western coasts of Iceland as well as all exposed coasts of the British Isles, including northwest Ireland, both sides of Scotland and northern England, Wales, and even in places on the south coast of England.

In Denmark S'ly gales were blowing that morning in most parts of the country and S or SW force 10 was reported in various places in southern Sweden.

A report from Ribe on the coast of Jutland says that the storm on that day was accompanied by snow and rain, and in the following night the storm grew stronger and turned more W'ly. This brought a serious storm of sand from the North Sea shore; and later, when the tide rose, a sea flood over all the lower meadows approached the town. The sea flood on the German coast at Emden near the Dutch border was more serious, with a peak level 50 cm higher than the 1825 storm flood. Farther east the levels were lower, presumably because of different phasing with the tide, though at Husum (9° E) the peak level was only 10 cm less than in 1825 (Petersen and Rohde, 1977). In England, too, on the coast of north Norfolk there was a sea flood which inundated houses in Cley though – unlike the 1897 case – there was no damage at Blakeney a little farther west.

It was reported in Douglas, Isle of Man that 'a series of strong NE and NW winds set in on the 11th and prevailed throughout the remainder of the month, causing extremely cold weather.'

*Meteorology*

A cyclone centre with lowest pressure about 980 mb approached the coast of northwest Ireland from the Atlantic on the morning of 11 March, as an anticyclone that had dominated southern and central Europe for some days gave way. Very cold air from the Norwegian Sea and Iceland was drawn into its rear side and the cyclone deepened rapidly over the North Sea. On the morning of the 12th barometric pressures were down to below 960 mb as the system moved northeast through the Skagerrak. It later filled a little as it moved farther on across Sweden to the Gulf of Bothnia: the lowest pressure was about 965 mb in the centre near 62½° N north of Stockholm on the morning of the 13th, but no lower than about 978 mb the following day when the centre had reached 64° to 65° N, still over the Gulf.

Because of the strong curvature of the air trajectories in the central part of this Low, despite the strong pressure gradient there, it turns out that the strongest gradient winds in this system were in the belt of nearly straight isobars giving N'ly winds blowing over a long fetch of open water from the Norwegian Sea into the North Sea. Gradient wind strengths up to about 75 knots were measured there, and this distribution doubtless explains the build-up of water level and the sea floods that occurred on the low-lying coasts of the southern North Sea. There were few reports of wind damage inland in the British Isles, though the sandstorm near the Danish North Sea coast is evidence of a noteworthy gale.

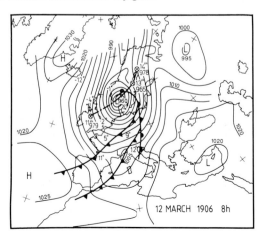

12 MARCH 1906 8h

## 3 December 1909

*Area*

British Isles, Irish Sea, southern and central latitudes of the North Sea and its coasts, by the latter half of the day the southern Baltic as well; also the European plain and adjacent areas from north France to north Germany.

*Observations*

There were many shipwrecks and many lives lost as a very deep depression crossed Ireland and then northern England, with central pressures down to about 948 mb at the deepest phase. It then continued northeast across the North Sea to be centred over south Norway on the morning of the 4th.

Gales force 8 from between WSW and WNW were blowing over much of England on the 3rd and force 10 from these directions was reported over the southern North Sea, the Channel and the Bay of Biscay. The gale on the 3rd began as a SE'ly generally force 8 to 9 over Denmark and force 10 SE'ly at the coast of southwest Norway.

There was a considerable sea flood in southwestern Jutland in the night of 3–4 December, the peak water level reaching 3.95 m above normal near (west of) Tønder, near the German frontier, and nearly 5 m above normal over the shallows of the bays and coastal lowland at Farup near Esbjerg. In these areas the flooding continued, or was repeated daily, until the 6th. The water level in Esbjerg (4½ ft over the quays) was the highest for over 50 years.

On the 4th the area experiencing gale force winds, generally force 8 to 9, was much smaller, mostly in Denmark but also some places in north Germany and south Sweden. NW force 10 was still being widely reported on the Atlantic between about 45–55° N 10–30° W and the Bay of Biscay.

*Meteorology*

Pressures had been low over the British Isles–North Sea–Norwegian Sea region for about a week beforehand. On the afternoon of 2 December an already deep cyclonic centre (below 970 mb) approached the west coast of Ire-

land near the River Shannon from the Atlantic. It crossed England to the North Sea during the following night (the depth reaching about 948 mb) and had reached the Skagerrak by the evening of the 3rd. By the morning of the 4th it was centred over south Norway west of Oslo and with another centre close to the west coast near Bergen (lowest pressures in both about 959 mb). The system continued moving farther north during the 4th to reach a point off the coast near Trondheim that evening. Its path then recurved southwestwards towards Shetland.

Meanwhile another, secondary, low pressure centre (which deepened to about 970 mb) crossed Wales and England from west to east on the 4th, emerging onto the southern North Sea near the Wash that evening. And on the 5th and 6th this was followed by yet another one (pressures down to 980–985 mb) which crossed England from near the Bristol Channel. Thus, each of these centres followed a path a bit farther south than the one before and the air over Britain became generally colder (−4 °C to +4 °C over England on the morning of the 6th), and it was presumably the belts of strong west winds over the southern North Sea with these later depressions which continued the sea flooding of the Danish coast near Esbjerg.

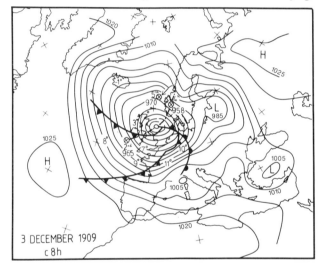

3 DECEMBER 1909
c 8h

## 16 February 1916

*Area*

Ireland, England, the Channel, the southern North Sea and Baltic, and the Danish and German coasts.

*Observations*

Accounts are rather sparse because the war inhibited publication at the time.

There was an important sea flood on the German and Danish North Sea coasts. Near the Dutch frontier the flood level was lower than in the 1906 storm, but on the coast of Schleswig-Holstein this flood was almost as high as that in the 1825 storm (at Husum just 10 cm lower).

There was flooding of the Jutland coast farther north, including the island Fanø off Esbjerg, and some damage to dykes.

Strong W'ly gales were reported from early on the 16th, with force 9 at 8h already on the Dutch and Belgian coast and in northern Ireland. By evening (19h) on that day the wind was WSW force 9 in Hamburg and Beaufort force 10 from about W on the North Sea coast of Friesland and at Esbjerg in Jutland. On the 17th at 8h the wind from about W was force 11 at Bornholm in the southern Baltic and force 9 all along the Baltic coast of Germany at least as far as 20° E. Stockholm reported ENE force 9 with snow. Force 8 gales were still blowing at exposed points off the coasts of northern Ireland, Wales and southwest and southeast England.

*Meteorology*

Winds over the British Isles–North Sea region were generally W'ly all through that winter.

On the 15th the main low pressure centre was over Iceland (about 970 mb) – roughly where it had been already for three days – and a small secondary depression (lowest pressure about 990 mb) derived from a frontal wave disturbance which had formed in mid-Atlantic near 50° N on the 13th–14th passed east across England and Europe near 51° N during that day, while a trough of low pressure extending from Iceland gradually produced a centre over southern Sweden and later the Baltic near 58° N (985 mb).

On the 16th a frontal cyclone with a well-marked warm sector over the North Sea and its centre (970 mb) at 8h near the northeast coast of Scotland – either the same centre or derived from the one over Iceland the day before – was deepening as it moved eastsoutheast across the North Sea. That evening it was close to Jutland (959 mb) and moved in across Denmark and south Sweden overnight with little change of depth and accompanied by strong gale to storm force winds on its southern side.

The measured geostrophic wind across southern Jutland and the German Bight seems to have been about 150 knots, and the gradient wind indicated after allowing for the cyclonic curvature of the wind's path is close to 100 knots.

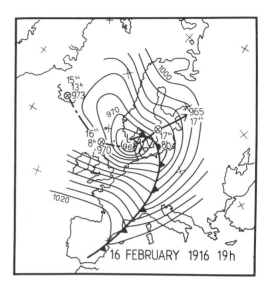

16 FEBRUARY 1916 19h

## 26–27 January 1920

*Area*

Mainly the British Isles, and at times the North Sea, also the eastern Atlantic between 45° and 55° N, and the Bay of Biscay. Later on the 27th also Danish coasts, notably Skagen in the north and Bornholm in the east.

*Observations*

A S'ly gale, Beaufort force 8 in the Hebrides, force 9 in Shetland and with SSE force 9 at the southwest coast of Norway, was reported on the evening of the 26th. Gale force from SW was reached that night also on the coasts of Ireland, and on the morning of the 27th S force 8 was reported on the coasts of northeast England and East Anglia. During the 27th the gale was very severe in Ireland, Scotland, and the west of England, particularly violent near the west and south coasts of Ireland. At Queenstown Harbour (Cobh) near Cork great damage was done to the quays and wharves. The strongest gust measured in that area was about 70 knots at Weaver Point at 8h on the 27th, but at Quilty near Spanish Point in Co. Clare, on the west coast near 52°50′ N, a gust of 97 knots was measured.

On the 27th also the SE'ly wind in the Baltic, being drawn into the circulation of the same cyclone, reached force 9 at Bornholm.

*Meteorology*

The review of this month of January in the *Meteorological Magazine* (vol. 55, p. 11) says that 'weather of a . . . southwesterly type prevailed . . . Depressions, which were often of great size and intensity, followed one another in rapid succession . . . Gales were frequent and widespread.' This

gale of the 27th was the most serious. The cyclone which caused it formed as a wave disturbance on the well-marked Atlantic polar front near 40° N 70° W on the 25th and travelled quickly east across the ocean. By midday on the 26th it was centred near 52° N 33° W, having covered over 1600 sea miles in 24 hours, a speed of nearly 70 knots and having deepened from 1019 to 984 mb. From then on the direction of movement of the centre curved towards the north and progress slowed to 600 nautical miles in the next 24 hours (25 knots) and to only half that rate as it approached south Iceland on the 28th with a central pressure as low as 958 mb.

This had become a very big depression, which dominated the whole Atlantic between latitude 30° N and the pole or beyond and from Labrador to Europe at 10–20° E. An older depression centre (970 mb) over Iceland on the 26th with a big circulation had been adsorbed and replaced, leaving no trace of its separate identity. But the

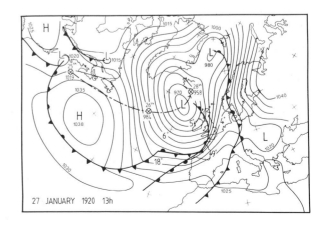

27 JANUARY 1920 13h

overall pressure range (of about 70 mb) on the Atlantic was not exceptional. The noteworthy intensity of the development was marked just in the shear zone along the front which crossed the British Isles on the 27th. Between latitudes 45° and 55° N the strongest gradient wind developed in this zone measured 80 knots and, in view of the apparent gradient of the shear, it probably reached 90 knots at its most extreme.

The strongest thermal contrast in the surface air over the Atlantic had developed between about 40° and 48° N.

## 28 January 1927

### Area

The British Isles and neighbouring seas between latitudes 50° and 60° N.

### Observations

'Gales in some districts of exceptional violence', particularly in western and northern districts of the British Isles, were reported in the *Monthly weather report* of the Meteorological Office. There was 'considerable structural damage, involving loss of human life in some cases', many of the deaths through buildings being blown down. Damage was confined to places south of the path of the depression centre. Fishing boats were swamped by the seas and dashed on the rocks of the coasts of western and northwest Ireland in Co. Galway and Co. Mayo.

The strongest winds recorded were at Dunfanaghy (northwest Ireland) about 70 knots 'over a brief period' and an average of 64.3 knots over a complete hour, with a strongest gust of 95 knots. At Tiree (off the west coast of Scotland) there was a strongest gust of 94 knots; and at Paisley and Renfrew (inland, near Glasgow) gusts of 90 and 89 knots respectively. At Lerwick in Shetland the strongest gust measured was rather less at 80 knots.

The storm wind came on suddenly and passed quite quickly across the British Isles, striking first at Valentia in southwest Ireland about 16.40h and Dunfanaghy at 22.50h (strongest gust at 23.35h) and reaching Cranwell, Lincolnshire (in eastern England) about midnight. The duration of the gale force winds was generally about 3 to 4 hours but lasted over 9 hours at Southport, Lancashire (strongest gust 83 knots). Around Morecambe Bay on the Lancashire coast of northwest England the sea wall gave way at Fleetwood and the town and surrounding country were flooded for some days.

### Meteorology

As in the previous winter, there was a great predominance of W'ly and SW'ly winds over Britain and northern Europe with a continual succession of depressions often of great size and considerable intensity. This storm is illustrated here by very wide-area maps of the meteorological situation over eastern North America, the Atlantic and Europe, to show the commanding size of the circulation systems.

In keeping with the size of the systems, this case seems to have been of classic simplicity for the development of a noteworthy storm. Two days before the storm struck the British Isles a remarkable thermal contrast was built up in a frontogenetic and cyclogenetic situation over the western Atlantic south of Newfoundland, as our map for the 26 January shows. Cold air brought swiftly from northern Canada by strong W to NW winds was still producing surface temperatures as low as −2 °C a hundred miles or more out onto the Atlantic, over the waters of the Labrador Current, in longitudes where subtropical air with temperatures 20–21 °C was streaming towards the frontal zone from the south: at some points the distance separating these temperatures was only 100 to 200 nautical miles.

The depression forming in that area deepened from 1014 to about 990 mb as it moved towards mid-Atlantic, where for some time on the 27th the pattern became more complex, with two low pressure centres between 30° and 40° W and still two separate frontal systems. By the 28th, however, this had been reduced to a single intense cyclone with central pressure about 950 mb and one main front approaching the west coasts of the British Isles, with squally winds widely reported as force 7 to 10 from about SW veering W. The strongest gradient wind indicated seems to be about 75 knots.

There were S'ly gales on the west coast of Norway as the system approached in the early part of the 29th and presumably a strong S'ly gale during the preceding 6 to 10 hours over the North Sea.

On the 29th the depression again split into two centres as it entered the Norwegian Sea, one centre staying between Shetland and the Faeroe Islands while the other proceeded quickly northeast to the Barents Sea.

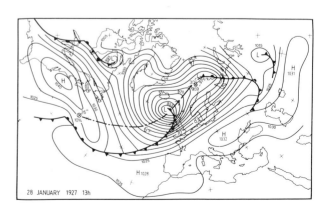

26 JANUARY 1927 13h

28 JANUARY 1927 13h

## 28–29 October 1927

*Area*

Southern half of the British Isles, especially the coasts of Wales, northwest England and western Ireland. On the 29th storm force winds across Denmark and the southern Baltic.

*Observations*

SSW'ly winds had been blowing over the Bay of Biscay during the 27th and freshened to a S'ly gale over the South-Western Approaches and western parts of England and Wales during the day on the 28th, later veering SW and W'ly. Gusts then reached 83 knots at Southport on the Lancashire coast, and the wind at Southport averaged about 61 knots (force 11–12) over a period of 1 hour around midnight on the 28th–29th. Later, when the storm crossed Denmark and moved on into the southern Baltic gales force 10 were reported at many exposed points on most coasts and force 11 at several points near 55° N east of 10° E, including the Bornholm area where the storm was at its strongest in the evening and night of the 29th–30th.

Damage associated with the storm seems mainly to have been caused by the sea surge. The evening tide on the 28th came in as a wall of water on the Welsh coasts about the northern end of Cardigan Bay. Coastal cottages were tumbled down by the water at Criccieth, railway lines were washed away; the sea flood in North Wales spread inland to Aberglaslyn, drowning many cattle. The harbour wall broke at Portmadoc and later the sea wall was demolished at Fleetwood, Lancashire. Flooding occurred on the Lancashire plain and about the Mersey. Five people were drowned in Fleetwood, trees were felled and houses damaged in that area.

On the 28th, early in the history of this storm, there was a disaster off the coast of Co. Mayo in western Ireland when the wind, which had been SE'ly there, first died down and then sprang up as a strong NW'ly gale overturning two fishing boats with a loss of ten lives.

*Meteorology*

A cyclone with central pressure rather below 980 mb approached Ireland from the southwest, moved up the Irish west coast during the 28th and then crossed Scotland to the North Sea.

The strongest S'ly and SW'ly winds were at some distance from the centre of low pressure. Later a strong rise of pressure coming in from the west sharpened the trough to the southern side of the cyclone centre and intensified the W'ly to NW'ly winds on the rear side of the system.

This is a somewhat surprising synoptic weather pattern to be associated with an important storm. Although the whole North Atlantic, particularly between latitudes 40° and 65° N was dominated by a single low pressure area, the individual circulation systems were fairly small and not particularly intense. Frontal wave disturbances were continually arising at distances less than 800 nautical miles apart over a period of many days duration (here illustrated by maps for the 25th and 29th) and even in the mature stage the cyclones were separated by barely more than 1200 miles. This repeating pattern was established by the 29th all the way from the western Atlantic across Europe and the USSR to northeast Asia. But on the 28th–29th one of the cyclonic systems drew in Arctic cold air from the Barents Sea and beyond to its rear side and so deepened for a time almost to 970 mb.

The jetstream was evidently becoming stronger over Europe with the meeting of Arctic and subtropical air, and the progress of this storm was speeding up from 700 nautical miles in 24 hours on the 27th–28th to 900 nautical miles from the 29th to the 30th.

The strongest gradient wind measured was in the cold air flow over the British Isles and North Sea on the 29th but seems not to have been much more than 60 knots. It seems therefore that the sea flooding must be attributed to the extra water driven into the eastern Atlantic by long continuance of strong SW'ly winds.

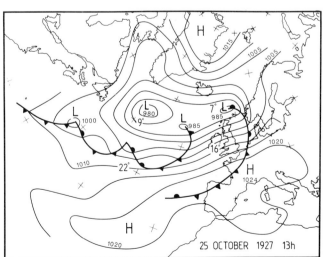

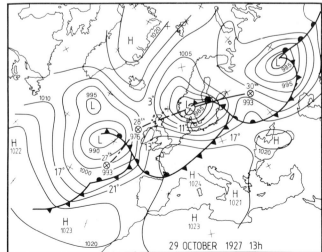

## 6–7 January 1928

*Area*

North Sea.

*Observations*

This was a brief, though severe, NW'ly gale at the rear side of a fast-moving storm depression. The cyclone centre, which had appeared as an open wave on the main Atlantic cold front at 13h on 5 January (lowest pressure 1009 mb) was over the Inner Hebrides near Skye at 7h on the 6th, pressure about 978 mb, and crossed Denmark in the evening of that day (about 981 mb), to lie over eastern Poland in longitude 25° E (985 mb) by 7h next day. This gale, following the cold front, was preceded by 6–9 hours of rising SW to W winds: the strongest wind reported on land was WNW Beaufort force 10 at Spurn Head on the Yorkshire coast at 13h on the 6th. The gale only lasted 5–10 hours, and by 1h on the 7th W by S Beaufort force 3 was reported at Spurn Head.

A storm surge was built up by the NW'ly wind in the North Sea raising the height of the tide about 50 cm above expectation at Dunbar at 15h on 6th and 1.5 m in the Thames Estuary (Southend), where it nearly coincided with the high spring tide around midnight. The surge continued south to the Strait of Dover and also affected the continental shore. Serious flooding occurred in inner London as the tide came over the embankments from the city and Southwark up to Putney and Hammersmith. Fourteen people were drowned, trapped in the basements in which they were living. Though this storm passed quickly, it came in the midst of a very stormy sequence. A thaw of the great snowfall of Christmas 1927 in England set in on 31 December and 1 January. The first week of the New Year brought nearly twice the normal rainfall in England. Another, more moderate, gale passed across England and the North Sea on 4–5 January, and yet another crossed southern Scotland on the 7th with WSW wind reaching Beaufort force 9 in the Firth of Forth.

*Meteorology*

With the gradual disappearance eastwards of the blocking anticyclone over continental Europe, where the frost gave way to westerly winds sweeping in across the German plain on the 4th, vigorous Atlantic depressions advanced eastwards from near Newfoundland to Scandinavia and the Baltic in quick succession.

The sequence was remarkable for the speed of travel of these occluding depressions and the brevity of the interval between each major depression and the next. The centre of the cyclone which produced this Thames flood and the severest of the gales in this sequence on the North Sea and surrounding coasts passed along a track which was generally some 5° to 10° of latitude farther south than the others.

Another noteworthy feature was the high temperatures prevailing over the Atlantic and Europe's Atlantic coasts as far north as Stornoway in the Outer Hebrides (even after the warmest airmass had passed on the 6th: 9.5 °C at

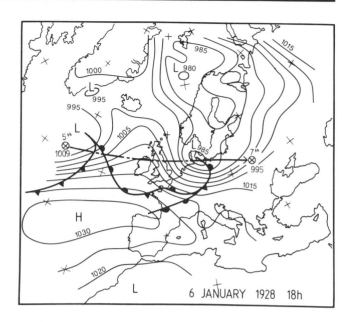

6 JANUARY 1928 18h

Stornoway with NW force 6 at 7h). Air temperatures +13 °C to +14 °C were reported by ships near 50° N in mid-Atlantic on the 6th. All this seems to point to abnormally high ocean surface temperatures.

Meridional patterns of air flow redeveloped over the northern part of the East Greenland Sea–Norwegian Sea on 4–5 January and broadly lasted until the 7th–8th. Hence a great temperature contrast came to exist between latitudes about 57° and 65° N. The warm air entering the storm depressions crossing the British Isles was also very humid, dew-points in the west on the 6th exceeding 10 °C.

*Maximum wind strengths*

The 24-hour movement of the depression, which produced the North Sea storm and flood, along a track from WNW to ESE of 1100 nautical miles from 7h on the 6th to 7h on the 7th allows an estimation of the jetstream strength by the same method as we have used for the storms in 1588 (Palmén, 1928; Douglas *et al.*, 1978). For a frontal wave or young cyclone such a movement would indicate a statistical probability of jetstream winds about 50 m/sec (100–110 knots); with partly occluded cyclones, as in this case, such disturbances are more rarely covered so that the statistical basis is weaker. The jetstream was presumably stronger, probably in the range 60–70 m/sec (120–150 knots). Between the 5th and the 6th at 13h the 24-hour movement of the system was 1600 sea miles (67 knots). This was the fastest stage, and, as the system was occluding, probably implies that the jetstream was then over 150 knots.

The strongest gusts of the surface wind reported in this storm were 73 knots at Spurn Head near Hull, 72 knots at Southport, Lancashire, and 71 knots at Fleetwood (also on the Lancashire coast). The mean hourly wind reached 52 knots (Beaufort force 10 to 11) at Fleetwood in the afternoon of the 6th. The strongest pressure gradients measured in this storm, over East Anglia and the southernmost part of the North Sea, in the evening of 6 January, indicate geostrophic winds of 90–100 knots, corresponding to gradient winds of about 70–75 knots. Gradient winds of this same strength were measured over the German Bight in the evening of the 6th. With this gradient wind, the strongest gusts would probably be in the range 75–95 knots or thereabouts.

As in the case of the sea flood which caused the Thames to flood Westminster on the night of 1–2 March 1791, the greatest wind strengths, although among the strongest found in the twentieth century, seem not to have been up to those in the most extreme North Sea gales in this catalogue. Evidently both in 1791 and 1928, the flood must be attributed to a great extent to the direction in which the winds impelled the tidal surge into the Thames and probably to some progressive build-up of water in the sea areas around Britain and the near-continent by the strong W'ly winds over the Atlantic during the preceding 3 to 6 days.

*Data sources*

Information about this storm was obtained from a number of sources, notably Geophysical Memoir No. 47 of the Meteorological Office (M.O.207g), entitled *Report on Thames floods* by A.T. Doodson and *Meteorological conditions associated with high tides in the Thames* by J.S. Dines. This memoir was published by His Majesty's Stationery Office, 1929. Further details were found in a Note by S.T.A. Mirlees 'The Thames floods of January 7th' in the *Meteorological Magazine* (1928, vol. 63, pp. 17–19). The weather observation reports used were published in the British and International Sections of the *Daily weather report* by the Meteorological Office.

# 16–17 November 1928

*Area*

The British Isles, Channel and north France, also eastern Atlantic near 50° N; later the North Sea, Denmark and southern Baltic.

*Observations*

There was widespread gale damage in southern Britain from this W'ly gale. The strongest winds measured were gusts of 75 knots at Cahirciveen (Valentia Observatory) near the southwest tip of Ireland and 81 knots at a 50-metre mast-head at Cardington airship field in Bedfordshire, England. Croydon in south London and Lympne, Kent recorded gusts of 70 and 69 knots respectively, for both places the strongest gusts ever measured there till that time.

W'ly winds of Beaufort force 8 to 10 were widely reported in Denmark, especially in southern and eastern parts of the country, and in the southern Baltic, when the gale reached there on the 17th.

With the frequent gales between this date and the end of the month cross-Channel steamship services were often interrupted.

*Meteorology*

Westerly winds prevailed in the British Isles and neighbouring regions from the 10th until well beyond the end of the month. There was a long succession of cyclonic activity from the western Atlantic to the shores of Europe and from there on, sometimes into the Baltic and sometimes to the Barents Sea and the north coast of Asia.

The depression which led to this storm on the 16th–17th appeared as an open wave disturbance on the main Atlantic front near 39° N 49° W on the 14th with lowest pressure 1008 mb. It moved quickly east across the ocean, covering 1100 miles (45 knots) from the 15th–16th and deepening to about 968 mb, centred near North Wales at 13h on the 16th and developing a particularly strong pressure gradient on its southern side between about 48° and 52° N, measured as 85–90 knots in the strongest part.

The centre crossed the British Isles and the North Sea and proceeded through the Skagerrak to reach south Sweden by the afternoon of the 17th, with lowest pressure then about 967 mb.

The strongest winds continued to affect just a rather narrow 'corridor' south of the centre over southern Denmark and then southernmost parts of the Baltic on the 17th as the system moved away farther east and began to fill up. By the 19th the lowest pressure in the system was 993 mb over the Baltic States and western Russia.

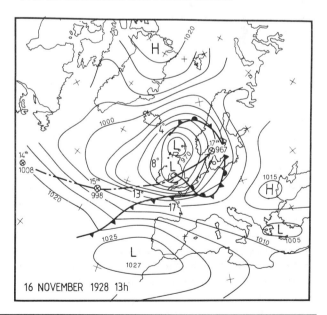

16 NOVEMBER 1928 13h

# 23–25 November 1928

*Area*

The British Isles and neighbouring sea areas, the North Sea, and then Denmark and parts of south Sweden and the Baltic.

*Observations*

Associated with a storm cyclone that crossed southern Scotland on the 23rd gales were felt in all districts, a severe gale in the W'ly winds everywhere south of the track of the depression centre. The strongest gusts reported were

94 knots at St. Ann's Head near Pembroke and 79 knots inland in the heart of East Anglia at Mildenhall, 75.5 knots at Eskdalemuir in the Southern Uplands of Scotland, 73 knots at Aldergrove in northern Ireland, and 72 knots at St. Mary's in the Scilly Islands.

The W'ly winds were reported as force 11 at points on the west coast of Jutland in the night of the 23rd–24th and force 9 to 10 at many other points in Denmark on the 24th. Force 11 was reported again on the east (i.e. the Copenhagen) side of Sjaelland and on the island of Bornholm, farther east in the Baltic, on the afternoon of the 24th. The gale in the area around Bornholm was SW'ly force 11 until midnight on the 24th–25th and continued there and spread to other eastern (and later northern) parts of the Baltic on the 25th, becoming gradually S'ly and then SSE force 10.

There was a sea flood all along Jutland's North Sea coast on the 24th, which broke the dykes in the southern part and caused serious damage in some places behind the dykes. The water rose 4.5 m above its average level at Ribe and 3.4–3.5 m at Esbjerg harbour, these being higher than any previously experienced levels at these places (Peders-sen, 1977).

*Meteorology*

The development and history of the storm was in many respects similar to the previous one, a week before. It first appeared as a wave on the main front over the western Atlantic on the 21st near 40° N 64° W, lowest pressure 1005 mb. On the 22nd it was approaching mid ocean near 48° N 42° W, 995 mb and from there it crossed to between northern Ireland and south Scotland by 13h on the 23rd, 1500 sea miles in 24 hours (62 knots), i.e. faster than the previous storm, and having deepened further. At Edinburgh the barometric pressure fell to 950.7 mb in the afternoon of the 23rd. That was the deepest stage. The depression proceeded to cross the North Sea and to pass through the Skagerrak on the 24th, when the lowest pres-

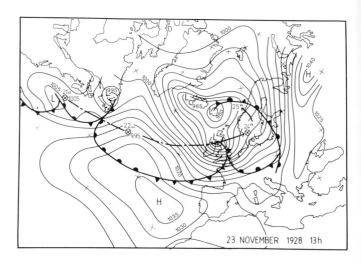

23 NOVEMBER 1928 13h

sure was about 963 mb. From that point on, however, this storm cyclone differed from its forerunner in curving off to the north, to a position off the Norwegian coast just south of Trondheim on the 25th. A new depression associated with the next front, arriving quickly from the Atlantic and passing north of Scotland, and then developing southeastwards over the North Sea, was centred near the coast of northeast England and southern Scotland at 13h on the 25th. It then appeared as part of a large complex cyclonic system, which together with the previous storm centre covered the entire North Sea and most of the Norwegian Sea. At that time the lowest pressure in both centres was between 961 and 964 mb, and the NW'ly winds over Britain and British coasts again strengthened to gale force. The strongest gusts reported in this renewed gale were 76.4 knots in northeast Wales near the Dee estuary and 73 knots at Dunfanaghy in Co. Donegal.

Renewed sea floods on the 26th and 27th had 'catastrophic effects', bursting the sea defence dykes etc. farther south along the continental coasts in Germany, Holland, Belgium and France.

## 5–7 December 1929

*Area*

The British Isles and seas near southwestern districts.

*Observations*

This was a month of excessive rainfall and repeated or persistent gales, the winds being generally from SW or W. The gales were associated with the frequent passage of depressions and secondary depressions and their fronts: they were felt most severely in coastal districts of the south and west of England and Wales and the south of Ireland, particularly between the 5th and 7th but again on the 12th and from the 20th to the 29th.

The strongest gusts reported were over 96 knots in the Scilly Isles and 89.5 knots at Falmouth on the 7th, 73 knots was recorded at Liverpool. On the morning of the 5th some very high hourly mean wind velocities were recorded: 61 knots at Pendennis Castle near Falmouth,

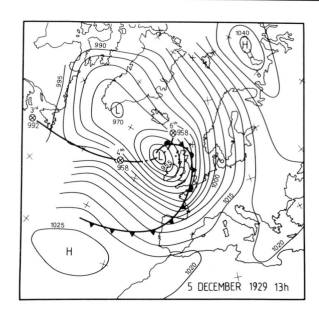

5 DECEMBER 1929 13h

where a gust of 82 knots was reported. The 96-knot gust at Scilly on the 7th was the strongest reported there.

*Meteorology*
The cyclone shown near the British Isles on the 5th developed from a centre (992 mb) between Nova Scotian waters and Newfoundland on the 3rd. It had advanced 1300 sea miles (54 knots) to $52\frac{1}{2}°$ N $27\frac{1}{2}°$ W by 13h on the 4th and deepened to 958 mb. After that its progress was quite slow,

and it contributed to the maintenance of a complex of deep centres between south Greenland and south Norway from the 6–7 December onwards. This great cyclonic system was rejuvenated by a succession of new centres developing on cold fronts daily crossing the Atlantic.

Barometric pressure fell to 949.5 mb at Cahirciveen (Valentia Observatory) near the southwest tip of Ireland on the evening of the 6th.

## 9–11 November 1931

*Area*
The British Isles, especially the Channel coast.

*Observations*
This was a month with abnormal prevalence of S'ly to SE'ly winds in the British Isles, mild and with excessive rainfall except in the north of Scotland and east of England.

The S'ly and SW'ly gales on the 10th–11th coincided with exceptionally high tides along the south coast of England and therefore caused much damage by coastal flooding and battering of houses along the shore by the waves. At Shoreham, Sussex, bungalows were considerably damaged. In Littlehampton streets were flooded to a depth of 2 ft (60 cm). A great lagoon covered the low-lying ground between Winchelsea beach and Rye harbour and the road was under 4 ft (1.20 m) of water, and part of the west of the Isle of Wight west of the river Yar and Freshwater Bay was cut off from the rest of the island at high tide.

At Blacksod point on the west coast of Ireland pressure fell to 961 mb at 13h on the 10th, but the maximum strength of the winds seems not to have been remarkable. NW force 10 was reported on the Atlantic near 48° N 22° W.

*Meteorology*
The month was characterized by persistent high pressure over Russia and depressions halted over the Atlantic.

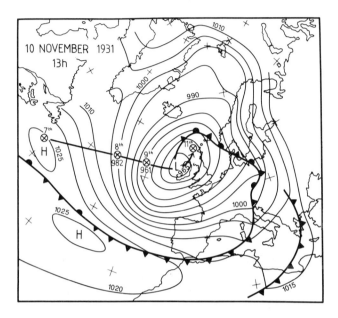

Although the monthly mean pressure was below normal in all parts of the British Isles, especially in the west and north of Ireland, there were not many gales and the strongest wind mentioned in the *Monthly weather report* of the Meteorological Office was 72 knots in the Outer Hebrides in another gale.

## 9 April 1933

*Area*
East Greenland Sea–Norwegian Sea

*Observations*
According to H.C. Shellard in an internal memorandum of the UK Meteorological Office Climatology Division dated 1963 (approx.), the strongest wind ever thus far measured anywhere in the world was a gust of 188 m.p.h. (163 knots) at Jan Mayen on this date.

*Meteorology*
The observation was in a N'ly storm which developed between northeast Greenland and the fronts of a depression (central pressure about 980 mb) which had moved northeastwards from mid-Atlantic, passing over Iceland on the 9th to near Bear Island and the Barents Sea on the 10th.

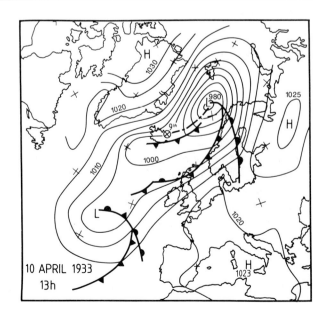

*Maximum wind strength*

The strongest pressure gradient as drawn on the usual small-scale meteorological charts indicated a gradient wind from the N of only about 70–75 knots. The much greater speed of the strongest gust measured is surely to be explained by convergence effects and turbulence caused as the wind skirted the slopes of the great mountain Beerenberg on Jan Mayen island. The pressure gradient may also have been less uniformly distributed between Greenland and Jan Mayen than the small-scale maps indicated. But pressure gradients develop and pass over very quickly in the regions about Greenland and the Greenland Sea, and it is likely that our map for the 10 April at 13h, although presenting a stronger N'ly current near Jan Mayen than 24 hours earlier, does not capture the strongest pressure gradient developed in the region of this storm about Jan Mayen.

As is common with N'ly outbreaks over this sea, bringing air from near the North Pole rapidly southward, the air's path seems to have been nearly straight north to south (great circle) and there was little cyclostrophic effect to reduce the gradient wind below the geostrophic wind value.

*Data source*

H.C. Shellard (*op. cit.*) and the observations reported in the British and international sections of the *Daily weather report* published by the Meteorological Office, London.

# 17–19 October 1936

*Area*

The British Isles, North Sea and its coasts, later Denmark and southern Baltic.

*Observations*

Strong W'ly gales force 8 to 10 about the northern coasts of Scotland from midday on the 17th spread to the west coast of south Norway by that evening. On the 18th the gale had ceased in the British Isles, but force 10 to 11 from between W and NW was reported in Denmark and north Germany, including the Baltic coast and widely over central and eastern parts of the North Sea. On the Norwegian North Sea coast from about Kristiansund southwards N force 8 to 9 prevailed all afternoon on the 18th. Winds of gale force from NW or N were again prevailing on British coasts from the Hebrides and Irish Sea to the North Sea on the 19th.

The water in the German Bight and southeastern North Sea reached severe flood levels on the 18th; but the recently raised dykes protected the coast of Schleswig-Holstein, although the dykes were damaged in some places.

*Meteorology*

This gale was caused by a deep cyclone which developed from a frontal wave (approx. 993 mb) in an almost classically frontogenetic and cyclogenetic situation southwest of Iceland on the 16th. The system deepened rapidly to 965 mb as it moved about 900 miles to a position just north of Shetland by the evening of the 17th. On the 18th this centre crossed southern Norway and Sweden to the mid-Baltic (975 mb) and then its path curved northeast-wards towards the White Sea. Meanwhile another, already occluded, depression, which had crossed Canada and reached a position southwest of Iceland on the 18th moved quickly 1200 miles southeast, crossing Scotland, to the southern North Sea by the 19th and later on, as a weakening feature, into central and eastern Europe.

The strongest gradient winds measured in this storm over the southeastern North Sea, Denmark and the German Baltic coast on the 18th seem to have been about 70–75 knots and perhaps 80–90 knots over Schleswig-Holstein. Gusts of 75–76 knots were measured in the Shetland and Orkney Islands, at Lerwick and Kirkwall on the 18th.

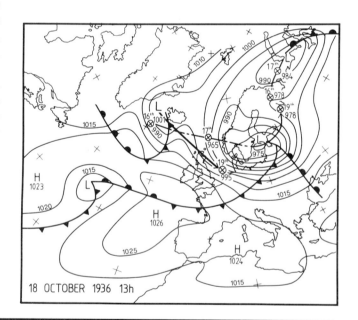

18 OCTOBER 1936 13h

# 26–27 October 1936

*Area*

The British Isles, North Sea and its coasts, Denmark and southwest Baltic.

*Observations*

This development somewhat resembled that of the previous storm nine days earlier, but this one was a more severe gale. Gusts of over 90 knots were reported at Tiree and 82 knots at Bell Rock in the North Sea east of Dundee on the 26th, as well as 82.5 knots at Paisley near Glasgow and 75.5 knots as far south as Bidston Observatory near Liverpool on the 27th. Aeroplanes and hangars were wrecked by the W'ly gale at Abbotsinch airfield and a tramcar weighing 15 tons was blown off the rails in Glasgow.

The SSW wind freshened to gale force on the coasts of Holland and Germany in the evening of the 26th and SSW force 10 was reported inland in the lower Rhineland. On the 27th W'ly gales force 10 to 12 were reported over the southern North Sea, the German Bight and on the Jutland coast. That evening force 11 was reported in the southern Baltic at Bornholm and force 9 on the Swedish south and southeast coasts in the extreme south of the country.

There was a sea flood with water levels up to 4.5 m above normal on the southern part of the Danish coast near Ribe with damage to the dykes. On the German Schleswig-Holstein coast the recently raised dykes again protected the land from serious flooding.

### Meteorology

The depression which brought this storm developed from a small Low centre (1009 mb) on an open wave on the main Atlantic front near Newfoundland on the 25th. This feature crossed the Atlantic very quickly, covering 1500 miles in 24 hours (62 knots) to a position west of the Hebrides (974 mb) on the 26th. It then moved on east much more slowly to south Norway (965 mb) by the 27th before curving northeast to the Gulf of Bothnia and beginning to fill up (988 mb on the 28th).

In this case the thermal gradients indicated by the reported surface temperatures in the frontogenetic and cyclogenetic col over the western Atlantic east of Newfoundland were stronger than those noted in connection with the previous storm. The greater strength of the winds measured in this storm is in harmony with that, although the strongest gradient winds measured on the maps available to us seem hardly distinguishable from the previous storm.

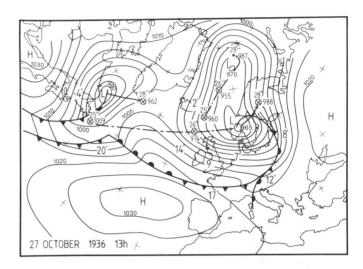

27 OCTOBER 1936 13h

---

## 16–22 January 1937

### Area

Continuous SE'ly gale to storm force·winds around the south and southwest coast of Norway, also associated gales affecting the British Isles generally on 17th–18th and whole North Sea on 18 January.

### Observations

Ships in distress over an unusually wide area on 17–18 January. Severe and long continued gale over waters near the coast of Norway. At Utsire lighthouse off the Norwegian coast the wind, always from between SSE and ESE was reported as Beaufort force 7, 8 or 9 at the morning observation every day for a week from 16 to 22 January, being force 9 on four of those days and force 8 on two of the other days.

Gales were reported to have 'lashed the coasts of Britain during the weekend' of 16–17 January. In Shetland the mail steamers were in port stormbound and fishing was at a standstill, with 'tremendous seas running all round the island' making roadways near the shore impassable. Vessels broke from their moorings in ports on the Yorkshire coast. Others ran ashore on the coast of East Anglia. A coasting steamer was damaged in Strangford Lough near Belfast on the evening of the 17th and crew members taken off. Conditions in the Strait of Dover were 'exceptionally bad' with the Channel ferries altering course to reduce the battering of the seas and running many hours late; the train ferries were cancelled. The leading French Atlantic liner, Ile de France, was driven against the quay and damaged in harbour at Le Havre, one propellor being destroyed.

On the evening of the 18th the Norwegian vessel S/S *Trym* (1909 tons) bound with a cargo of iron ore for Middlesbrough radioed that she was sinking near 58°40' N 3°20' E. The mailship *Venus* (Bergen–Newcastle line) sailed to her aid but could not achieve a rescue of the crew until the morning of the 20th.

A Russian steamer *Ilmen* (2369 tons) was in difficulties at the same time, with her steering broken, drifting towards Shetland. A German naval ship *Welle* sank with all hands lost in the gale in the southern Baltic near Fehmarn. The Aberdeen trawler *Strathebrie* sent an SOS as late as the 21st from the central North Sea area east of Aberdeen, but succeeded in limping home by late on the 22nd.

Winds of force 9 from the S or SE were reported at various points on the British coasts from Pembroke to Shetland, and also at Brest in northwest France, on the 17th and 18th. On the Danish North Sea coast SE force 10 was reported at two places on the 19th, and on the 20th force 9 was reported briefly as far north as $67\frac{1}{2}°$ N on the coast of Norway. Force 8–10 from the SE was also reported in the Baltic region, e.g. at Copenhagen and Härnösand ($62\frac{1}{2}°$ N), on the 19th–20th.

### Meteorology

The westerlies were most strongly developed in a rather low latitude zone, mainly between about 40° and 50° N (and, when the zone was broadest, sometimes extending farther south to near 30° N) throughout this sequence. The main surface low pressure system was slow-moving, with two centres about 975 mb between latitude 62° and 73° N, broadly near 20° W, off southwest Iceland and east Greenland on the 15th, being gradually replaced by a

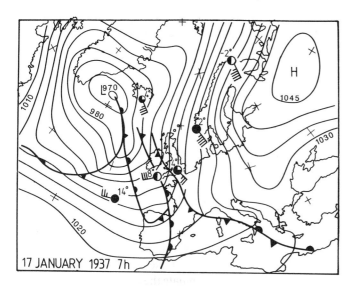

17 JANUARY 1937 7h

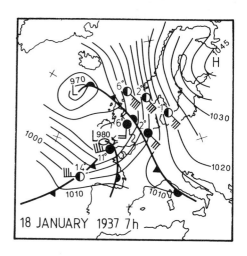

18 JANUARY 1937 7h

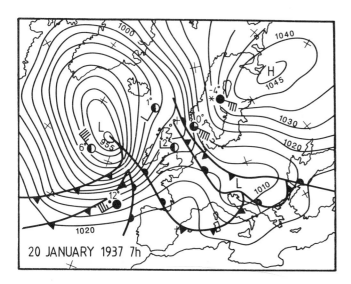

20 JANUARY 1937 7h

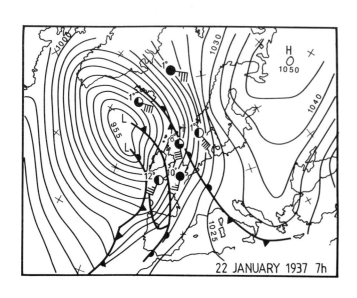

22 JANUARY 1937 7h

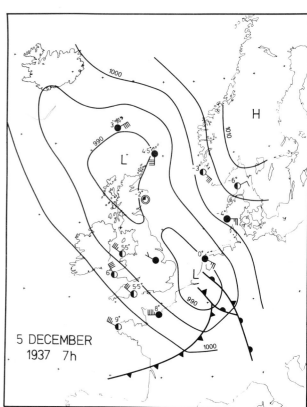

5 DECEMBER 1937 7h

single centre with depth about 950 mb near 55° N 25° W on the 20th. This centre then moved north again and later northwest, deepening further, and was centred near 62° N 35° W on the 23rd.

The vigour of the system seems to have been maintained, at least in part, by deepening frontal wave depressions approaching Ireland from the southwest on the 17th and 20th, their paths later curving to the north and their centres becoming absorbed in the very large main area of low pressure. On the 17th–19th a cyclonic centre with pressure about 985 mb crossed the British Isles from Co. Kerry in southwest Ireland to reach the central North Sea near 55° N 3° E before losing its separate identity. It was the passage of this secondary system which produced the strongest winds reported in the British Isles and later in the North Sea, on the Norwegian and Danish coasts, and in the Baltic region.

The thermal contrast between the airmasses in these systems was great, with surface air temperatures of 11 °C to 13 °C being reported in latitude 49° to 51° N between the Bay of Biscay and the Scilly Isles (and by the 22nd as far north as Glasgow) and temperatures between +1 °C and −3 °C over southern Britain in the colder air behind the cold front on the morning of the 19th.

There was a stationary, blocking anticyclone centred over Russia near 60° N, highest pressure 1045 mb on the 15th, rising to nearly 1055 mb by the 23 January. The cold continental air reaching central Germany from the SE brought temperatures generally in the region of −8 °C there, though on the coasts of Scandinavia readings between about −5 °C and +1 °C were general in the strong winds.

*Maximum wind strengths*

The strongest gradient winds on the morning of the 19th before the advancing cyclone centre and its occlusion over the eastern North Sea were in the region of 100 knots from over Denmark to near the Norwegian coast and over 70 knots over southernmost Sweden and south Norway. It seems likely that the strongest actual surface winds were in gusts in this same range of speeds in the convergence of the wind flow in the narrowing corridor between the occluded front over the eastern North Sea (aligned from near 56° N 5° E to 55° N 8° E to near Hamburg at 7h on the 19th) and the mountains of Norway.

*Note.* This type of storm, characterized by a SE windstream bringing cold air from the continental interior and producing high wind speeds as it converges between the mountains of south Norway and the front of warmer airmasses advancing from the Atlantic, is not so rare as its appearance in this catalogue might suggest. However, it must be most characteristic of climatic periods when winter anticyclones constituting a nearly stationary blocking situation frequently dominate the atmospheric circulation over northern Europe and Russia. At other times its occurrences are likely to be much rarer and short-lived.

Somewhat comparable storm situations were noted on:

4–6 December 1937 (map p. 164)
22–23 December 1937
14–17 January 1938
5 December 1938
14–19 December 1938 (see p. 167)
23–25 February 1939

The January and December 1937 storms are here illustrated by maps. None of the storms listed above caused so much or such widespread damage as the January 1937 storm here analysed. Only that over the period 14–19 December 1938, however, approached the duration of the 1937 January storm and the situation was in many respects similar. This was a run of years noted for the frequency of north European blocking anticyclones.

## 10–13 February 1938

*Area*

The North Sea.

*Observations*

This is a case of a N'ly gale which developed almost simultaneously over the whole width of the North Sea. On the 11th the N'ly windstream, with reported winds of Beaufort force 7 and 8 on the coasts, extended all the way from the area between Spitsbergen and Jan Mayen to the Mediterranean coast of France and the Alps. Force 9 was reported briefly in a few places, at the coast of northern Norway on the 11th and in north Holland, as early as the evening of the 10th.

There was an interruption to the development between the evening of the 11th and various times on the 12th as a further warm sector cyclone formed near Iceland and travelled southeast to reach north Germany by the evening of the 12th.

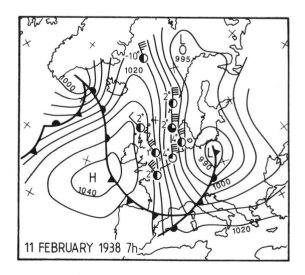

11 FEBRUARY 1938 7h

Gale force was also reported at a few places inland in France on the 13th, when the mistral brought the N'ly wind up to similar strength in the Rhône Valley and on the Mediterranean coast, and force 7 on Irish Sea as well as North Sea coasts.

*Meteorology*

A cyclone centred near 70° N 10° E – not far from Lofoten – in the Norwegian Sea on the morning of the 10th moved southsoutheast to Prussia on the 11th and reached the heel of Italy by the 13th, while pressure remained low in a long trough to north Norway and later the Barents Sea.

Warm air streaming from the W to NW over the British Isles on the 10th was soon limited to an occluding warm sector that passed south over France and reached the western Mediterranean by the 11th. At the same time the anticyclone over the Atlantic west of Biscay intensified and was transferred north, to be centred near 56° N 18–20° W on the 13th. The N'ly winds became less strong over the northern North Sea, where on the 13th gale force only occurred either in squally showers or in regions with local convergence effects near the northern English and Norwegian coasts.

## 1–2 June 1938

*Area*

The British Isles, especially southern England, the Channel and neighbouring coasts.

*Observations*

This summer gale in the night of 1–2 June was described (*Monthly weather report* of the Meteorological Office) as attaining 'a violence unprecedented for the time of year since systematic wind measurements began'. Gusts of 76.4 knots were measured at Calshot on the south coast of England, 69.5 knots at the Lizard (Cornwall), and 62.5 knots at Shoeburyness in the Thames Estuary. The wind averaged 53 knots over a whole hour at the Lizard and 47 knots at Shoeburyness.

This gale was described in the *Meteorological Magazine* (1938, vol. 73, pp. 139–42) as the 'climax of three very disturbed days, at a season when the average gale frequency is at about its minimum', though gale force had also been experienced in eastern England three days earlier, at Spurn Head and Yarmouth and in Lincolnshire and the London area where gusts reached 48 knots, and again in the southwest on 31 May.

*Meteorology*

A small, but intense, depression moved about 600 miles in the 24 hours from 13h on the 1 June northeast across England from the mouth of the Channel to the northern North Sea, where it lingered on the 2nd–3rd while filling up. There were gales near the centre, S force 8 being reported on the Brittany coast already at 13h on the 1st and over a rather wider area on the 2nd, when SW by S force 9 was also reported on the coast of Holland. The N'ly winds on the rear side of the depression also became strong to gale for a time on the 2nd.

This intense development in late spring/early summer clearly owed its occurrence to the unusual temperature contrasts existing. There was an exceptional extent of Arctic sea ice flowing fast out from the polar basin through

the Fram Strait west of Spitsbergen to the Greenland Sea and near Iceland. The winds reaching Scotland from the north were so cold that snow covered the low ground near the Moray Firth in the early hours of 2 June and some snow fell as far south as the Peterborough area in eastern England. Meanwhile, very warm air was flowing from the S and SW in the warm sector over central Europe with temperatures reaching 26 °C in south Germany on the 1st and 28 °C in central Italy. Similar temperatures were reached in Poland and Latvia on the 2nd and 3rd.

The suddenness of onset of the gale on 1 June was attributed in part in the *Meteorological Magazine* account (ibid.) to distortion of the isobars caused by development off the French west coast of a new cold front (roughly corresponding to the one shown on our map) ahead of the original cold front of the depression. There were two distinct belts of heavy rain over northern France as the cold air swept in from the Bay of Biscay in the evening of the 1st and the temperatures fell in two stages from 25 °C to 10 °C to 11 °C.

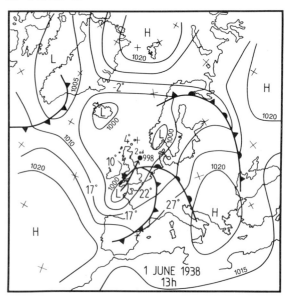

1 JUNE 1938 13h

## 23–24 November 1938

*Area*

The British Isles, North Sea and its coasts, Denmark, southernmost Sweden, and the south Baltic.

*Observations*

Severe gales reported over Ireland, England, and southern Scotland on the 23rd, and later on the North Sea, and then Denmark where the storm continued until past midday on the 24th. The storm also affected southernmost parts of the coasts of Norway and Sweden. The wind blew strongly (force 9–10) in the Skagerrak and in the south, and later central, Baltic.

The storm was unusually strong inland in England. The average wind speed over an hour reached 51 knots at Cardington (Bedfordshire, east midlands of England) and 49 knots at the Lizard and Scilly Isles in southwest England. Strongest gusts reported were 94 knots at St Ann's Head, Pembroke in southwest Wales and 73 knots at Birmingham and near the southwest point of Ireland.

There was a storm flood in the night of 23rd–24th in the German Bight and other coasts close to the track of the depression. The surge rose 10–30 cm higher than the 1936 floods.

This storm surge severely tested the sea defences. Dykes were seriously damaged in southern Jutland. The harbours at Esbjerg and at Helsingør (Elsinore) on the east coast of Denmark (Øresund) were flooded.

Force 11 from WSW or SW was recorded at several points (Lyngvig, Hantsholm and Hirtshals) on the exposed west and north coasts of Jutland in the early part of the 24th and on Bornholm in the south Baltic. The S'ly to SSE wind had risen to force 9 at a number of places as the storm approached.

## Meteorology

The frontal cyclone which caused these gales had appeared as an open warm sector near 50° N 28° W (993 mb) in mid-Atlantic on the 22nd but was already drawing into its rear very cold air from Greenland and Davis Strait. It

deepened rapidly to cross the British Isles as a 960 mb centre on the 23rd, when temperatures as low as − 2 °C to − 4 °C were observed in mid-Atlantic between 52° and 56° N near 30° W. Its advance in 24 hours from the 22nd to 23rd was about 1000 nautical miles (42.5 knots), slowing to about half that speed in the next 24 hours to the neighbourhood of Oslofjord.

The centre then turned north, to reach the Arctic coast of Norway quite quickly, as other deep cyclonic systems reached the waters near the British Isles and Iceland, bringing further gales (though not so strong as this one).

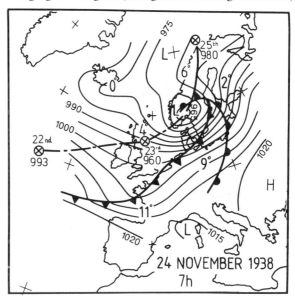

24 NOVEMBER 1938 7h

---

# 14–19 December 1938

## Area

Prolonged SE'ly storm, strongest around the south and southwest coast of Norway, but gales from between S and SE, later becoming more easterly, affected the whole North Sea on the 16th–19th and most of the British Isles at various times, particularly on the 15th.

## Observations

There were fewer reports of serious difficulties than in the storm of 16 to 21 January 1937, though the situations were meteorologically similar. The crew of the Eddystone lighthouse off the south coast of southwest England were relieved by breeches buoy after being marooned for a week in the gales which preceded those dates. The situation ended on this occasion with continental cold air and frost spreading over western Europe and the British Isles.

## Meteorology

In this case the advancing front of mild Atlantic air and the cyclonic activity impelling it were brought to a halt as the anticyclone over northern European Russia and western Siberia intensified, its central pressure reaching 1065 mb, and causing cold fronts to advance westwards over the German plain to the North Sea and later across England.

## Maximum wind strengths

SE'ly winds of Beaufort force 9 were reported at Lindesnes and Utsire on the coast of Norway on the 16th and force 10 was reported a little farther off the coast. S'ly and SE'ly winds force 8 were widely reported over the western half of the British Isles on the 15th and force 9 from SSE or S at Shetland and the Faeroe Islands in the night of 15th–16th. Easterly winds up to force 8 affected the southern and eastern coasts of England on the 19th. The gradient

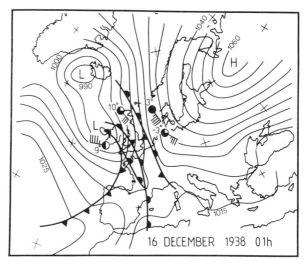

16 DECEMBER 1938 01h

winds, though long continuing at values appropriate to gale force, seem never to have been quite as strong as in the January 1937 storm. This time the strongest gradient was clearly developed over the eastern, central, and north-ern North Sea, including close to the coast of Norway rather than over the coast itself, and cannot be fixed exactly from the available pressure observations. Maximum values around 80–90 knots seem likely.

## 9–10 February 1949

### Area

The British Isles south of latitude 55° N, southern North Sea, German Bight and continental coasts, including most Danish coasts and more briefly the southwest Baltic.

### Observations

A W'ly gale force 9 was first reported at the Scilly Isles on the morning of the 9th, and force 7–9 was widely reported later that day on the coasts of England and Wales from Anglesey and Yorkshire southwards as well as in a major Low between Iceland and Greenland. This new Low Inland in England the strongest wind reported seems to have been a gust of 55.6 knots at Mildenhall, Suffolk.

The winds over the eastern North Sea between the Netherlands and Denmark veered NW to NNW overnight, and 'a powerful storm' was reported bringing a storm flood tide in the Wadden Sea and German Bight which broke some dykes and flooded a village on the border of Oldenburg and East Friesland. At Husum on the Schleswig-Holstein coast the water rose 4.09 m above its average datum level in the early hours of the 10th.

### Meteorology

This cannot rank as a very severe storm, but the sea flood threat was serious on the German Bight coasts.

The very high water levels must be attributable to the long duration of stormy W'ly winds over the Atlantic between latitudes 45° or 50° and 60° N from the 7th onwards, with the collapse of a large blocking anticyclone frost situation dominating northern and central Europe in the first week of the month. A small secondary low pressure centre (lowest pressure 990–995 mb) formed in the early hours of the 9th between southern Ireland and the coasts of Wales, at the point of occlusion of the front of a major Low between Iceland and Greenland. This new Low moved across England to the central North Sea by the evening of the 9th and then deepened very rapidly to somewhat below 980 mb as it neared the Jutland coast. It then crossed Denmark to lie near Copenhagen next morning at much the same depth, but from that point on it filled rapidly, and the winds eased during the 10th as the Low continued on its path near the southern coasts of the Baltic towards Poland and Lithuania.

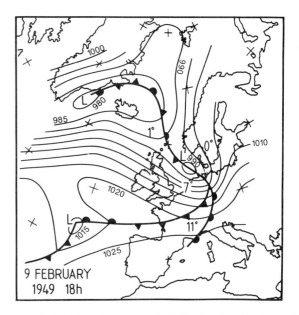

9 FEBRUARY 1949 18h

## 23–26 October 1949

### Area

The British Isles, North Sea and Baltic.

### Observations

Gales from SW, W, and later N or NW in most parts of the British Isles and North Sea apart from the extreme north, on the 23rd to 26th and over Denmark, south Norway and the southern Baltic on the 24th–25th. There was a renewal of gale force winds (up to force 9) from SSW over the southern and eastern North Sea on the morning of the 26th, veering N by the evening of that day. The gale on the 23rd caused considerable damage to the sea wall at Folkestone and other property along the coast in East Sussex and Kent.

The highest wind speeds reported in Britain were in gusts of 74 knots at Dover and 71 knots at Manston, Kent on the 26th. On the 25th Bell Rock lighthouse in the North Sea east of Dundee reported a gust of 68 knots, presumably from the N or NE as a low pressure centre crossing the British Isles from south of Ireland to reach the Skagerrak on the 26th passed near.

S to SW gales force 8–9 were reported in most parts of Denmark on the 23rd and force 8–10, locally 11, on the 24th. There was some storm damage from flooding around Ribe on the North Sea coast and also at Copenhagen on the evening of the 24th. The new North Sea gale on the 26th produced SSW force 10 in eastern Denmark lasting into the evening and followed by W force 8–9 on the 27th.

The swing of the gale to NW on the North Sea coast on the evening of the 26th brought sea water over the dykes on the coast of Jutland, causing fresh damage in the Ribe area, and flooded Esbjerg harbour, where the highest level was 3 m above normal, from 17 to 19.30h. A ship was cast ashore on the coast farther north, near Hanstholm.

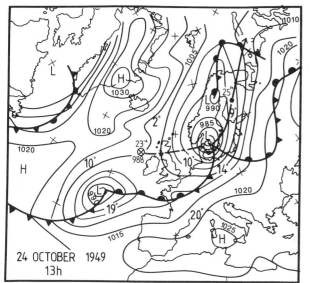

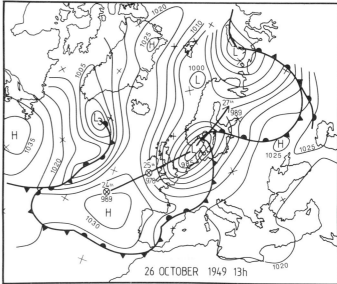

## Meteorology

This series of gales was caused by two main storm cyclones which crossed the North Sea on the 24th and 26th (see maps). A small secondary low pressure centre on a frontal wave at 48° N 25° W at 13h on the 22nd moved quickly northeast, covering over 700 nautical miles to near North Wales 24 hours later, and was absorbed into the North Sea cyclone seen on the map for the 24th. It was this feature that produced the damaging winds on the 23rd. No reports have been found of damaging winds north of the paths of the low pressure centres.

The strongest gradient winds measured in these systems seem to have been no more than about 70 knots.

The heights reached by the storm tides and the damage occasioned on the Danish coasts must be largely attributable to excess water driven into the European coastal seas by the long spell of strong W'ly winds over the Atlantic – also perhaps to very rough seas produced by the N'ly outbreak over a very long fetch of sea from the central Arctic to the North Sea from the 24th onwards.

## 30 December 1951

### Area

Northern and eastern Scotland and northern Ireland and the neighbouring sea areas about the Hebrides, Pentland Firth, and the northern North Sea.

### Observations

Violent gale caused extensive and widespread damage, also some flooding in coastal districts. Considered the most extensive and severe gale in Scotland since 1927 (report in *Weather* 1952, 7(1), p. 21). There were several deaths caused.

The strongest gusts reported in this storm were: 94 knots at Millport on the island of Bute, 87 at Bell Rock lighthouse in the North Sea 20 km east of Dundee, 85 at Benbecula and Tiree in the Hebrides, and 88 knots at Edinburgh airport (Turnhouse). The wind at Benbecula averaged 73 knots (force 12) over a one hour period in the forenoon.

### Meteorology

There were several severe gales affecting the British Isles during the last five days of December 1951, this one on the 30th being the severest. A series of very deep depressions approached the British Isles, crossing the Atlantic on eastnortheast or northeastward tracks from a zone off the American seaboard south of Newfoundland. Most of the

activity was south of 60° N. At least two of the storm cyclones over the central to eastern Atlantic during those days – on the 26th and 30th – were characterized by somewhat unusual, strong deepening of the occluded system, leading to very strong pressure gradients close to, and nearly all around, the centre. There were strong thermal contrasts between the main airstreams feeding into the cyclone, as is usual in intense systems in these latitudes, but the unusually strong pressure gradients near the

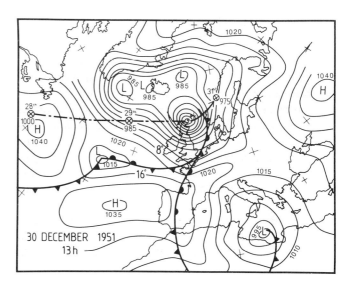

centre surely betoken the release of energy from some latent instability in the upper air.

The 500 mb level contour map in both cases shows a very strong W'ly jetstream over the Atlantic in latitudes 45° to 55° N debouching into a very strong diffluence near the British Isles.

It is hard to measure accurately the gradient winds in such strong vortices and likely that the actual winds present and passing through the system had not time to attain balance with the pressure gradient. Geostrophic winds of 300–400 knots are indicated in this case, leading to gradient winds of the order of 130–150 knots over northeast Scotland and the sea areas between Wick and Aberdeen. It is not surprising therefore that actual winds of around 100 knots were experienced in gusts.

## 17 December 1952

### Area

Scotland and northern and eastern England, with nearby parts of the North Sea.

### Observations

Severe NW'ly and N'ly gale. Similar winds up to gale force associated with the same storm also affected Ireland, the southwest coasts of England and some coasts of the Bay of Biscay for a time, and there were SE'ly winds of gale force for a while in most parts of Denmark.

The strongest gusts noted were 96.4 knots at Cranwell, Lincolnshire, 91 at Liverpool airport, 87 at Tiree in the Inner Hebrides, 80 at Stornoway, 79 at Fleetwood, Lancashire and 76 at Sellafield on the coast at the southwest side of the Lake District. The gust at Cranwell was reputedly the strongest ever measured anywhere inland in Britain to that time.

### Meteorology

The depression in the North Sea seen on our map for 18h on the 17th had been very deep (below 960 mb) when it passed between northeast Scotland and the Northern Isles in the morning of that day. The system travelled southeast at about 27 knots from a position near the Orkney Islands at 6h to the central North Sea by 18h. The strongest gales developed over Britain as the Low moved southeast over the North Sea, but it was beginning to fill

up and proceeded to decay quickly over the Low Countries. By the 18th a W'ly wind pattern was established over the British Isles and lasted for a week. It was, however, mainly a cold month with blocking patterns on most of the other days.

In the afternoon of the 17th winds of gale force 8 from NW or N were reported at many British coastal stations and force 7 on the west coast of Ireland. That morning Stornoway in the Hebrides had reported NW force 9.

Geostrophic winds up to 100 knots and gradient winds up to 74 knots are indicated over central and eastern Britain by our map for 18h on the 17th.

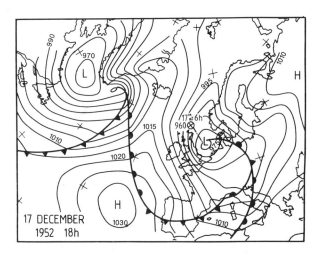

17 DECEMBER 1952 18h

## 31 January to February 1953

### Area

The North Sea, especially western and central parts from east of Shetland and the whole east coast of Scotland and England to the coast of Holland where the severest effects were felt. Also most of the British Isles.

### Observations

Great sea floods occurred on the coasts of the Netherlands and Lincolnshire, East Anglia and the Thames Estuary, where the tidal surge raised by the storm peaked within a couple of hours of the predicted high spring tide about midnight of the 31st–1st. At Vlaardingen in the Netherlands the surge raised the water about 3 m above the level of the normal tide, at King's Lynn the greatest deviation was 2.5 m and in the Lower Thames (Sheerness) about 2.2 m higher than the normal tide. Altogether, about 2000 people were drowned by the sea floods, some 1600 in the Netherlands and 350 in England, where the greatest death toll was on Canvey Island in the Thames.

The ferry S/S *Princess Victoria* foundered in this storm on the 31st in the North Channel area between Scotland and northern Ireland.

There was also very great damage to forests in Scotland. Winds of Beaufort force 10 from NW to NNW were reported on the most exposed parts of the British coast especially in northeast England and force 11 in northeast Scotland. A strongest gust of 109 knots (56.1 m/sec) was measured between 9 and 10h on the 31st by an anemometer (at the standard 10 m above the ground) on Costa Hill, Orkney; 98 knots was measured on that day at the RAF base at Kinloss just south of the Moray Firth and 93 knots elsewhere in Orkney.

### Meteorology

The strong winds over the North Sea region on these dates developed in the cold air from the Arctic – a direct northerly airstream reaching Shetland – at the rear side of a depression, which had approached the Faeroes from the west to westsouthwest on 30th January and then plunged southeastwards to the German Bight. The central pressure was no deeper than 996 mb near 60° N 17° N at 0h on the 30th. It had deepened to 980 mb near the Faeroes by 18h on that day and 968 mb at a point midway between Aberdeenshire and south Norway at 12h on the 31st. At the same time an anticyclone was building up over the Atlantic and extending in a north–south orientation. On the 1st this High was centred near 56° N 15° W, central pressure about 1034 mb. By 12h on that day the depression was moving slowly farther to the southeast over Germany and filling, its central pressure already only 984 mb.

The sequence of developments in which this storm was involved marked a very sharp change of weather in Britain and on the European plain. On 30 January, the air over these areas was of maritime tropical origin, steered from the south between Madeira and the Azores and then around a great anticyclonic system covering Spain and the Alps, to the plains between northern France, the Low Countries, north Germany and Poland. Temperatures ranged from 7 to 12 °C in the British Isles and were as high as 7 to 8 °C in central Germany and Poland. Behind the cold front which is seen reaching the Alps on our map for midday on the 31st temperatures ranged from 0 to +5 °C over Britain in the N'ly gale which was reported as force 12 in Shetland.

*Maximum wind strengths*

The strongest pressure gradient was measured over the central North Sea, from the N to NNW, around midnight on the 31st–1st: geostrophic winds were 120–150 knots over a limited zone, corresponding to gradient wind in the range 100–130 knots, regarded as 'phenomenal' at the time.

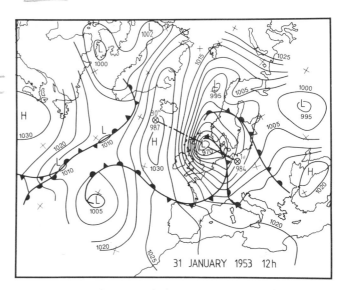

31 JANUARY 1953 12h

## 21–23 December 1954

*Area*

The North Sea, Scotland, eastern England, the Low Countries and Germany.

*Observations*

Two storms affected the region named, on 21 and 23 December respectively.

On the 21st W to NW winds of Beaufort force 9 and 10 were reported at many places, chiefly in Scotland but also in northern and eastern England as far south as Felixstowe.

Gusts of between 60 and 70 knots occurred widely. Notable extremes on the 21st were 90 knots at Kinloss just south of the Moray Firth about midday, 82 knots at

Wick and 78 at Middleton St. George in Co. Durham – all these at the time of passage of the cold front. Similarly, 84 knots from WNW to NW was recorded at Nordeney on the coast of Germany and Beaufort force 10 with gusts to between 80 and 88 knots at other points on the Dutch and German coasts.

The storm on the 23rd followed a similar path and pattern, notable gust strengths reported being 75 knots at Stornoway in the northern Hebrides, 76 knots at Middleton St. George and 86 knots at Bremerhaven on the German coast. 78 Knots was recorded as far away as Exeter and several places on the continental coast reported Beaufort force 10, 11 and even 12 on the afternoon of the 23rd.

The sea reached flood levels once more in the storm on the 23rd in the German Bight, notably in the Ems, Jade

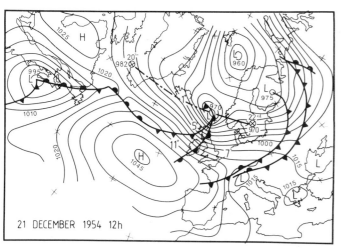

21 DECEMBER 1954 12h

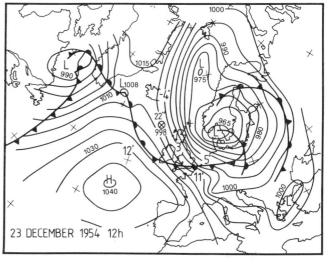

23 DECEMBER 1954 12h

and Weser mouths, and despite the strengthened sea defences one dyke was breached (Petersen and Rohde, 1977).

*Meteorology*

These gales, and the severe squalls accompanying the cold fronts, and the cold air outbreaks at the rear of two depressions, occurred to the right of the path of the centres of the depressions, which were being steered by a NW'ly jetstream (from about 310°).

The thermal contrast was great between the warm air reaching mid-Atlantic and forming the warm sectors passing over or near the British Isles (especially the south-western portion of these islands) and the Arctic air predominating over the region, which was drawn from both sides of Greenland and northern Canada.

R. Murray and C.P.W. Marshall published a note ('The storms and associated storm surges of December 21–23, 1954', *Meteorological Magazine*, 1955, vol. 84, pp. 333–41) reporting these two NW'ly storms over the North Sea which produced severe squalls and considerable storm surges, the highest up to that date since the storm of 31 January–1 February 1953. There was no significant coastal flooding as the storms did not coincide with spring tides. The meteorological circumstances were similar to other cases here reported, but the pressure gradients near the surface seem to have been less strong than in the 1953 case. The gradient winds of about 70–75 knots were reported as 'comparable with the highest observed this century, excluding the January 1953 storm'.

*Maximum wind strengths*

The jetstream at heights of about 10 km and above over Scotland was of exceptional strength, close to 200 knots (100 m/sec): 198 knots was measured at about 11 km over Leuchars on the east coast on the evening of the 22nd and 196 knots over Stornoway the previous evening. These speeds were described as 'among the highest figures that have (ever) been observed over the British Isles'.

The strongest gusts reported exceeded the gradient wind speed by a margin of 15 knots or some 20 per cent.

---

## 29 July 1956

*Area*

Southern England and Wales and the Channel.

*Observations*

This summer gale produced a pressure of 976.6 mb at Yeovilton, Somerset (in southwest England), the lowest pressure known in July over the British Isles apart from one observation of 976 mb at Tynemouth on 6 July 1920. There was great damage to trees, bushes and foliage near the south coast – at least 500 fallen trees and branches blocking roads in Brighton. Vegetation was also severely 'burnt' by salt spray carried by the wind up to 25 km inland.

The strongest gust measured on this occasion in 1956, namely 76.4 knots at Culdrose, Cornwall, was the greatest wind speed measured in July over the British Isles since before 1920. Gusts up to 74 knots also occurred in Brighton.

This was described as a violent gale which cost many lives in shipwrecks and caused widespread damage inland in southern Britain. In Brighton it came at first from S but turned finally to NE.

*Meteorology*

The storm was caused by a depression, which had been centred near 47° N 28° W (north of the Azores) at 12h on the 27th (1002 mb) and advanced about 700 sea miles in the next 24 hours (nearly 30 knots). Next it deepened

to below 980 mb and curved somewhat to the northeast as it moved to a position centrally over England on the 29th. From there it passed on northeast to the northern North Sea on the 30th, filling up very gradually and its rate of advance slowing gently.

The system presumably owed its developing vigour between the 28th and 29th to its drawing in a long stream of direct Arctic air from the northeast Greenland–Spitsbergen–Barents Sea area, although the temperature differences apparent at the surface were not extreme.

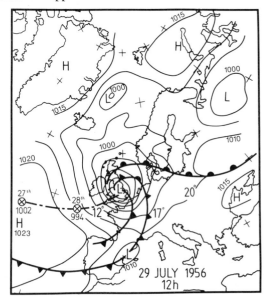

29 JULY 1956
12h

---

## 23–25 August 1957

*Area*

The British Isles and most near-lying waters, and ultimately the whole southern North Sea, extending for a time to parts of southern Danish waters and the southwest Baltic.

*Observations*

Gales unusually widespread, prolonged, and strong, for the time of year, causing widespread damage.

Gusts to over 70 knots were reported in the Orkney Islands. Beaufort force 7 or 8 was reported widely in the Hebrides, and by ships at sea between longitude 20° W and

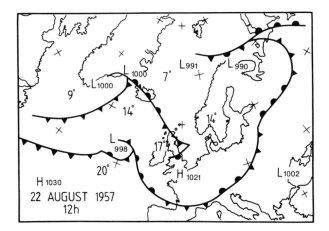

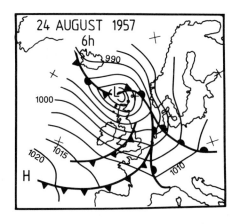

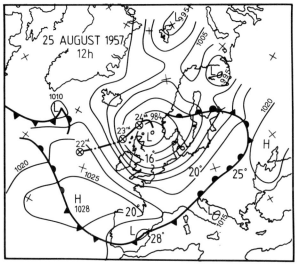

the Hebrides, all day on the 23rd and 24th. Later on the 24th and on the 25th these strengths also occurred in W'ly gales along the west coast of Ireland, in the Irish Sea too, and on the coast of Wales. On the 25th force 7 to force 9 was reported by all ships in the southern North Sea and German Bight as well as on some of the Danish islands in the Baltic south of Copenhagen.

Barometric pressure fell to 966 mb at Cape Wrath (northwest Scotland), reported at the time as a new low record for the British Isles in August.

*Meteorology*

The circulation pattern over the Atlantic and northern Europe during August had been quite variable and on the 22nd, as shown by our outline map of the fronts and centres, was still quite complex. It was, however, generally W'ly between latitudes 40° and 70° N apart from a small anticyclone over the Channel and western and central Europe. A long airstream of cool air was spreading over more of the ocean from Canada and over the next two days gradually displaced the warmer airmasses from the eastern Atlantic, the British Isles, and Biscay. The small, rather complex, depression (centre 998 mb at noon) west of Ireland on the 22nd deepened rapidly to about 965 mb as it moved quite slowly (at 25 knots slowing to about 10 knots) northeast to near Cape Wrath by the 24th. It absorbed the remains of the low pressure southeast of Greenland, and as the trough west of the intensifying cyclone centre shortened, simplifying the pattern, a more direct supply of Arctic cold air was drawn south from the area about Spitsbergen and northeast Greenland to mid-Atlantic and over the British Isles on the 25th.

As this thrust of cold air pushed farther south the strongest W'ly gales in this system developed farther south than before in latitudes 52° to 55° N across England and the southern North Sea.

---

## 16–17 September 1961

*Area*

Atlantic fringe of Ireland, Scotland, and later Norway from about 60° to 70° N.

*Observations*

This S'ly storm over the western half of Ireland resulted in communications and electric power supplies being cut over wide areas and very great damage was done, particularly to woodlands. The storm caused at least 16 deaths in that country.

*Meteorology*

This had been an Atlantic tropical hurricane first spotted near 16° N 30° W on 7 September, when it was drifting slowly west along the southern edge of the Azores anticyclone. It later curved to the north and was near 30° N 46° W at 12h on the 12th with strongest surface winds estimated at 100 knots close to the centre. It then turned east, passing along the northern side of the Atlantic anticyclone which was in a rather southerly position. The storm centre passed over the Azores during the morning of the 15th. Ships were reporting surface winds of 50 to 70 knots at points near 35° N 35°–40° W on the 14th as the storm drifted eastwards across the ocean. Its strength seems to have been declining until some time on the 15th when it became linked to the polar front a little northeast of the Azores. Its motion then speeded up, and steered by the southwesterly jetstream, it covered rather over 1000 nautical miles in the 24 hours from noon 15th to noon 16th. Its centre then lay near 55° N 10° W, close to the west coast of Ireland, and had deepened in the previous 18–20 hours from about 980 to 956 mb according to the indications of Irish observing stations close to the centre. The system continued to the northeast, as our map shows, but began to fill again during the 17th as the supply of fresh Arctic cold air ceased, and on the 18th was near North Cape with central pressure 975–980 mb.

In the most intense phase of the storm on the 16th some new records for strongest gusts occurred in Ireland, notably 93 knots at the Shannon airport in Co. Clare and 98 knots at Malin Head, Ireland's northernmost tip. The strongest hourly mean winds also reached hurricane force at some points in Ireland: 64 knots at Malin Head and 56 knots at Claremorris in Co. Mayo.

SW'ly gales affected northern Scotland and much of the coast of Norway north of 60–62° N on the 17th and 18th. Mean winds up to 55 knots were reported at a number of stations on the Norwegian coast and rather over 50 knots near the Scottish coast during the passage of the storm.

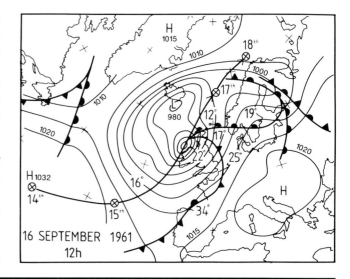

16 SEPTEMBER 1961 12h

---

## 16–17 February 1962

*Area*

North Sea and neighbouring lands.

*Observations*

Winds of strong gale force (Beaufort force 9) from SE were observed already at midday on 15 February at the weather ship position 66° N 1° E in the Norwegian Sea and a S'ly gale of mean speed 55 knots (force 11) at the angle of the Norwegian coast in 62° N, at both places with snow falling. Twenty-four hours later Beaufort force 10 or 11 was observed in a widespread NW'ly storm from northern Scotland and the Hebrides to the coast of Denmark, force 8 or more everywhere in the North Sea region from the Faeroe Islands and the Hebrides to the German Bight as well as WSW force 9 in the southern Baltic. At the following midnight (17th 0h) WNW or NW force 9 to 10 was occurring widely from the central North Sea to the sea and islands off the SE coast of Sweden, locally force 11 near the Dogger Bank and in the south Baltic.

A disastrous storm surge was built up in the German Bight, reaching its peak around the time of the normal high tide about midnight on the 16–17 February, the water level then being about 3.9 m above the predicted tide at Cuxhaven on the German coast. The storm flooding drowned 340 people in the districts about Hamburg and Oldenburg in northwest Germany and destroyed the dykes over a great distance. Flooding and breaches of the dykes also occurred in Denmark and Holland.

There was widespread storm damage all over Britain on the 16th, especially in Sheffield where in the early hours of the day an unusually exaggerated wave motion throughout a 3 km depth of the air which had been lifted over the Pennines produced a spatial pattern of intermittent concentrations of wind downstream from the hills. A mean hourly wind speed of 65 knots (force 12) occurred at one stage in Sheffield with gusts to 83 knots. There was much structural damage.

*Meteorology*

This gale was produced by a deep depression which had approached the coast of Norway in latitude 66° N from the west, central pressure 959–960 mb in two centres near longitudes 7° E and 8° W at 0h on 16 February when the W winds in the wide warm sector were reaching gale force over the northern two-thirds of the North Sea. Our maps show that at its deepest phase the cyclone centre produced a pressure of 953 mb over central Sweden around midday on the 16th. The outbreak of NW to N winds which covered the North Sea on the 17th, after the cold front had crossed the area during the previous day, ended a long sequence of westerly winds over the British Isles and North Sea which had started on the 30 January.

The temperature contrasts between the different air masses of Atlantic and Arctic origin involved in this storm were quite great but by no means exceptional at the surface. However, the temperature at the 500 mb level over the Faeroe Islands fell from −18 °C on the 15th to −35 °C on the 16th and the 500 mb height fell to about 4.9 km over southern Norway.

This prolonged spell of strong westerly winds in latitudes between 50° and 65° N is considered to have built up the water level in the North Sea to higher than normal before the additional surge impelled by the NW and N winds which followed this depression.

*Maximum wind strengths*

The strongest pressure gradients measured at 0h on 17 February at the height of the sea flood episodes indicated NW'ly geostrophic winds up to 90 knots from the eastern North Sea off SW Norway to over the German Bight and 110 knots over the north German Baltic coast. The radius of curvature of the air's path seems to have been 600 sea miles or slightly over, so that the strongest gradient winds were about 70 to 72 and 85 knots over these respective areas. The pattern was giving gradient winds up to 85 knots over the region of the German Bight three to six hours earlier.

In the jetstream higher up WNW'ly to NW'ly winds of over 100 knots were measured at the 500 mb (5–6 km height) in a zone extending from 60° N 25° W across Nor-

thern England to Denmark and the German Bight. The highest measured values at that height were about 130 knots, and no doubt greater speeds occurred at heights of around 10 to 12 km.

Strongest gusts at the surface would be expected to have been of the same order of strength as the gradient wind. In fact, however, a gust of 154 knots – 'by far the highest wind speed ever measured in the British Isles'* – was recorded in Shetland (*Weather*, *17*(3), p. 114, 1962, reporting 'The weather of February 1962') and one of 103 knots on Lowther Hill, Dumfriesshire (55°23′ N 3°45′ W 736 m above sea level) in this storm on 16 February, presumably just when the cold front passed.

*The highest figure ever reported in the world has been, according to an internal memorandum of the UK Meteorological Office Climatology Division compiled by H.C. Shellard about that time, a gust of 163 knots at Jan Mayen (71° N 8° W) in a N'ly storm on 9 April 1933.

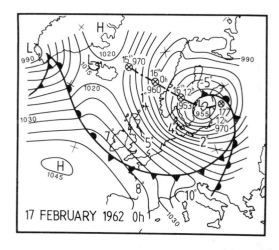

16 FEBRUARY 1962 0h                 17 FEBRUARY 1962 0h

## 23 February 1967

*Area*

Southern North Sea.

*Observations*

This storm, in which the German rescue vessel *Adolph Bermpohl* was lost, was reported by the *Deutsche Seewarte*, Hamburg to have produced the strongest winds ever actually measured over the North Sea. The average wind speed was over 140 km/hr or about 75 knots over a period of some hours duration. There was no serious flooding of the coastlands, because the storm surge came about the time of low tide.

*Meteorology*

After some days of very strong winds from between SW and WNW over the British Isles between 18 and 21 February, mostly in cold air from west of Greenland and the Canadian Arctic, a depression first reported near the Azores with central pressure 1010 mb on the 21st moved northeast across the central districts of the British Isles on the 22nd and the North Sea on the 23rd. During this time the centre deepened and lay near the west coast of Jutland with a pressure of about 965 mb at 12h on the 23rd. It

later passed over the Baltic into Russia near latitude 56° N on the 25th, filling to about 1005 mb.

Pressure gradients over the North Sea on the 23rd were strong, but the air's path was strongly curved and the cyclostrophic effect presumably reduced the strength of the

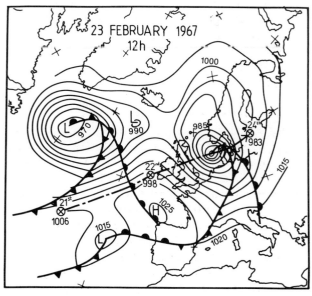

gradient wind to well below that of the geostrophic wind.

Upper winds of 100 knots from about WSW were measured over western Germany between latitudes about 49° and 53° N around midday on the 23rd at the 500 mb (about 5 km) level as a sharp cold trough developed southwards over the North Sea, distorting what had been a rather smooth, long stretch of W'ly flow.

*Maximum wind strengths*

No information about the strongest gusts measured in this storm has been found. The pressure gradients for N'ly and W'ly winds on the western and southern sides of the storm centre over the central and eastern North Sea may have produced geostrophic winds of up to 130 knots at the height of the storm, but the curvature effect presumably reduced the gradient wind speed to just 70–85 knots.

Named in Germany the '*Adolph Bermpohl* hurricane', it was considered remarkable chiefly for the duration of gale force winds, especially over and near the German Bight. Nevertheless, it is clear that some of the other storms reviewed in this catalogue, particularly those of 21–22 March 1791 and 3–4 February 1825 produced stronger winds over the North Sea over substantially longer periods than this case in 1967.

*Data source*

M. Petersen and H. Rohde *Sturmflut – die Grossen Fluten an den Küsten Schleswig-Holsteins und in der Elbe*, published in Neumünster (Karl Wachholtz Verlag) 1977.

---

## 6 March 1967

*Area*

Atlantic fringe of northwestern and northern Ireland, and all Scotland and, later, Norway coast. Sea areas off these coasts were affected by a notable SW'ly gale, with fierce gusts recorded.

*Observations*

First reports of Beaufort force 9 SW'ly came from Blacksod Point and Malin Head on the coasts of Ireland already at 0h on the 6th. The gale then swept across the whole of Scotland, producing gusts of 79 knots at Tiree in the Hebrides (56½° N) soon after midnight and 78 knots at Stornoway (58° N) in the Outer Isles at 2.30h, both of these from about S by W, and at 8.35h a gust of 124 knots from W by S was measured on the Cairngorm summit (1245 m at 57° N) in eastern Scotland. The greatest hourly mean wind speed noted was 61 knots between 8 and 9h, also on the top of Cairngorm. Lerwick in Shetland reported WSW winds of gale force from 9 to 18h on the 6th, and from about 18h to the early hours of the 7th SW'ly Beaufort force 8 or 9 was reported on the coast of Norway about 62° N.

*Meteorology*

A great current of strong SW'ly winds was maintained from mid-Atlantic 45° to 55° N, north of the Azores, to past North Cape over a period of many days, by a series of Lows moving northeast and passing between Scotland and Iceland, through the Norwegian Sea to the Barents Sea. Temperatures were generally from 8 °C to 10 °C or 12 °C in the British Isles on the 5–7 March and from + 1 °C to 6 °C on the north coast of Norway. Meanwhile Arctic cold air was streaming south, with temperatures mostly between − 2 °C and − 6 °C in Iceland. The strongest gradient winds measured over the British Isles and northern North Sea were from about 72 to 85 knots around midday 6 March, from about WSW to W. The duration of the winds of damaging strength (Beaufort force 10 or over) seems to have been no more than a few hours anywhere on low ground, but force 9 was maintained for 24 hours or more on the top of Cairngorm. The width of the stream of winds of damaging strength across Scotland and the North Sea at the height of the storm was about 100–120 sea miles and its length only about 300.

---

## 4 September 1967

*Area*

Northern England

*Observations*

Noteworthy W'ly gale caused structural damage and deaths along a narrow zone in northern England (Swaledale and near Teesmouth). The sea was also driven inland in Morecambe Bay, causing a sea flood about Blackpool.

*Meteorology*

This was a rather narrowly localized, but intense, storm, brought by a cyclone travelling northeastwards very close to the British Isles in a W'ly to SW'ly sequence with successive centres at rather different latitudes. This one drew Arctic cold air from the region of northeast Greenland and Spitsbergen south at a time when the sea ice belt was more prominent than had been general up to 1964.

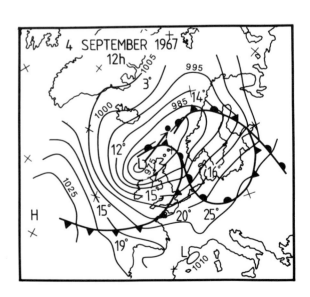

## 14–15 January 1968

*Area*

From the eastern Atlantic about 48° to 53° N around midday on the 14th this SW'ly gale affected the northwestern fringe of Ireland between about 18h and midnight. It then swept the whole of Scotland and northern England, producing great gusts, gradually turning to WSW to WNW, between 0h and 9h on the 15th, and passing on across the central North Sea it reached Denmark about the middle of that day.

*Observations*

A SW'ly gale force 9 was reported at Blacksod Point, on the coast of Ireland, by 18h on the 14th and Malin Head and Tiree in the Hebrides at midnight. Tiree had gusts of 89 knots at 23.50h and 102 knots at 2.25h, while gusts of over 93 knots from about W were recorded on the top of Cairngorm (1245 m in 57° N) in eastern Scotland at 1.15h and 3.30h on the 15th. Great Dun Fell (54°41′ N 2°27′ W 847 m) on the high Pennines in northern England had a gust of 116 knots from SW about 3h and Lowther Hill (55°23′ N 3°45′ W 754 m) one of 108 knots from W at 4.25h. A gust of 94 knots was measured on the Forth Road Bridge at 5.05h, and Bell Rock lighthouse in the North Sea, off Dundee, had 96 knots at 4.10h. Gusts of 90 knots were also measured at Prestwick and Edinburgh (Turnhouse) airports.

The strongest hourly mean winds measured were 86 knots on Great Dun Fell between 4 and 5h, while 76 knots was measured on Lowther Hill and 68 knots on the Forth Road Bridge between 5 and 6h. 65 knots was measured at Bell Rock between 4 and 5h, and when the storm reached the North Sea coast of Denmark later Beaufort force 11 was reported at one point near 55° N (Esbjerg).

*Meteorology*

This gale was a feature of a westerly sequence, lasting about a week, with the main depression centres in a belt about 58° to 63° N, and vigorous secondary Lows developing from waves on a trailing cold front between the Azores region and the North Sea, being ultimately absorbed into and supplying energy to the main system. A 967 mb centre near 53° N 20° W at noon on the 14th deepened to about 957 mb just off the Hebrides 12 hours later and was still about 960–962 mb in the Skagerrak at 12h on the 15th. From that time on the pressure gradients and winds were weakening. By evening the centre had passed into the Baltic.

Gradient winds up to about 140 knots probably affected a limited area near the depression centre 50–52° N over the Atlantic on the 14th, and later were slightly over 100 knots in the W'ly cold airstream across Scotland and the North Sea between about 55° and 57° to 58° N.

The greatest duration of winds of damaging strength at any one place was only about 4 to 5 hours at low levels and 5 to 10 hours at the level of the summits mentioned, although strong gale force continued on the top of Great Dun Fell for about 18 hours. The width of the zone of damaging winds was only about 100 sea miles over a length of 250 to 300.

## 7 February 1969

*Area*

The British Isles and North Sea, especially northern North Sea.

*Observations*

At 9.15h on 7 February a gust of 118 knots, 'the second highest wind speed ever recorded in the British Isles', was measured by a Meteorological Office trial anemometer of non-standard design, intended for use in winds of exceptional strength, at Kirkwall airport in the Orkney Islands. Verbal reports from Meteorological Office staff at the airport confirm that driving on the roads was impossible about the time of the gust because of the density and height to which snow was being blown up from the surface and that staff inside the airport building dived for safety as they expected the building to collapse.

Reports of gale force winds were not, however, very widespread.

*Meteorology*

This was an incident in a N'ly gale bringing direct Arctic air from the polar basin south over the British Isles behind the cold front of a depression which travelled slowly southeast from east of Iceland at 12h on the 6th to about 60° N 0° near Shetland at 12h on the 7th. The central pressure had been about 985 mb near Iceland on the 6th but had risen to 998 mb by the 7th. Very strong pressure gradients existed on the 6th within the ambit of this depression, particularly west and north of the centre, in the form of a long sweep of strong N'ly winds bringing exceptionally cold air rapidly south through the Fram Strait between Greenland and Spitsbergen from near the north pole. On the 7th the diversion of this current around the low pressure system in the Iceland region was straightened out as the system moved east away from Iceland. Temperatures in the N'ly gale were −20 °C at Jan Mayen (71° N) and −17 °C in Iceland at midday on the 7th.

The Arctic airstream spread south over the British Isles during the 7th. Temperatures fell below freezing point in central London that evening with the strong wind lifting frozen snow from the streets in spirals of drift up to, and over, the dome of St Paul's Cathedral. Most main roads were blocked by drifted snow and strewn with hundreds of abandoned vehicles.

*Maximum wind strengths*

The strongest gradient wind indicated by the isobars over the Norwegian Sea and the region of the Orkney and Shetland Islands was N'ly 80–90 knots. (Curvature of the air's path was negligible, so that the geostrophic and gradient wind values were alike.)

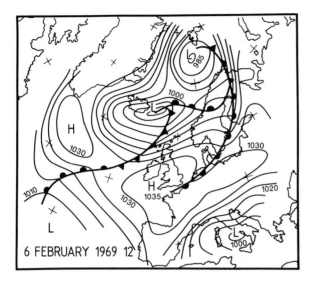

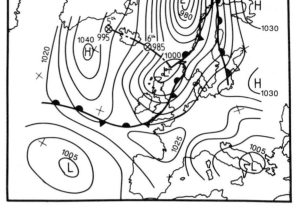

The strongest gust measured at the surface (in Orkney in this storm) exceeded the gradient wind by 30% to 35%. It was reported that very active convective motions of the air in the vertical plane in association with a cumulonimbus cloud overhead were rendered visible by the powdery snow lifted at the time of the extreme gust.

By the 8th western Iceland had come within the central region of an anticyclone (pressure 1036 mb) and the low pressure system extended north and south as a trough before the coast of Norway, from Spitsbergen to the German Bight and was extending farther south as the cold air spread towards the Mediterranean.

*Data source*

I am indebted to Miss M. Roy, Superintendent of the Meteorological Office, Edinburgh, for information from the data files in that office.

## 12–13 November 1972

*Area*

England, southern North Sea, Holland and Germany.

*Observations*

An intense storm passed quickly across the British Isles and North Sea into the continent in the 18 to 24 hours between about 15h on the 12th and 12h on 13 November. Ireland was affected in the afternoon of the 12th, Wales and the southern half of England in the evening of that day, the southern North Sea during the night and forenoon of the 13th, and the Netherlands during the middle part of the 13th. Enormous numbers of trees were blown down and buildings damaged, especially in East Anglia (particularly Norfolk) and the east Midlands of England, but also in the Netherlands and on into central Europe. On the evening of the 13th it was said to be unsafe out on the streets in the cities of eastern Germany (Dresden) because of falling masonry and flying roof tiles.

The storm caused 50 deaths in the course of its passage across Britain, the Netherlands and Germany.

*Meteorology*

This storm was produced by a rather small cyclone with an open warm sector, formed in a sharpening frontal zone across the Atlantic – seen on our map for the 13th near 47° N with a strong temperature contrast. A jetstream was formed and maintained near 50–53° N with winds of over 100 knots down to the 500 mb (5.4 km) level. The small Low centre travelled over 1000 nautical miles in 24 hours (43 knots) in latitude 53° to 55° N from west of Ireland on the 12th to the southwest Baltic at noon on the 13th and deepened rapidly from 983 mb to 959 mb when it reached the area of north Holland and the German Bight coast around 6h on the 13th.

Temperatures in the warm sector over southern England and northern France on the 12th reached 13 °C. Temperatures in the colder air from the Arctic fringe of northern Europe advancing south behind the cold front on the afternoon of the 13th were generally between 2 °C and 6 °C over Britain and no lower than 7 °C to 8 °C over Germany, though sub-zero temperatures extended south in the Norwegian Sea to near latitude 60° N.

The situation was noteworthy for the narrowness of the long straight jetstream. Another very marked, and not very common, characteristic was the development of an

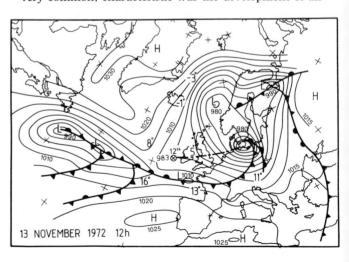

intense surface pressure gradient close to the cyclone centre. This suggests that there may also have been some thermal instability for vertical motion (probably associated with a high lapse rate of temperature in the cold air) and may have produced some tornado-like activity in the central region.

The storm was intense, and may have included some tornado activity, but the area affected by the damaging winds was at no stage very great.

*Maximum wind strengths*

The geostrophic wind indicated by the isobars over Norfolk, the southern North Sea and Holland in the early hours

of the 13th and over Denmark later may have reached about 220 knots. But because of the strong cyclonic curvature of the air paths in the central region of the storm the strongest gradient wind was probably no more than 70–80 knots. Strongest gusts of wind at the surface were doubtless of the order of 70 to 80 knots or possibly a little higher.

*Data sources*

Routine daily weather maps of the British and Danish meteorological services; English and continental newspapers and personal correspondence.

---

## 2–3 April 1973

*Area*

England, southern North Sea, Holland and north and east Germany.

*Observations*

Another intense storm which passed quickly east across the region named, this storm had noticeable common characteristics with the previous one of 12–13 November 1972. Again enormous numbers of trees were uprooted: 150 000 trees were reported blown down in the Netherlands and many houses there were unroofed during the evening of the 2 April. The rather narrow trail of damage largely repeated that of the previous case. Similar damage was again extensive in Norfolk.

*Meteorology*

The storm was again produced by a depression with an open warm sector which travelled east across the region near latitude 53–54° N at a steady speed of about 40 knots, deepening rapidly until it reached the continental landmass and then beginning to fill up. Again there was a straight and rather narrow jetstream from about W across the Atlantic and much of Europe between about 50° and 53° to 55° N. At the 500 mb (5.5 km) level winds up to 90–95 knots were measured. The central pressure of the cyclone near 54° N 21° W at 12h on the 1st was about 1006 mb. Over the North Sea near 53° N 3° E at 12h on the 2nd the central pressure was close to 970 mb, but 12 hours later over northern Germany it had filled to about 978 mb.

*Maximum wind strengths*

Mean wind speeds up to Beaufort force 12 were reported

over Holland on the evening of the 2 April. As in the November 1972 storm, an intense pressure gradient was again developed close to the centre of the depression.

The geostrophic winds measured by the strongest pressure gradient close to, and at the rear of, the centre of the storm depression over the southernmost North Sea reached 190 to 200 knots. Because of the strong curvature of the air's path in this region the strongest gradient winds were probably no more than about 80 to 90 knots. The strongest gusts at the surface were probably of this order.

*Data source*

Weather maps published by the British and Danish meteorological services; the Eastern Daily Press, Norwich and Dutch newspapers.

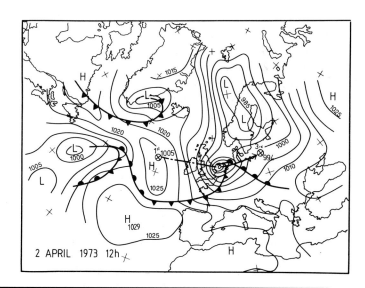

---

## 6 November to 17 December 1973

*Area*

The North Sea and especially the German Bight and neighbouring coasts.

*Observations*

This was a remarkable series of North Sea storm surge flood tides on the coasts about the German Bight produced by a series of N'ly and NW'ly storms, recorded by Petersen

and Rohde (1977), who found no other comparable sequence since the winter of 1792–3 (in which the storms and floods were more severe). The five cases which produced the highest water levels were on 13, 16 and 19 November and 6 and 14 December 1973. The tide levels reached in the highest surges on these dates were only 30 to 60 cm lower than in the 1962 storm.

Ships in the central North Sea and coastal stations in the German Bight registered these storms as:

NW to N force 7 to 9 on 6 November
WNW to NW force 8 to 9 on 12 and 13 November
WNW to NW force 9 on 15 and 16 November
W to NW force 9 on 19 November
WSW force 8 to 9 on 24 November
N force 7 on 26 and 27 November
NW force 8 to 9 on 13 and 14 December
NW force 9 on 17 December

*Meteorology and interpretation*

This long run of similar events is illustrated here by maps of the barometric pressure pattern for the dates of the five highest surges mentioned above.

The sequence of daily circulation patterns over the British Isles during these six weeks shows an unbroken spell of W'ly, NW'ly and N'ly situations, modified on a few days by the passage of a depression centre or anticyclone across the country. Most of the depressions were passing southeastwards from near Iceland to the Baltic, and in those cases the centres missed the British Isles.

There was a jetstream from between W and NW over the eastern Atlantic in about the latitude of the British Isles on most days, but winds reported at the 500 mb level were not often over 100 knots. The extremes at that level seem to have been 110 to 140 knots (on 13 and 14 November and 3 December).

The frequent high surges of the sea in the German Bight during these weeks surely owed a great deal to the continual transport of water towards the German Bight, since the gales were seldom, if ever, exceptionally strong.

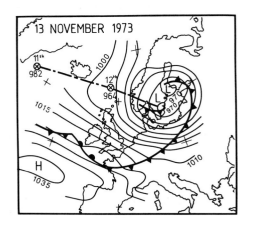

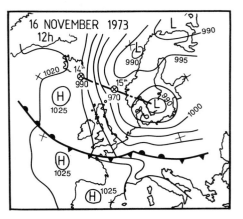

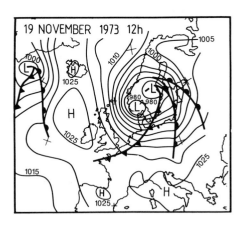

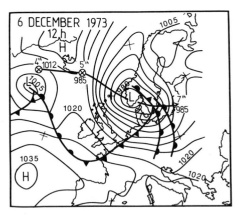

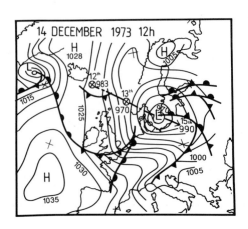

180

# January 1974

### Area

Ireland and neighbouring parts of the Atlantic and at times all the British Isles.

### Observations

The vigour and persistence of the storm activity near the Atlantic fringe of the British Isles was remarkable, especially in two storms on the 11th–12th and 27th–28th. Malin Head at the extreme north of Ireland experienced gusts of 48 knots or more on 21 days of the month, Roches Point near Cork in the extreme south on 17 days, and many other places on half the days of the month.

The extreme gusts recorded were 96 knots from 210° in Co. Mayo, near the northwest coast, on the 27th and 94 knots from 210° at Cork airport on the 12th.

With the S'ly and SW'ly storm winds, there was severe flooding with the exceptional tides in the area of Cork city on the 11th–12th and in Sligo, in the northwest of the country, where the tides reached 'new record' heights. Sea flooding also occurred about Waterford, in the southeast. Many harbours and boats in harbour were damaged by the wave action. Very large numbers of trees were felled all over Ireland, blocking roads and causing further damage.

### Meteorology

The main focus of cyclonic activity during the month seems to have been in latitudes between 50° N and south Iceland, frequently in the latitude of Ireland, over the central and eastern Atlantic. A wide area around 50° N 40° W was much colder than usual, with Arctic cold air supplied from the Davis Strait and northern Canada flowing towards mid-Atlantic, while it was a mild month in the British Isles.

*Note.* The author is indebted for information on these storms to D. Fitzgerald of the Irish Meteorological Service which produced a special memorandum on the storms of January 1974.

# 2–3 January 1976

### Area

The North Sea, especially the southern half, also neighbouring lands from Ireland and England to central Europe.

### Observations

This storm is believed to have caused more damage and loss of life on land over the region named than any since the storm of 31 January 1953, 23 years earlier. The extreme tide level at Husum on the northwest German (Schleswig) coast at 13.40h on 3 January was 4.12 m above normal, 39 cm higher than with the disastrous flood in that area on 17 February 1962. 10 000 people were evacuated from their homes in the region of the German Bight coastland and the Elbe estuary, but on this occasion no lives were lost from the sea flood. Farther north in Denmark at Højer in southwest Jutland, the tide reached 4.92 m above normal on the 3rd, 10 cm below the top of the dykes. In Holland the extreme levels reached on this occasion were lower (4.06 m above normal at Vlissingen).

The worst losses and damage with this storm were reported on land. About 60 people died: four in Ireland, 24 in Britain, 11 from a vessel which sank off the Dutch coast, two in Holland and 16 in Germany, besides two in Belgium and one in France. Enormous numbers of trees were blown down or snapped off at the trunk: over 10% of the trees in a Norfolk forest and 670 in Norwich city alone. The insurance cost of the damage of all kinds in Britain alone was finally assessed at about £126 million (at 1976 prices).

At Birmingham Airport the wind was of storm force (Beaufort force 9, over 38 knots) for 3½ hours continuously on the evening of 2 January and at Coltishall, Norfolk for 5 hours from late evening on the 2nd to the early hours of the 3rd.

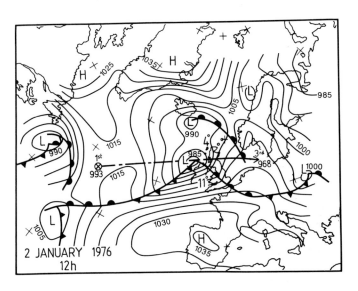

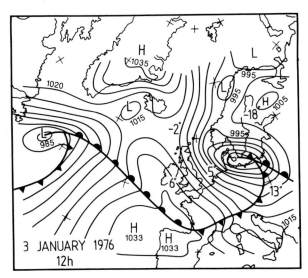

In Berlin on 3 January the wind was of gale force for the 24 hours, with gusts to violent storm (over 52 knots) over a period of 17 hours. (Both these durations were, however, exceeded in the February 1962 storm.)

*Meteorology*

This storm was produced by a mobile depression which broke away in a very strong thermal gradient from a low pressure area which covered the central Atlantic between latitudes 20° and 50° N in the last days of December 1975. Very warm, humid subtropical air was streaming north on the eastern side of this sluggish low pressure area. Dew-points were up to 11° over southern England and Ireland.

A great temperature contrast built up between this air, with 16 °C at the Azores and 12 °C in England and Ireland, and a notably cold outbreak of Arctic air from the north over Greenland and Iceland with temperatures as low as − 24 °C in southeast Greenland and − 11 °C in southwest Iceland.

A series of warm front wave depressions broke away and travelled quickly eastwards from the Atlantic system to central Europe in the last days of December: the one which produced this storm was the first of the series to engage the main Arctic air outbreak, and it deepened rapidly as it crossed the British Isles, travelling at 35–40 knots, on 2 January. The central pressure was a little below 1000 mb on the Atlantic early on the 2nd, 972 mb as it passed over the Scottish coast onto the North Sea that evening, and reached 968 mb over Denmark next morning. Filling of the centre began as it passed over the continent: the lowest pressure was 976 mb over the southern Baltic on the evening of 3 January and 988 mb over western Russia/Ukraine on the morning of the 4th.

*Maximum wind strengths*

The strongest gradient winds seem to have reached almost 100 knots from the NW over the central and eastern parts of the North Sea in the morning hours of 3 January. Farther west over the North Sea the maximum was probably in the range 87–97 knots.

The strongest gusts reported were at Birmingham Airport, 73 knots on the 2nd evening; at Coventry Airport and another airfield (Wittering) in eastern England, 91 knots; at Gorleston on the North Sea coast of Norfolk, 89 knots; and on the island of Sylt, off the coast near the German/Danish border, 88 knots on the 4th. This was the second highest speed ever recorded on Sylt.

*Data source*

Many details of this storm are given by C. Loader in an article 'The storm of 2–3 January 1976' in *Journal of Meteorology*, vol. 1 (9), pp. 273–83, Trowbridge (England), June 1976.

---

## 11–12 January 1978

*Area*

Western, central and southern North Sea.

*Observations*

This was a strong N'ly storm which lasted some hours in the night of the 11th–12th on the coasts of Norfolk and in the Fenland, the districts where it did its worst damage. Dykes were breached and the coastal lowland flooded, notably at Wells on the north coast of Norfolk where the sea bank and harbour-side road were badly breached. A small coasting freight ship was lifted from the quay in the town and deposited on the street. The flood tide (spring tide) rose somewhat higher than in the great storm in January 1953, but because of the strengthened and raised sea defences only one person lost her life – in a basement in the little port of Wisbech in the fenland.

There was also damage on the north coast of Kent, where the 150-year-old pier at Margate was demolished utterly as the wind and tide beat against the shore.

*Meteorology*

The storm occurred with a nearly straight Arctic outbreak, a 'corridor' of N to NNE'ly winds from a thousand miles beyond Spitsbergen across the polar basin to latitude 50° N in the Atlantic, which then as it progressed slowly east across the British Isles extended farther south over the Bay of Biscay and France. This development took place at the rear of depressions over the Barents Sea and Biscay–southern North Sea; this latter centre later passed east over Germany. The lowest central pressure was about 978 mb

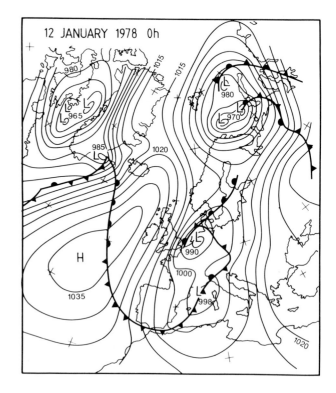

over the southern North Sea about 12h on the 11th and 968–970 mb over the Barents Sea near the north coast of Norway around the same time.

An anticyclone over the central North Atlantic near 41° N 29° W at 12h on the 11th and 44° N 25° W at 0h on the 12th had an intense ridge with pressures over 1030 mb extending north to 60° N and pressures over 1020 mb far north in the Norwegian Sea.

Geostrophic winds up to about 115 knots were indicated over the southern North Sea on the night of the 11th–12th, corresponding to gradient winds of about 90 knots. According to one assessment published by the Meteorological Office, the wind at low levels in the free air above the surface at its strongest stage exceeded 100 knots over eastern and southern England and may have reached 125 knots locally. The strongest gusts observed on that night reached 74 knots at Gorleston near Yarmouth and 70 knots at Kew (London). The hourly mean wind speed at Gorleston reached 53 knots at the height of the storm.

*Data source*

An article by E.G.E. King on 'The northerly gales of 11–12 January 1978' (*Meteorological Magazine*, 1979, vol. 108, pp. 135–46) gives the details here used as well as some further information. This article is also of interest for its reporting of successive revisions made shortening the estimated return periods of very high wind speeds over the British Isles as a result of the increased frequency of storms since about 1970.

## 13–14 February 1979

*Area*

Channel and south coast of England.

*Observations*

Fierce E'ly storm in the Channel and wave battering of the south coast of England breached the Chesil Bank, isolating Portland as an island.

*Meteorology*

The long E'ly wind stream of continental polar air indicated by our map of the situation on the 13th, with temperatures near the freezing point ($-1$ °C to $-5$ °C in Holland and Germany, $+1$ °C to 3 °C in England with fairly widespread snow cover) strengthened during the later part of the 13th as the Low advanced eastwards along the Channel and into the continent. The gradient wind strengthened over the English south coast to about 100 knots and was still of that strength over Kent till noon on the 14th.

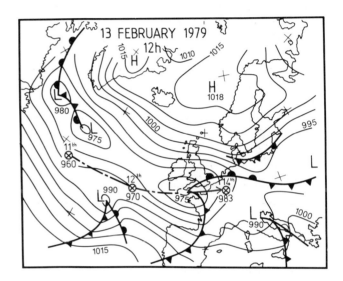

## 13–14 August 1979

*Area*

Southwestern approaches to the British Isles. (The Fastnet race disaster.)

*Observations*

Mid-August is well after the normal seasonal minimum of wind strengths in the British Isles which falls between some time in May and early July. Even so, this storm in August 1979 must rank as a notably disastrous one for high summer.

The biennial Fastnet yacht race saw 303 yachts leaving Cowes, Isle of Wight, about midday on the 11th to sail to Fastnet Rock off southwest Ireland and back to Plymouth via Bishop's Rock near 49½° N 6½° W. On the night of the 13th–14th a great storm struck the competitors between Fastnet and the Scilly Isles, 24 yachts were abandoned, 15 of their crew were drowned, and only 85 of the starters were able to finish the race.

*Meteorology*

A depression which had been a rather small feature near Nova Scotia on the 11th (central pressure 1002 mb) moved towards mid-Atlantic in latitude 47° N and filled to 1006 mb on the 12th, but then deepened and turned more towards the northeast as it approached the British Isles, drawing in Arctic cold air at its rear side from the circulation of an older depression near Iceland. The centre was near southwest Ireland, pressure 978 mb, at 0h on the 14th.

This was a notably rapid deepening for the area and time of year. And it was, of course, the quick increase of wind and sea that underlay the upset to the sailing race.

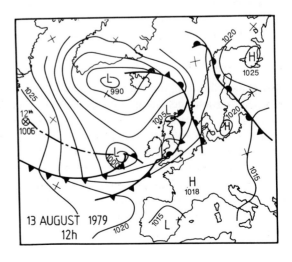

*Maximum wind strengths*

Mean wind speeds over the sea area Fastnet–Scilly were estimated reaching 50–55 knots with gusts to 68 knots at the time of the tragedy on the night of the 13th–14th.

The geostrophic wind is estimated to have reached 110 knots, but owing to the strong cyclonic curvature of the wind flow the strongest gradient wind was probably no more than about 65 knots.

Strongest gusts up to and somewhat in excess of this were measured in southwest England and Wales: 74 knots at Hartland Point, 65 knots at Milford Haven. At several places the strongest gusts were the highest speeds ever measured in August. The Hartland Point value was 'treated with some reserve' officially, probably because it exceeded the gradient wind strength – though examples of this seem to be established in a number of the storms in this catalogue.

The source used in this account was: A. Woodroffe 'The Fastnet storm – a forecaster's viewpoint' (*Meteorological Magazine*, 1981, vol. 110, pp. 271–87).

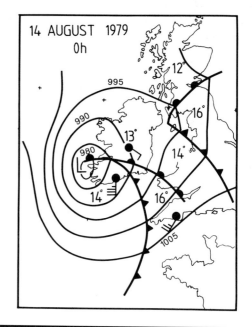

## 4–5 December 1979

*Area*

Coasts of Norway, especially between latitudes 60° and 67° N, also coasts of northern Scotland.

*Observations*

Hurricane force winds from SW, and later W, produced a sea flood (storm surge up to 1.2 m above the predicted tide) in the night of the 4–5 December washing over the quays, washing away barns in the coastal fields, and causing damage totalling millions of kroner in the counties of Møre and Romsdal, Trøndelag and Nordland. There were also storm winds on the Skagerrak coast.

Gusts up to 111 knots were measured by anemometers at Haramøy and up to 107 knots at Svinøy, near the most exposed places at the angle of the coast near Ålesund. At Kråkenes Light in the same area the mean wind somewhat exceeded 70 knots at times in the early hours of the 5th.

Such wind speeds may have occurred once or twice a decade since 1970 in similar situations on this part of the coast, but the combination with the tide made this case an outstanding one.

*Meteorology*

The storm developed from an open wave disturbance seen on the polar front near 47° N 40° W at noon on the 3rd. The airmass thermal contrast in that region was strong between maritime tropical air giving temperatures of 16 °C in the warm sector and Canadian cold air with surface air temperatures close to the freezing point. The disturbance travelled 1200 miles in the next 24 hours (50 knots) steered by a jetstream giving observed winds up to 100 knots as low as the 5 km level. The central pressure fell to about 959 mb in this time, and deepened further to about 941 mb in the night of 4–5 December as the Low entered the Norwegian Sea, drawing fresh cold air south in its rear.

Over Europe temperatures as high as 9 °C were measured near the Gulf of Riga and 7 °C in the Gulf of Finland at noon on the 5th, with 13 °C to 15 °C in northern Germany, while values of 3 °C to 5 °C ruled on the Norwegian coast.

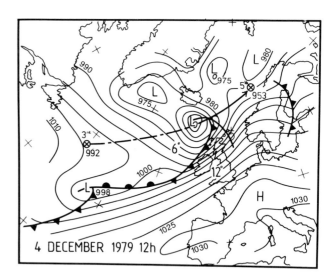

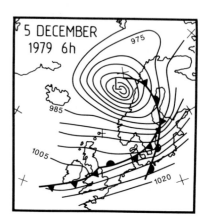

*Maximum gradient winds*

The strongest winds indicated by the pressure gradients measured in this storm were:

   (i) about 80 knots in the warm sector over northern Ireland and Scotland around midday on the 4th

  (ii) near 65° N 3° to 6° W, between Iceland and Norway, 90–100 knots around 0h on the 5th

 (iii) between about 63° and 68° N over the Norwegian Sea, close to the coast of Norway, possibly over 100 knots.

*Data source*

Details were obtained from the article 'Orkanen 4.-5. desember 1979: kan stormflo varsles?' by E. Martinsen, B.G. Harsson, P. Skjøthaug, L. Andersen, J.D. Jenssen and M. Lystad in *Vaeret*, 4 (3), Oslo, Universitetsforlaget, 1980.

---

# 27 March 1980

*Area*

Central North Sea. The Alexander Kielland catastrophe.

*Observations*

'From 21 March onwards, for nearly a week before the 10 000-ton hotel-platform Alexander Kielland collapsed and overturned with the loss of 123 human lives, SE'ly winds had been blowing in the North Sea. Their strength varied but was mostly gale force. On the 27th by 12h GMT the SE wind had increased to strong gale (Bft 9) and wave heights of 4 to 6 metres were reported at the platform. During the afternoon the wind increased further to Bft 10 in the early evening, when the catastrophe occurred, helicopters reported 50 to 60 knots and the wave heights may have been up to 7 to 9 metres.' A few hours later the wind weakened, as a low pressure centre passed.

'It is clear that the weather was hard when Alexander Kielland overturned, and that there had been a long period with fairly strong wind over several days before the accident happened. But the development of the weather followed a familiar pattern for the North Sea in winter time and such situations may be said to be relatively normal at the time of year. There is also ground for believing that the waves were in no way extreme.'

'Even if the storm may have been the trigger which set off the collapse in the construction of the platform, there is no reason to believe that either the weather or the waves were extraordinary. The Alexander Kielland platform was built to withstand the conditions it was exposed to on 27 March 1980.'

*Meteorology*

For some days around 24–25 March a rather deep depression (central pressure at 12h on the 24th 984 mb near 54° N 13° W) had been centred west to northwest of Ireland, with S'ly winds over most of Britain and SE winds over the North Sea between the depression and a large anticyclone (central pressure 1033 to over 1040 mb) over Russia with an extension to Scandinavia. On the 26th a further depression series (centres 980–985 mb), which had crossed the Atlantic in latitudes 40° to 50° N, rather further south than the previous ones, began to push northeastwards across England into the North Sea. By midday on the 27th the leading depression was centred over northeast England, depth 987 mb, and a complex frontal belt including an old warm front and an occlusion was approaching the platform site. The wind probably remained generally SE force 8 to 9, perhaps gusting to Beaufort force 10 and fluctuating a little in direction, as each front passed. The wind probably did not turn to SW and slacken off until the cold front finally ended the sequence, passing about 20h GMT.

The position of the rig at the time of the disaster was near $56\frac{1}{2}$° N 3° E in the North Sea.

*Maximum wind strengths*

The strongest gradient winds indicated in the SE'ly windstream over the central North Sea on the 27th were about

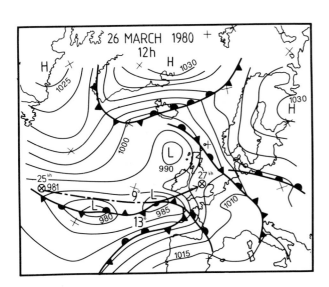

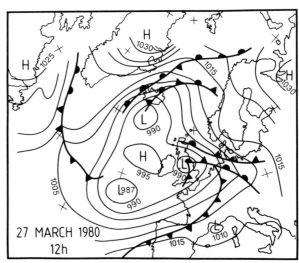

75 knots, possibly briefly up to 90 knots at the extreme. There was little curvature of the wind's path; geostrophic and gradient wind strengths were similar.

*Note.* This was hardly a great North Sea storm but is included because it caused the greatest disaster to an oil or gas rig to that time.

*Information source*

Ørnulf Fremming 'Katastrofe i Nordsjøen' in *Vaeret*, 4 (3), pp. 90–3, published by Universitetsforlaget, Oslo, for the Norwegian Meteorological Institute, 1980.

## 23–25 November 1981

*Area*

The whole North Sea, with all sea areas between the Faeroe Islands, the Hebrides, and the coasts of Holland, Germany, Denmark and west Norway. Later also the southern Baltic.

*Observations*

NW'ly winds of 50 knots were reported in the north-western sector of the North Sea from the evening of the 23rd; and by 6h on the 24th 50 and 60 knot winds from about N were reported east of Shetland, while all wind reports in the central and southern North Sea were of either about 50 or 60 knots from between NW and W. At 12h on the 24th 80 knots from NNW was reported over the sea west of Stavanger, Norway. NW 75 knots (with squally showers of hail) was reported about a 100 sea miles west of the coast of Denmark. Most places in Denmark experienced gusts of 50 knots or more, many had gusts over 70 knots. At one place in northern Jutland a gust to 83 knots was reported. Helicopters engaged in rescue operations at the rigs (Transworld 58 and the Sedco/ Phillips service platform) which dragged, and broke loose from, their anchors in the central North Sea reported gusts up to 100 knots.

The highest waves in the area where the rigs were in difficulties were reported to be of the order of 20 m high.

The storm in Denmark was described by the Danish Meteorological Institute as the 'storm of the century'. On the west coast of Jutland a sea flood developed with the high tide. At Esbjerg and Ribe the water levels reached the highest levels ever recorded there. The port of Esbjerg was closed and in parts of the town the people were evacuated from their houses. At Ribe at one time during the storm the sea rose to 5.02 m above the predicted (normal) tide level.

Thanks to the building up of the dykes after the January 1976 sea flood, and thanks also to the long warning given by the Meteorological Service, and despite losses among cattle, no human lives were lost in this greatest ever flood tide in Denmark.

In the early hours of the 25th winds over the northern part of the North Sea and over Denmark were still blowing from the NW at between 40 and 65 knots. Over the southern Baltic WSW to W 50 knots was widely reported.

Enormous damage was done to woodlands in Denmark, some of which will take 50 to 100 years to recover. The overall cost may be £70 million. The insurance companies estimated the cost of damage to buildings etc. in Denmark at the equivalent of £45 to 50 million sterling.

*Meteorology*

The slow and steady development of the situation meant that the prospect of a severe NW'ly storm began to be apparent 3 to 4 days beforehand, as the depression centre approached Shetland from Iceland with sharpening of the airmass contrast between the mild, humid air at that time over Britain, Denmark and Germany and Arctic cold air being drawn south over Iceland. The cyclone centre near Shetland had deepened to 963 mb by 18h on the 23rd and maintained about that same depth as it was transferred across southern Scandinavia to the Baltic on the 25th.

*Maximum wind strengths*

Geostrophic wind strength approached or exceeded 100 knots over all parts of the North Sea north of about $52\frac{1}{2}°$ N at some time on the 23rd and 24th. Over the western North Sea, and later over a corridor of W'ly wind stretching from the central North Sea to Lithuania and Latvia,

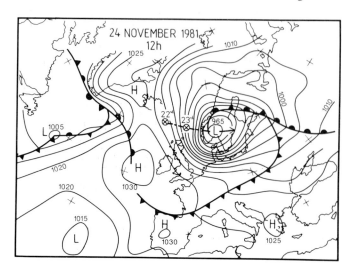

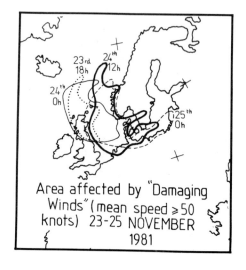

Area affected by "Damaging Winds" (mean speed ≥ 50 knots) 23-25 NOVEMBER 1981

where there was least curvature of the air's path, the gradient wind also probably reached this strength at some stage. At 12h on the 24th the gradient wind strength was 120–130 knots over the eastern North Sea between 55° and 58° N and was 100 knots over Denmark and from east of Shetland to the coast of Norway. The strongest gusts at the surface must be supposed to have been in the range 100–120 knots in the region of the strongest gradient winds.

Our map of the development of damaging surface winds as this storm advanced slowly across the region indicates the areas where most reports at the times named indicated surface winds of more than 50 knots. Data for this compilation were unusually accessible in this case.

The winds at jetstream level were also very strong: at heights between the 500 and 200 mb level (between 5 and 12 km) winds exceeding 100 knots were measured at many points over the region between the North German plain and southern-central Scandinavia on all three days, the strongest winds observed in this storm at those heights being between 130 and 135 knots. Geostrophic winds at the 500 and 300 mb levels were up to 227 knots near 53° N over the Dutch–German coast of the North Sea on 24 November 1981, gradient winds at that height of about 170 knots being implied. It seems likely therefore that in the core of the jetstream an actual wind maximum of about 200 knots (100 m/sec) existed.

*Data sources*

Details of this storm and observation reports available were obtained from the articles 'Stormflodsorkanen den 24.11.1981' by S. Lund and H. Faurby, pp. 5–16, and 'Stormfloden den 24. november 1982', pp. 17–20, in *Vejret*, 4. årgang (No. 1), Copenhagen (Dansk Meteorologisk Selskab), January 1982, and 'Stormen den 24 november 1981' by N. W. Nielsen, G. Jensen and O. Christensen, pp. 27–40, in *Vejret*, 5, February 1983. Further information was kindly supplied specially for this study by the Danish Meteorological Institute, Copenhagen and by Mr Michael Hunt of Anglia Television, Norwich. An article 'The severe storm in Denmark on 24 November 1981' by E. Skjødt in *Journal of Meteorology*, 1982, vol. 7 (No. 66), pp. 43–5, Trowbridge (England), gives a few further details and mentions some other comparable storms in Denmark, including two others in November 1981.

# 18 January 1983

*Area*

Northern, central and eastern North Sea, Denmark and north Germany.

*Observations*

Force 8 to 10 gales from W to NW over the North Sea between latitudes 53° and 57° N, and similar storm force winds from between W and WNW across Denmark, caused widespread destruction. Trees were uprooted and telegraph lines brought down in many places on the 18th.

In Copenhagen on that day 10 tons of roofing were torn off the Riksdag (parliament) building – part of the historic Christiansborg Palace – killing passers-by on the street below.

*Meteorology*

This storm developed as a depression, which was a small feature centred near the Faeroe Islands (985 mb) on the 17th, deepened and moved east at about 30 knots, to reach the Baltic near Stockholm (959 mb) around noon on the 18th. On the 19th, as a direct Arctic air outbreak continued in the rear of the depression, force 8 and 9 N'ly gales were widely reported over central and eastern parts of the North Sea, from outside Trondheim to the German Bight.

Temperatures in and around the British Isles in the warmer air on the 17th were generally from 9 °C to 11 °C, in Denmark 6 °C to 7 °C. In the gales in the rear of the system over the sea areas they were 3 °C to 5 °C, though slight frosts occurred in sheltered places inland. No very sharp airmass contrast was obvious on the surface weather maps, but doubtless a much greater contrast was seen in the upper air.

The 1982–83 winter, remarkable for its mildness in many sectors of the northern hemisphere, was also remarkable for the frequency of severe storms in the North Atlantic and over Europe, which were marked by unusual depth of a number of cyclone centres, namely:

931 mb on 19 December 1982 about 57° N 18° W
934 mb on 20 December near north Scotland, about 59° N 5° W
930 mb on 5 January 1983 near 62½° N 21° W

The maps of the circulation development over the North Atlantic were characterized by the very great vigour displayed throughout this winter, particularly between mid-December and mid-February. The cyclone which produced the storm of 18 January first appeared on the 15th as a small wave (central pressure 1004 mb) on the western Atlantic on the sharp front between the warm air from south of 30° N being steered almost due north in longitudes 40° to 50° W around a very big Azores anticyclone and a strengthening stream of direct Arctic air

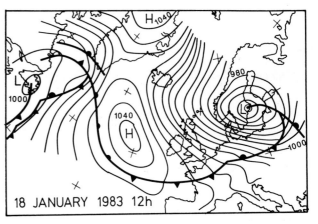

18 JANUARY 1983 12h

from the N. The latter was flowing south over both sides of Greenland, from near the pole to subtropical latitudes off the American seaboard. As the storm developed over the 17 to 19 January, this polar air current in its rear was switched temporarily to the easternmost Atlantic and Europe.

The developments at the end of January, which led to the next historic storm in this compilation, could to some extent be regarded as a repetition of this sequence of development.

## 1 February 1983

### Area

All the North Sea, Britain, and parts of the continental coast.

### Observations

Exceptionally high tide levels occurred on the coast of England, up to 6 m above normal, causing sea flooding in parts of Yorkshire (Filey, Scarborough), Lincolnshire (Mablethorpe) and to a lesser extent in East Anglia (Lowestoft); though the sea defences, which had been improved after the 1953 and 1978 storm floods, mainly held the water back and no lives were lost on the coast. The W to NW gale in England killed 4 or 5 people inland and 30 old people were blown off their feet in northwest England.

### Meteorology

This gale system, which produced storm force winds, particularly the W'ly to NW'ly winds on its southern side but also at some stage an E'ly storm near 60° to 63° N in the North Sea and a N'ly storm over and near the coast of Scotland, was associated with a depression which crossed the Atlantic on an eastnortheastward track, deepening steadily. It passed across northern Scotland near Inverness, its central pressure 950 mb as it reached the North Sea. It was then steered E by S, to cross central Jutland and the Danish islands into the Baltic, filling slowly.

The gradient winds were about 100 knots over the North Sea between latitudes 53° and 57° to 58° N as the centre of the depression passed over. They seem to have reached about the same strength over Denmark as the centre crossed the country later on the 1st.

There was a good deal of similarity in the way this storm developed over the Atlantic from the 30 January onwards to the sequence studied in connexion with the previous storm of 18 January. But this later storm appeared as a rather bigger system, with its circulation already more developed, on the western Atlantic near longitude 45° W (central pressure 1000 mb) on the 30th.

At the time of our map, at midday on the 1st, winds of force 9 and 10 were reported very extensively: in the region of the Shetland Isles, the central and northern North Sea, near the coast of northeast Scotland, as well as off Hull, in Dover Strait and the Channel, and both off the northern tip of Ireland and off the south coast of Ireland. At the same time force 7 to 8 was generally reported elsewhere in England, in the northern part of the Bay of Biscay, on the coasts of the Low Countries, Germany, and Denmark, as well as the southeast coast of Sweden and elsewhere in the southern Baltic.

Some similarities were noticed between this storm and that on the same date in 1953, which caused a great sea flood with the loss of many lives in England and the Netherlands. The 1953 storm had changed course sharply after reaching mid-Atlantic on a course towards the NNE, west of the Azores. From that point on the tracks were similar, both storms being steered along the southern edge of a developing outbreak of Arctic cold air from the Norwegian Sea. The central pressure was lower in this 1983 case (965 mb in 1953), and the reduced loss of life and less flooding clearly owed much to improved sea defences.

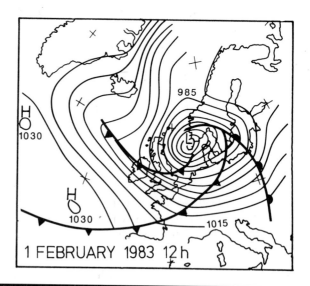

1 FEBRUARY 1983 12h

## 15 December 1986

### Area

Central North Atlantic, near and eastsoutheast of Cape Farewell, Greenland.

### Observations

The deepest cyclone ever recorded over the North Atlantic.

### Meteorology

Namias (1987) used the map here illustrating the 916 mb

centre (from Burt, 1987) to reiterate his conviction that such intense synoptic developments 'are usually foreshadowed by large abnormalities in the general circulation for periods of a month or more beforehand'. It is as if the large-scale anomalous circulation patterns have predisposed the atmosphere and ocean surface to favour singular events.

In this case the average mid-troposphere flow pattern represented by the mean height of the 700 mb pressure surface over the period 11 November to 11 December 1986, preceding development of this record storm centre, showed great departures from normal. An intense low pressure centre for this month-long period just where the storm reached its greatest intensity showed a 700-mb height 2.5 standard deviations below normal; a positive anomaly centre over central Europe, near 50° N 20° E, was 2.6 standard deviations above normal and another positive centre over the Atlantic near 33° N 48° W was also of unusual intensity. Furthermore statistical studies show that anomalies in the same senses are strongly correlated with each other – in other words, such a pattern when it occurs is a very stable one. Hence, it is reasonable to maintain, as Namias does, that sizeable departures from normal in the sea surface temperatures, the overlying air temperatures, the static stability and moisture content of the air masses, and other properties, must have been built up.

The abnormal wind pattern for the 30-day period before this storm studied by Namias favoured strong cyclonic intensification (cyclogenesis). The distribution of airmasses encouraged a strongly baroclinic zone along the path of the abnormally strong jetstream.

Gradient wind strengths up to 100–120 knots were indicated on the southern, western and northern sides of the low pressure centre in the Atlantic on the 15th. And, far away from that region, S'ly surface winds up to force 12 were reported ahead of the fronts as they advanced east across the North Sea on that day. Gradient wind strength in that area too seems to have been 100 to 120 knots but in a more narrowly restricted belt of strong wind.

N.B. The isobars on the synoptic map for 15 December 1986 illustrating this storm are at 10 mb intervals.

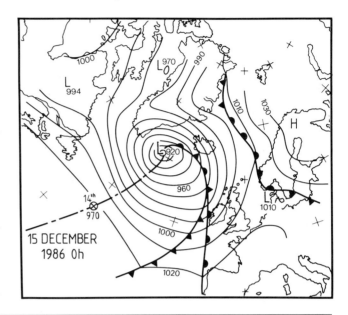

15 DECEMBER 1986 0h

---

## 16 October 1987

*Area*

England, especially the southeastern sector between London and the coasts of Sussex, Kent and East Anglia, also all parts of the Channel and Channel coasts from Cornwall east. In northern France, especially all of Brittany, and then coastal places from Normandy to the Netherlands and south Norway. A few hours after the westernmost places mentioned, all the southeastern half of the North Sea was affected and the Skagerrak. In northern Jutland (Denmark) a new storm beach platform was added to the north side of the Skagen promontory. Elsewhere in northern and northwestern Jutland, near Hurup and Thisted, many roofs were blown off and walls collapsed, but in other parts of Denmark little or no significant damage was reported. Around Tønsberg, on the south coast of Norway, there was sea flooding of the lowest grounds.

*Observations*

After generally moderate, or lighter, winds from variable directions over the British Isles on the 15th, winds rapidly strengthened over southern England after midnight. The strongest winds, attained between 2h and 6h, produced gusts of probably 100 knots at Shoreham on the Sussex coast and over 90 knots at at least six places with anemometers between Thorney Island, Herstmonceux, Ashford (Kent), and Sheerness in the Thames estuary. Gusts over 80 knots occurred in central London and at places from Jersey in the Channel Islands (85 knots) to Gorleston on the east coast of Norfolk (85 knots). Gusts somewhat exceeding 95 knots were measured in the Gorm field in the eastern North Sea. The strongest winds in most places were from about SSW.

The highest reported gust speed anywhere in this storm was an estimated 119 knots at Quimper coastguard station on the west coast of Brittany (48°02′ N 4°44′ W) and the strongest measured gust (117 knots) was also in France, on the west-facing coast of Normandy, at Pointe du Roc (48°51′ N 1°37′ W), near Granville.

There was enormous damage to buildings, trees, electricity and telephone lines. It is estimated that about 15 million trees were lost (broken, felled or uprooted) in southern England, including valued specimen trees in Kew Gardens and other parks. About 90 000 trees on the streets of London were lost. Electric power lines were brought down over a wide area. The Central Electricity Generating Board's national grid was severely affected: hundreds of thousands of homes in southeast England were still without power the following night and 2000 had not been reconnected after two weeks. Telephone lines were similarly disrupted. Road and railway links were widely disrupted by fallen trees and other debris of the storm. Almost no trains were running on British Rail's Southern Region before noon on the 16th and London Midland Region had no trains running south of Rugby, much of the difficulty due to staff being unable to get to work. Buildings and cars were smashed by falling trees and masonry. Chichester cathedral was damaged by a pinnacle from the tower falling through the roof, and several valuable stained-glass windows were blown in. The bell tower of the medieval abbey at Caen in northern France was blown down, crushing four cars.

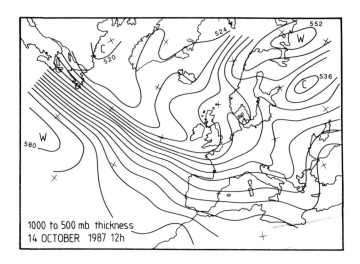

1000 to 500 mb thickness
14 OCTOBER 1987 12h

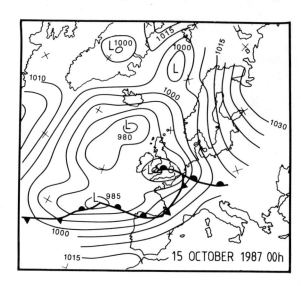

15 OCTOBER 1987 00h

Hundreds of small boats were smashed or blown away. Numerous caravans were similarly wrecked as well as innumerable glasshouses. A Channel ferry was driven aground on the coast outside Folkestone and others were unable to get into port. And a bulk carrier was capsized in Dover harbour. At Harwich a ship being used as a detention centre for illegal immigrants broke adrift, and in Felixstowe ships loaded with toxic chemicals also came adrift and threatened to break up. In the North Sea off Lincolnshire a drifting diving vessel threatened an oil rig with 80 on board, which was also adrift.

Eighteen people lost their lives in the storm in England and at least four more in northern France. Death rolls would doubtless have been bigger if the storm had struck in the daytime.

Later in the forenoon of the 16th the storm struck south Norway causing flooding by the sea on parts of the south coast near Tønsberg. Tide levels were 1.0 to 1.9 m above normal around the coast between Stavanger and Oslo, the highest since 1914. Hundreds of parked cars were destroyed by the salt water. Innumerable buildings near the coast and harbour installations were badly damaged. Forests were blown down, the loss in timber alone being calculated at about £20 million, much of this destruction of forests being inland in southeastern and eastern Norway, about Oslo and in Hedmark. At least one person was killed by the storm in Norway.

The total cost to the insurance industry due to losses and damage in England was estimated at £1000 million at 1988 prices. This figure cannot be regarded as the total cost of the storm since it leaves out losses that were not insured.

Newspapers and radio media were quick to suggest comparability of this storm with the great storm in 1703, since such events are rare in the extreme south and southeast of England. The 1987 storm produced its damaging effects in much of the same area of England as in 1703. Return periods of over 200 years were suggested on the basis of twentieth century wind measurements for the extreme gusts produced by this storm in England southeast of a line from Southampton through London to Norwich.

A report on the storm by Burt and Mansfield (1988) adds that 'it should be borne in mind that north and west of a line from Cornwall to Durham mean winds in excess of 55 knots and gusts in excess of 90 knots can be expected more frequently than once in 50 years, and 100 knot gusts can be expected at this frequency over the north of Northern Ireland and over western and northern Scotland.' What was remarkable about the 1987 storm was the occurrence of such gusts over southern England.

*Meteorology*

In the early hours of the 15th before the development of the storm cyclone over the ocean southwest of the British Isles there was a strong jetstream across the Atlantic near latitude 40° N with wind speeds up to at least 170 knots and a continuous cloud band embedded in it marking the frontal zone (Monk and Bader, 1988). The associated polar front at the surface lay across the Atlantic in latitudes 38° to 45° N, with temperatures 19 °C to 21 °C in the warm air near the Bay of Biscay and 10 °C to 12 °C in the cold air southwest of Ireland. Surface pressure was low over a wide area from England to mid-Atlantic and from northwest of Scotland to near the Azores.

A small, intense Low had formed close to the British Isles on the 14th and was changing the pattern of air flow, bringing a very warm, moist, cloud-laden airmass closer to the British Isles from the southwest and accompanied by deepening of the cold air from the north in mid-Atlantic, with deepening convection there. The cloud development in the jetstream frontal zone became more irregular, indicating a number of pulses of cold air and incipient wave disturbances developing along its length.

Ultimately one frontal wave over Biscay developed very rapidly into a cyclone centre with pressures down to 952 to 955 mb, which crossed England from Devon to Yorkshire between 1 and 5h to 6h on the 16th, moving northeast at about 50 knots and reaching the Norwegian Sea near 70° N 7° E, west of north Norway, with a depth of 975 mb by evening of the 17th.

The thermal pattern over the Atlantic and Europe at midday on the 14th, expressed as 1000 to 500 millibars

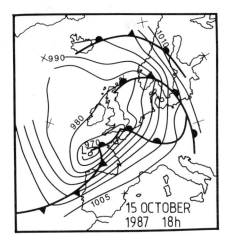

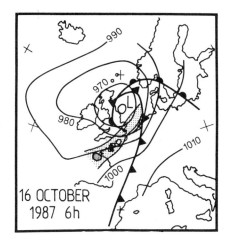

thickness or the separation between the height of the 1000 and 500 mb pressure levels, is here illustrated, and it indicates already at that time a very strong jetstream fanning out into a diffluence over France and close to the south of the British Isles where the curvature of the upper wind flow becomes very sharp. This is a classic situation for strong cyclogenesis (cyclonic development).

The thermal constrasts between the airmasses involved in the storm cyclone were vividly demonstrated by the surface temperature changes registered at places in southern England (see, for example, Templeman *et al.* (1988)). Temperatures were about 8 °C until 21h or after on the evening of the 15th and rose rapidly to 16 °C to 18 °C (18.2 °C at South Farnborough) between 22h and midnight as the warm front passed, only to fall back again to about 8 °C about the time of the strongest winds with the cold front, between about 3 and 5h. A little-noticed feature which probably also testified to the origin of the warmest air was the muddiness, due to high dust content, of the rain that accompanied this front: by that time the strong winds were from nearly due S and dust from the Sahara or Spain may

have been carried by the wind. In many places south of the low pressure centre there was a second climax of high wind speed from about SW or WSW in the cold air.

Our map of the strongest wind speeds reported in gusts with this storm shows that the zone of strongest wind was developed some 150 nautical miles (300 km) to the right (southeast) of the path of the low pressure centre. Along that path the greatest wind speeds were little more than half the highest values reported in the areas that took the brunt of the storm.

On the map for 6h on 16 October 1987:

> *the lightly shaded area* marks the belt within which gusts of 80 knots or more were experienced
> *the dark shading* marks areas where some gusts exceeded 100 knots
> *the outer lines*, roughly paralleling the belt of strongest winds, mark approximately the northern limit over England and the southern and eastern limit over the continent of the zone where gusts exceeding 50 knots were experienced

---

## 9–10 February 1988

### Area

The British Isles and ultimately most of the North Sea, especially the southern part, Dutch and German coasts.

### Observations

This was another great storm in terms of buildings damaged and trees down, especially in northwest England. At least six people died, one swept overboard from a ship near Britain's southwest coasts, two or three deaths inland in Ireland on the 9th and more in the following night near the coasts of Lancashire and Cumbria.

Strongest gusts reported included one of 93 knots at Belmullet on the northwest coast of Ireland and 89 knots on the Cornish coast near Land's End.

### Meteorology

An intense cyclone, formed on the western Atlantic, appeared at 0h on the 8th as an open wave (centre 985 mb) on the front near 47° N 42° W between subtropical air from the southwest with temperatures about 18 °C

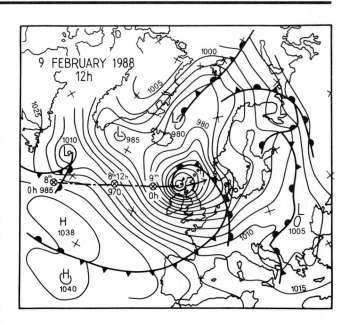

and air from the Canadian Arctic giving temperatures of −2 °C not much more than 100 nautical miles west of the front in latitudes 42° to 48° N 48° W over the ocean. It then deepened to 955 mb as it travelled about 1500 sea miles in the next 24 hours (61 knots) to a position near 56° N 17° W. Its speed of advance then slowed to 25 knots as it approached Scotland, deepening further, and was centred in the Hebridean Sea (945 mb) at 12h on the 9th.

The strongest winds occurred on the south side of the system as it crossed the British Isles. By 12h on the 10th the centre (964 mb) was over the North Sea at 58½° N 4° E near the coast of southwest Norway. Determination of the strongest gradient winds is difficult because of the strong cyclonic curvature, but they were probably about 70 knots over northern Ireland, southwest Scotland and the northern part of the Irish Sea during the night of 9–10 February.

---

### 3–4 March 1988

*Area*

Northern North Sea and the east coast of Britain.

*Observations*

This N'ly storm owes its importance to the fact that Spurn Head peninsula at the mouth of the Humber on the east coast of England (which had suffered some scathe in the storm of 31 January–1 February 1983) was breached by the action of the waves and swell. About 350 m of the concrete roadway and its foundations were washed away, and Spurn Point with the lighthouse was isolated.

*Meteorology*

The long track N'ly outbreak from the high Arctic to the North Sea developed behind a cyclone that had moved from the western Atlantic to the Denmark Strait between Greenland and northwest Iceland, where it showed a 1002 mb centre on the 1st. It then moved steadily, but slowly, southeast to reach south Sweden on the 5th.

The strongest N to NNE gradient winds over the northern North Sea on the 3rd–4th seem to have been in the range 70–75 knots.

The violent action of the sea must be related to the long fetch over open water of the strong wind, from about lati-

tude 80° N, which had persisted since midday on the 2nd (and continued till the forenoon of the 5th).

*Data source*

Note by P.C. Spink 'Further damage on the Spurn peninsula 4 March 1988' in *Journal of Meteorology*, 13 (131), pp. 281–3, (Bradford-on-Avon) 1988.

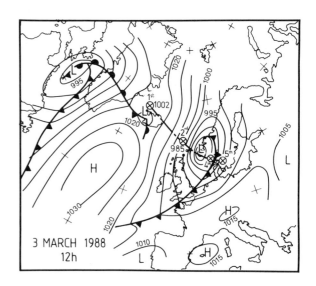

---

### 21–22 December 1988 and mid-January 1989

*Area*

Northern North Sea between the Faeroe Islands and the Norwegian coast about 61° to 64° N.

*Observations*

Strong W'ly storm, especially damaging in southern parts of the Faeroe Islands, associated with a frontal wave depression that had come up from the southwest. Mean wind speeds up to 82 knots were measured before the anemometer ceased action after a gust of 121 knots blew it

down. Later gusts were stronger and damaged 250 buildings. The storm proceeded to the west coast of Norway, where buildings, including the town hall in Ålesund were also damaged.

The central pressure of the cyclone at that time, near 65° N, was 940 to 945 mb.

Another storm on 14–15 January 1989 caused serious damage in northern parts of the Faeroe Islands, demolishing 14 of the 18 houses in one small place.

The total damage on the islands from these two storms was assessed as the equivalent of £7.5 to 9 million.

---

### 13 February 1989

*Area*

Northern Scotland and neighbouring sea areas.

*Observations*

Violent W'ly and NW'ly gale caused widespread damage on land and at sea. Large buildings, including hospitals, were unroofed.

The strongest gusts included one of 126 knots at Fraser-

burgh lighthouse, near Kinnairds Head at the northeast corner of the coast of Aberdeenshire and Buchan. This was the strongest gust ever measured in mainland Britain.

*Meteorology*

A cyclone which had advanced over 1500 miles from 50° N 43° W (995 mb) in about 24 hours reached Shetland waters (975 mb) in the afternoon of the 13th and turned eastsoutheast, to lie near Oslo next morning before being absorbed into an older system near north Norway.

The latter had driven Arctic cold air south with temperatures as low as 1 °C at the Faeroe Islands and 5 °C to 6 °C over Britain, contrasting with the values of 10 °C to 12 °C in the warm sector airmass.

A very strong pressure gradient developed near the bent-back occlusion coiled round on the western side of the low pressure centre, which had deepened 20 mb in 18 hours or less in the later part of the 12th.

This was an incident in a long and very disturbed Atlantic cyclonic sequence which produced another, deeper centre (below 960 mb) between Scotland and Iceland on the 14th. The exceptionally mild winter in northern Europe is believed to have led to higher than normal sea temperatures near 60° N although the airmasses transported south from near the pole were as cold as usual (−12 °C at Jan Mayen on the 14th).

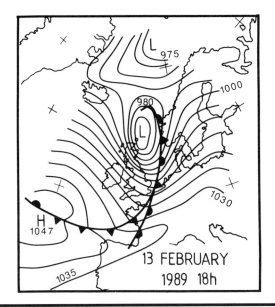

13 FEBRUARY 1989 18h

## 8 November 1989

*Area*

Southern England and the Channel, central and southern North Sea.

*Observations*

SW'ly and W'ly winds, estimated at up to 78 knots over the North Sea off the Norfolk coast, overturned and sank an oil rig minutes after all its crew had been rescued by helicopter. Wave heights in the area were estimated at 15 m.

*Meteorology*

A young open warm sector depression approached England fast from the southwest and crossed the country to Lincolnshire and the North Sea, occluding and intensifying rapidly as it did so. Strong W'ly gales developed over the east Midlands, the Wash and Norfolk and then on the North Sea. London had its strongest winds since the great storm in 1987.

## 16–17 December 1989

*Area*

British Isles and surrounding seas, Bay of Biscay and especially the northern North Sea.

*Observations*

Very strong gales, mainly S'ly and SW'ly but E or NE over northern Scotland.

Gusts to 104 knots on the Cornish coast and very heavy seas breaking over the land. At least nine lives were lost, mainly at sea and with people washed into the sea. A fishing boat was swamped and sunk in the Firth of Clyde on a crossing from the Holy Loch to Greenock, with the loss of all aboard. The crew of another ship in the Bay of Biscay was lifted off by helicopter.

*Meteorology*

A very deep, already largely occluded, cyclone, which had been centred (935 mb) near 49° N 14° W at midday on the 16th and had been moving northeast towards the British Isles during the 15th, arrived at the south coast of Ireland at 0h on the 17th and proceeded across Ireland and central Scotland to reach the North Sea off the northeast point of Aberdeenshire as a 953 mb centre on the evening of the 17th and the Norway coast near Trondheim (still below 960 mb) by the evening of the 18th. Temperatures in the S'ly gale in southwest Cornwall reached 13 °C at one stage, whereas temperatures in the cold air from the E and NE crossing Scotland before the centre arrived were between 3 °C and 5 °C. Colder air from the north was bringing temperatures as low as −2 °C to the Faeroe Islands on the 16th (and again later on the 18th) and passing on southwestwards towards mid-Atlantic.

The overall pressure pattern was tending to become stationary and later retrogressed westwards.

## Further notable storms in late January and February 1990

As this text neared completion, a remarkable succession of deep cyclones on the Atlantic, their central pressures well below 950 mb at the deepest phases, brought further destructive storms to northwest Europe. On 25 January 1990 a great SW'ly storm was severest over southern Britain and the nearer parts of the continent: gusts of wind exceeded 87 knots at a number of places. Unlike the great storm of 16 October 1987 this storm struck in the daytime, during working hours, and the death toll was therefore higher, chiefly through falling trees: 47 deaths were attributed to the storm in England and Wales, 10 each in

France and Belgium, 19 in the Netherlands and seven in Germany. It was estimated that about three million trees were felled in Britain. Another storm on 3 February produced by a smaller, secondary depression in the Channel caused up to 20 deaths in France and Germany and disastrous flooding in England in places north of the cyclone centre. And at least 14 deaths were caused in the British Isles and 12 more in neighbouring continental countries by yet another strong storm on 26 February 1990. Losses to be met by the insurance industry from the storms of the 1989–90 winter were provisionally assessed at £2.5 billion. A similar long succession of very deep cyclone centres arising at intervals of only a few days occurred on the North Atlantic in December 1942.

# References

## Consulted for storms before 1720

Bain, G. 1922, *The Culbin Sands or the story of a buried estate*. Nairn (Nairnshire Telegraph)

Bebber, W.J. van 1880, Die geographische Verteilung und Bewegung, das Entstehen und Verschwinden der barometrische Minima in den Jahren 1876 bis 1880. *Monatliche Übersicht der Witterung*, 5. Hamburg (Deutsche Seewarte)

Brooks, C.E.P. 1954, *The English climate*. London (English Universities Press)

Chromow, S.P. 1942, *Einführung in die synoptische Wetteranalyse*. Vienna (Springer), 532pp

Crawford, I.A. 1967, Hebridean settlement at Siabaidh, Berneray, Harris. *Post-Medieval Archaeology*, 1, 110–13

Crawford, I.A. 1973, Personal communication, 21 March 1973

Crawford, I.A. and Switsur, R. 1977, Sandscaping and C14: the Udal, North Uist. *Antiquity*, 51, 124–36

Defoe, Daniel, 1704, *The Storm*. London

Defoe, Daniel, 1724, *A Tour in Circuits through England and Wales*, vol. 1 of *A Tour through the Whole Island of Great Britain (1724–6)*. London

Defoe, Daniel, 1928, *A Tour through England and Wales*. London (Dent Everyman edition, in two vols. including the *Tour through Scotland*)

Derham, W. 1704, Observations concerning the late storm (a letter.). *Philosophical Transactions of the Royal Society, London*, 24, 1530–5 (see also pp. 1535–7, 1544–55)

Douglas, K.S., Lamb, H.H. and Loader, C. 1978, A meteorological study of July to October 1588: the Spanish Armada storms. *Climatic Research Unit Publication, CRU RP 6*. Norwich (University of East Anglia), 80pp

Douglas, K.S. and Lamb, H.H. 1979, Weather observations and a tentative meteorological analysis of the period May to July 1588. *Climatic Research Unit Publication CRU RP 6a*. Norwich (University of East Anglia), 39pp

Dyke Women's Rural Institute 1966, *A history of the parish of Dyke and Moy* (Culbin). Aberdeen (Waverley Press)

Edlin, H.L. 1976, The Culbin Sands. In *Environment and Man*, vol. 4, *Reclamation* (Edited by J. Lenihan and W.W. Fletcher), pp. 1–13. Glasgow and London (Blackie)

Evans, Thomas, 1698, Scheme of the wind and weather at Llanberis. (Twelve months of daily weather observations 1 March 1697 to 28 February 1698.). 12 manuscript sheets in the National Library of Wales, Cardiff

Frydendahl, K. 1986, Vejrhistorie. *1066 Tidsskrift for historisk forskning*, 16(5), 3–15. Copenhagen (Historisk Institut)

Gottschalk, M.K.E. 1975, *Stormvloeden en overstromingen in Nederland – Deel II: de periode 1400–1600*. (Storm floods and river floods in the Netherlands.) Assen (van Gorcum)

Gottschalk, M.K.E. 1977, *Stormvloeden en overstromingen in Nederland – Deel III: de periode 1600–1700*. Assen (van Gorcum)

Gram-Jensen, I.B. 1985, *Sea floods*. Copenhagen (Danish Meteorological Institute: Climatological Papers), 76pp

Hauerbach, P., Hansen, A.L. and Nielsen, H. 1983, *Havet, grenen, Skagen* and *Et supplerende kapitel til Skagen bys historie*. Skagen (Sjøbeck)

Hennig, R. 1904, Katalog bemerkenswerter Witterungsereignisse von den ältesten Zeiten bis zum Jahre 1800. *Abhandlungen des Preussischen Meteorologischen Instituts*, II(4). Berlin

Koster, E.A. 1978, De stuifzanden van de Veluwe: en fysisch-geografisch studie. Amsterdam (Doctoral thesis in the University), 195pp

La Cour, P. 1876, *Tyge Brahes meteorologiske Dagbog, holdt paa Uraniborg for Aarene 1582–1597*. Appendix to *Collectanea Meteorologica*. Copenhagen (Kgl. Danske Videnskabernes Selskab). 264 + lxxvpp

Lamb, H.H. 1977, *Climate: present, past and future*, vol. 2, *Climatic history and the future*. London (Methuen), 835pp

Lamb, H.H. 1979, Climatic variation and changes in the wind and ocean circulation: the Little Ice Age in the northeast Atlantic. *Quaternary Research*, 11(1), 1–20

Lamb, H.H. 1982, *Climate, history and the modern world*. London (Methuen), 387pp

Lamb, H.H. 1986, Ancient units used by the pioneers of meteorological measurements. *Weather*, 41(7), 230–4. London (Royal Meteorological Society)

Lamb, H.H. and Johnson, A.I. 1966, Secular variations of the atmospheric circulation since 1750. *Geophysical Memoirs* No. 110. London (H.M.S.O. for Meteorological Office)

Lamont, John, of Newton, Seventeenth century MSS farm diary from Fife in east Scotland. Held in St. Andrews University Library

Lenke, W. 1961, Bestimmung der alten Temperaturwerte von Tübingen und Ulm mit Hilfe von Haufigkertsverteilungen. *Berichte des deutschen Wetterdienstes*, No. 75, Band 10. Offenbach am Main. (Contains daily observations for 1691 to June 1694 and 1710–14.)

Ludlum, D.M. 1963, *Early American hurricanes 1492–1870*. Boston, Mass. (American Meteorological Society), 198pp

Manley, G. 1953, The mean temperature of central England, 1698–1952. *Quarterly Journal of the Royal Meteorological Society*, 79, 242–61. London

Manley, G. 1962, Early meteorological observations and the study of climatic fluctuation. *Endeavour*, 21, 43–50. London (Imperial Chemical Industries)

Manley, G. 1974, Central England temperatures: monthly means 1659 to 1973. *Quarterly Journal of the Royal Meteorological Society*, 100, 389–405. London

Middleton, W.E. Knowles, 1969, *Invention of the early meteorological instruments*. Baltimore (John Hopkins University Press), 362pp

Morrison, A. 1967–8, Report from the Harris estate papers in the *Transactions of the Gaelic Society of Inverness*, 45, 47–8

Mossman, R.C. 1896, The meteorology of Edinburgh, part I. *Transactions of the Royal Society of Edinburgh*, 38(iii), No. 20, 681–755

Mossman, R.C. 1897, The meteorology of Edinburgh, part II. *Transactions of the Royal Society of Edinburgh*, 39(i), 93–108

Mossman, R.C. 1898, Appendix of remarkable atmospheric and celestial phenomena. *Transactions of the Royal Society of Edinburgh*, 40(iii), No. 21, 476–8

Neumann, J. 1981, The cold and wet year 1695 – a contemporary German account. *Climatic Change*, 3(2), 173–87. Dordrecht and Boston, Mass. (Reidel)

Palmén, E. 1928, Zur Frage der Fortpflanzungsgeschwindigkeit der Zyklonen. *Meteorologische Zeitschrift*, 45, 96–9

Petersen, M. and Rohde H. 1977, *Sturmflut: die grossen Fluten an den Küsten Schleswig–Holsteins und in der Elbe*. Neumünster (Karl Wachholtz), 148pp.

Rohde, H. 1964, Nachrichten über Sturmfluten frühere Jahrhunderte nach Aufzeichnungen Tönninger Organisten. *Die*

*Küste*, **12**, 113–32. Heide i. H. (Boyens & Co. for Der Küstenausschuss Nord- und Ostsee)

Say, Revd. 1698–1724, Observations at Lowestoft by the Revd Mr Say. Original MSS in Bodleian Library, Oxford. Summary Catalogue 35448

Scheen, R. 1977, Fregatten *Lossen*'s historie 1684–1717. *Norsk Sjøfartsmuseum Årsberetning 1976*, pp. 41–110

Schröder, W. 1969, Zur 'Groszen Fluth' von 1717. *Acta Hydrophysica*, **14**(1–2), 237–9. Berlin (Institut für Physikalische Hydrographie der Deutschen Akademie der Wissenschaften zu Berlin). (This item refers to a number of earlier accounts, including newspaper accounts, of the Christmas 1717 storm, with a map of the extent of the sea flood, published in 1718. It also gives brief details of three seventeenth century floods included in this compendium of great storms.)

Schweckendieck, W. 1876, *Die Sturmfluten der Nordsee*. Lecture by Dr. Eilker, pp. 1–26 in *Jahresbericht des Königlichen Gymnasiums und der Höheren Bürgerschule zu Emden* (Dr W. Schweckendieck, Director). Emden (Th. Hahn Wittwe)

Short, Thomas, 1749, *A general chronological history of the air, weather, seasons, meteors . . .* London (T. Longman), in 2 volumes, 494 and 535pp

*Societas Meteorologica Palatina* 1781–92 *Ephemerides*. (Yearbooks of the daily meteorological observations collected by the Society from a network of observers across Europe and beyond.) Mannheim (in 12 volumes)

Steers, J.A. 1937, The Culbin Sands and Burghead Bay, *Geographical Journal*, **90**, 498–528

Tannehill, I.R. 1956, *Hurricanes, their nature and history*, 9th edn. Princeton, N.J. (Princeton University Press), 308pp

*Theatrum Europaeum*, 1702, *Theatri Europaei continuati: Vierzehender Theil*. Covering the years 1691 to 1695 in Europe and other parts of the world. By the heirs of the late Matthäus Merian. XIV Theil (part) of the series of 21 parts or volumes reporting events of the years mentioned, published in Germany between 1633 and 1738

Walton, K. 1956, Rattray head: a study in coastal evolution. *Scottish Geographical Magazine*, **72**, 85–96. Edinburgh

Weikinn, C. 1961, *Quellentexte zur Witterungsgeschichte Europas von der Zeitwende bis zum Jahre 1850*. I *Hydrographie*, **3** (1601–1700). Berlin (Akademische Verlag)

Willis, D.P. 1986, *Sand and silence – lost villages of the North (Forvie, Rattray, Culbin and Skara Brae)*. Aberdeen (University Centre for Scottish Studies)

Woebcken, C. 1924, *Deiche und Sturmfluten an der Deutschen Nordseeküste*. Bremen–Wilhlmshaven (Friesen Verlag)

Wright, Thomas, 1668, *Philosophical Transactions of the Royal Society*, vol III, No. 37, pp. 722–5

## Consulted for storms between 1720 and 1880

Brazell, J.H. 1968, *London weather*. Meteorological Office publication, Met. 0.783. London (Her Majesty's Stationary Office), 270pp

Carr Laughton, L.G. and Heddon, V. 1927, *Great storms*. London (Philip Allan: The Nautilus Library), 254pp

De Boer, G. 1968, *A history of the Spurn Lighthouses*. York (East Yorkshire Local History Society). 72pp

Dixon, F.E. 1959, Weather in old Dublin. *Dublin Historical Record*, **15**, 65–72

Dove, H.W. 1858, Storms produced by the lateral operation of aerial counter-currents on each other. In *The third number of meteorological papers, compiled by Rear-Admiral R. Fitzroy, F.R.S., published by authority of the Board of Trade*, pp. 34–5

Duphorn, K. 1976, Gibt es Zusammenhänge zwischen extremen Nordsee-Sturmfluten und globalen Klimaänderungen? (Are there connections between extreme North Sea floods and global climatic changes?) *Wasser und Boden*, **10**, 273–5

Espy, J.P. 1841, *Great Liverpool storm of the 6th and 7th January 1839*. In Espy, J.P. *The philosophy of storms*. Boston, Mass. (Little and Brown), pp. 294–300, 519–32

Fitzroy, R. 1860(a), Remarks on the late storms of October 25–26 and November 1, 1859. *Proceedings of the Royal Society*, **X**, 222–4. London

Fitzroy, R. 1860(b), Notice of 'The *Royal Charter* Storm' in October 1859. *Proceedings of the Royal Society*, **X**, 561–7

Fitzroy, R. 1861, Storms of the British Isles. *The tenth number of meteorological papers*, text and atlas. London (Board of Trade)

Geikie, A. 1901, *The scenery of Scotland*. New York (Macmillan: 1st edn. 1865)

Gram-Jensen, I.B. 1985, Sea floods. *Danish Meteorological Institute Climatological Papers No. 13*. Copenhagen, 76pp

Lauder, Sir Thomas D. 1830, *An Account of the Great Floods of August 1829 in the province of Moray and adjoining districts* (3rd edn. 1873). Elgin (M'Gillivray) 350pp. Also a review in *Blackwoods Magazine*, issue for August 1830

Lowe, E.J. 1870, *Natural phenomena and chronology of the seasons*. London

Lyell, C. 1830, *Principles of Geology*, (1st edn. p. 264; 5th edn. 1837, p. 399, but no longer mentioned in most later editions). London (John Murray)

Meteorological Office c. 1855 ff. Daily weather reports, weekly supplements to the daily weather report, and monthly weather reports. London (Meteorological Office)

*Meteorological Magazine*, 1939, January 6th 1839: the 'Great Wind.' *Meteorological Magazine*, **73**(876), 342. London (Meteorological Office)

Müller, W. 1825, *Beschreibung der Sturmfluthen an den Ufern der Nordsee und der sich darin ergiessenden Ströme und Flüsse am 3 und 4 Februar 1835*. Hannover

*New Statistical Account*, 1843, Statistical Account of Kincardineshire by the ministers of the respective parishes, pp. 274–5

Oliver, J. 1967, William Borlase's weather journal 1753–72. (Journal kept by the rector of Ludgvan, near Penzance, Cornwall.) *Journal of the Royal Institution of Cornwall*, **5** (New Series), Part 3, 267–90. Truro (Blackford)

Rohde, H. 1964, Nachrichten über Sturmfluten in frühere Jahrhunderte nach Aufzeichnungen Tönniger Organisten. *Die Küste*, **12**, 113–32. Heide i.H. (Westholstein Verlag Boyen)

Rohde, H. 1977, Sturmfluthöhen und säkularen Wasserstandsanstieg an der deutschen Nordseeküste (Storm flood heights and secular rise of sea level on the German North Sea coast) *Die Küste (Archiv für Forschung und Technik an der Nord- und Ostsee)*, **30**, 52–143

Russell Goddard, T. 1937, *Guide to the Farne Islands* (4th edn. 1956). Newcastle-upon-Tyne (Hindson and Andrew Reid)

Schröder, W. 1972, Zur grossen Sturmflut vom February 1825. *Acta Hydrophysica*, Band XVII, (Heft 1), 47–51. Berlin (Institut für Physikalische Hydrographie der Deutschen Akademie der Wissenschaften)

Sernander, R. 1936, Granskär och Fiby urskog: en studie över stormluckornas och marbuskarnas betydelse i den svenska granskogens regeneration. *Acta Phytographica Suecica*, VIII, section 4, 16–32. Uppsala (Almquist & Wiksell)

Shields, L. and Fitzgerald, D. 1989, The 'Night of the Big Wind' in Ireland, 6–7 January 1839. *Irish Geography*, **22**(1), 31–43

Short, Thomas. 1749, *A general chronological history of the air, weather, seasons, meteors . . .* London (T. Longman), in 2 volumes, 494 and 535pp

Traill, W. 1868, On submarine forests and other remains of indigenous woods in Orkney. *Transactions of the Botanical Society of Edinburgh*, **9**, 146–154

Watt, A. 1985, *Highways and byways round Kincardine*, Aberdeen (Gourdas House), 504pp

Wheeler, D.A. 1987, The Trafalgar storm 22–29 October 1805 *Meteorological Magazine*, **116**(7), 197–205. London (Meteorological Office)

White, G. 1789, *The Natural History and Antiquities of Selborne, in the County of Southampton: with Engravings and an Appendix*

White, G. 1795, *A Naturalist's Calendar with Observations in various Branches of Natural History; extracted from the papers of the late Rev Gilbert White, M.A.* (Ed. J. Aikin)

Willaume-Jantzen, V. 1896, *Meteorologiske Observationer i Kjöbenhavn*. Copenhagen (Danish Meteorological Institute)

Wishman, E. 1986 Gud bevare oss for 11 mars! *Vaeret*, 10(4), 115–23. Oslo (Meteorological Institute and Universitetsforlaget)

## References for storms after 1880

Burt, S.D. and Mansfield, D.A. 1988, The great storm of 15–16 October 1987 (with appendix on damage). *Weather*, 43(3), 90–114. London (Royal Meteorological Society)

Dines, J.S. 1929, Meteorological conditions associated with high tides in the Thames. In *Geographical Memoir No. 47*. London (HMSO for the Meteorological Office)

Doodson, A.T. 1929, Report on Thames floods. In *Geophysical Memoir No. 47*. London (HMSO for the Meteorological Office)

Fremming, Ø. 1980, Katastrofe i Nordsjøen. *Vaeret*, 4(3), 90–3. Oslo (Universitetsforlaget)

King, E.G.E. 1979, The northerly gales of 1–12 January 1978. *Meteorological Magazine*, 108, 135–46

Loader, C. 1976, The storm of 2–3 January 1976. *Journal of Meteorology*, 1(9), 273–83. Trowbridge (G.T. Meaden)

Lund, S. and Faurby, H. 1981, Stormflodsorkanen den 24.11.1981. *Vejret*, 4(1), 5–16

Martinsen, E., Harsson, B.G., Skjøthaug, P., Andersson, L., Jenssen, J.D. and Lystad, M. 1980, Orkanen 4–5 desember 1979; kan stormflo varsles? *Vaeret*, 4(3), 94–107. Oslo (Universitetsforlaget)

*Meteorological Office*, London. *Daily, weekly and monthly weather reports*. Also *Meteorological Magazine*

Mirrlees, S.T.A. 1928, The Thames floods of January 7th. *Meteorological Magazine*, 63(2), 17–19

Monk, G.A. and Baden, M.J. 1988, Satellite images of development of the storm. *Weather*, 43(3), 130–35. London (Royal Meteorological Society)

Murray, R. and Marshall, C.P.W. 1955, The storms and associated storm surges of December 21–23, 1954. *Meteorological Magazine*, 84, 333–41

Nielsen, N.W., Jensen, G. and Christensen, O. 1983, Stormen den 24 november 1981. *Vejret*, 5(2), 27–40. Copenhagen (Danish Meteorological Institute)

Oppermann, A. 1894, Stormen den 12ᵗᵉ Februar 1894 og dens Virkning i de danske Skove. *Tidsskrift for Skovvaesen*, 6A, 1–42. Copenhagen

Pedersen, K. 1977, *Det tabte land i vest. Sagn og beretninger om kystboernes kamp mod hav og stormflod*. Esbjerg (Vestkystens Forlag)

Petersen, M. and Rohde, H. 1977, *Sturmflut – die grossen Fluten an den Küsten Schleswig-Holsteins und in der Elbe*. Neumünster (Karl Wachholtz Verlag)

*Royal Meteorological Society*, London. 1988, *Weather*: special issue devoted to articles on the storm of 15–16 October 1987. *Weather*, 43(3), 66–142

Shellard, H.C. 1963, Meteorological Office Internal Memorandum, Climatology Division, on Extremes

Skjødt, E. 1982, The severe storm in Denmark on 24 November 1981. *Journal of Meteorology* 7(66), 43–45. Trowbridge (G.T. Meaden)

Spink, P.C. 1988, Further damage on the Spurn peninsula 4 March 1988. *Journal of Meteorology*, 13(131), 281–3. Bradford-on-Avon (G.T. Meaden)

Templeman, R.F., Oliver, H.R., Stroud, M.R., Walker, M.E., Altai, T. and Pike, S.L. 1988, The storm of 15–16 October 1987. Some wind speed and temperature observations. *Weather*, 43(3), 118–22. London (Royal Meteorological Society)

Weber, G.-R. 1990, Tropospheric temperature anomalies in the northern hemisphere 1977–1986. *International Journal of Climatology* 10, 3–19

Weismann, C. 1904, Julestormen 1902. *Tidsskrift for Skovvaesen*, 16B, 18–76. Copenhagen

Woodroffe, A. 1981, The Fastnet storm – a forecaster's viewpoint. *Meteorological Magazine*, 110(10), 271–87

## Other observations, sources

| | |
|---|---|
| Berlin 1697–1770 | Kirch family records in the possession of the Deutcher Wetterdienst, Frankfurt am Main |
| Edinburgh 1731–6 | *Medical Essays and Observations*. Edinburgh (W. and T. Ruddimans for Messrs Hamilton and Balfour), vol. I, 3rd edn, 1747 |
| Hamburg 1780–1808 | Daily observations published twice weekly in the newspaper *Hamburgische Adresse-Comptoir Nachrichten* |
| Kemnay (or Disblair), Aberdeenshire 1758–80 | Private diary of Janet Burnett, farmer's wife. Property of Mr M.G. Pearson of Meteorological Office, Edinburgh, an indirect descendant |
| London 1723–1811 | Daily weather with pressure and temperature values, worked from the original observers' records by the late Professor Gordon Manley. Available in the Meteorological Office Library, Bracknell, Berkshire |
| Lowestoft, Suffolk, England 1698–1724 | Observations by the Revd Mr Say. Bodleian Library, Oxford. Original MSS. 35448 |
| Ludgyan, Cornwall 1753–72 | Borlase's observations. See under J. Oliver |
| Perth, Scotland early nineteenth century | James Ramsay's observations. From *Perthshire Transactions and Proceedings*, available in the library of the Meteorological Office, Edinburgh |
| *Societas Meteorologica Palatina*, Mannheim 1781–92 | *Ephemerides*. (Yearbooks of daily weather observations at a network of places, mainly in Europe but extended for a time to Labrador, organized by the first meteorological society, in Mannheim.) Available in the Meteorological Office Library, Bracknell, Berkshire and some other meteorological libraries |

## Cited in Part I

Arends, F. 1833, *Physische Geschichte der Nordsee Küste und deren Veränderungen durch Sturmfluthen seit den Cymbrischen Fluth bis Jetzt* (in 2 vols). Emden (Woortman). Reprinted in Emden 1974 (Leer)

Bahnson, H. 1972, Spor af muldflugt; keltisk jernalder påvist; højmoseprofiler. *Geological Survey of Denmark. Yearbook 1972*

Bergeron T. 1928, Über die dreidimensionale verknupfende Wetteranalyse, Teil I. *Geofysiske Publikasjoner*, V, No. 6. Oslo

Bergeron, T. 1930, Richtlinien einer dynamischen Klimatologie. *Met Zeitschrift*, 47, 246–62. Braunschweig

Bilham, E.G. 1938, *The climate of the British Isles*. London (Macmillan), 347pp

Blüthgen, J. 1966, *Allgemeine Klimageographie*. Berlin (Walter de Gruyter), 720pp

Børresen, J.A. 1987, *Wind atlas for the North Sea and Norwegian Sea*. Oslo (Norwegian University Press and Norwegian Meteorological Institute)

Brazell, J.H. 1968, *London weather*. London (HMSO for the Meteorological Office, Met. O. 783), 249pp

Brooks, C.E.P. 1949, *Climate through the ages*. London (Ernest Benn, 2nd edn) 395pp. (also quoting from L.S. Higgins 'An investigation into the problem of the sand dune areas on the South Wales coast' in *Archaeologia Cambrensis*, June 1933)

Brooks, C.E.P. and Hunt, T.M. 1933, Variations of wind direction in the British Isles since 1341. *Quarterly Journal of the Royal*

*Meteorological Society*, **59**, 375–88. London

Chandler, T.J. and Gregory, S. 1976, *The climate of the British Isles.* London and New York (Longman), 390pp

Chromow, S.P. 1942, *Einführung in die synoptische Wetteranalyse.* Vienna (Springer), 532pp

Crawford, I. and Switsur, R. 1977, Sandscaping and C14: the Udal, N. Uist. *Antiquity*, **51**, 124–36

Defoe, Daniel 1704, *The Storm.* London

Defoe, Daniel 1724, *A Tour through the Whole Island of Great Britain* (1724–6), vol. 1. London (Dent Everyman edition 1928)

Douglas, K.S., Lamb, H.H. and Loader, C. 1978, A meteorological study of July to October 1588: the Spanish Armada storms. *Climatic Research Unit Research Publication No. 6 (CRU RP6).* Norwich (University of East Anglia)

Flohn, H. 1949, Klima und Witterungsverlauf in Zürich im 16. Jahrhundert. *Vierteljahresheft der Naturforschungsgesellschaft in Zürich*, **94**, 28–41

Flohn, H. and Kapela, A. 1989, Changes of tropical sea–air interaction processes over a 30-year period. *Nature*, **338**(6212), 244–6. London (16 March 1989)

Folland, C. and Parker, D.E. 1989, Comparison of corrected sea surface and air temperature for the globe and the hemispheres 1856 to 1988. In *Proceedings of the Thirteenth Annual Climate Diagnostics Workshop, October 31–November 4 1988*, held at Cambridge, Mass. Washington D.C. (National Oceanic and Atmospheric Administration), pp. 160–5

Frydendahl, K. 1971, The climate of Denmark: I. Wind (standard normals 1931–60). *Danish Meteorological Institute Climatological Papers* No. 1

Frydendahl, K, 1986, Vejrhistorie (Weather history). *1066 Tidsskrift for historisk forskning*, **16**(5), 3–15. Copenhagen

Frydendahl, K. 1990, Time-lag between long-term temperature changes in the oceans and over the continents. (As yet unpublished Memorandum, Danish Meteorological Institute. Personal Communication February 1990)

Funnell, B.M. 1979, History and prognosis of subsidence and sea-level change in the Lower Yare Valley, Norfolk. *Bulletin of the Geological Society of Norfolk*, **31**, 35–44. Norwich

Gottschalk, M.K.E. 1971, *Stormvloeden en rivieroverstromingen in Nederland: Deel I – De periode voor 1400.* Assen (van Gorcum)

Gottschalk, M.K.E. 1975, *Ibid. Deel II – De periode 1400–1600.* Assen (van Gorcum)

Gottschalk, M.K.E. 1977, *Ibid. Deel III – De period 1600–1700.* Assen and Amsterdam (van Gorcum)

Graham, H.G. 1899, *The social life of Scotland in the eighteenth century.* London (Black)

Gram-Jensen, I. 1985, Sea floods – contributions to the climatic history of Denmark. *Danish Meteorological Institute: Climatological Papers*, No. 13. Copenhagen, 76pp

Grove, J.M. 1988, *The Little Ice Age.* London and New York (Methuen), 498pp

Grude, M.A. 1914, *Jaederen 1814–1914.* Sandnes (Ingvald Dahles Forlag), pp. 492–511

Hauerbach, P., Hansen, A.L. and Nielsen, H. 1983, *Havet-grenen-Skagen.* Skagen (Sjøbeck)

Hennig, R. 1904, Katalog bemerkenswerter Witterungsereignisse von den ältesten Zeiten bis zum Jahre 1800. *Abhandlungen des Preussischen Met. Inst.*, **II**, No. 4. Berlin

Jelgersma, S. and van Regteren Altena, J.F. 1969, An outline of the geological history of the coastal dunes in the western Netherlands. *Geologie en Mijnbouw*, **48**(3), 335–42

Jelgersma, S., De Jong, J., Zagwijn, W.H. and van Regteren Altena, J.F. 1970, The coastal dunes of the western Netherlands: geology, vegetational history and archaeology. *Meddedelingen Geol Sticht*, Nieuwe Serie No. 21

Jenkinson, A.F. 1977, An initial climatology of gales over the North Sea. *Meteorological Office, Met. O. 13 Branch Memorandum No. 62* (unpublished). Bracknell

King, E.G.E. 1979, The northerly gales of 11–12th January 1978. *Meteorological Magazine*, **108**, 135–46

Kington, J. (editor) 1988, *The weather journals of a Rutland Squire: Thomas Barker of Lyndon Hall.* Oakham (Rutland Record Society, County Museum)

Knowles Middleton 1969, *Invention of the meteorological instruments.* Baltimore, (Johns Hopkins Press), 362pp

La Cour, P. 1876, *Tyge Brahes meteorologiske Dagbog.* (Collection of observations in Tycho Brahe's weather diary). Copenhagen (Kongelige Danske Videnskabernes Selskab: *Collectanea Meteorologica*). 264pp. + 75-page summary in English and French

Lamb, H.H. 1956, Meteorological results of the *Balaena* expedition 1946–7. *Geophysical Memoir*, **94.** London (HMSO for the Meteorological Office)

Lamb, H.H. 1957, Tornadoes in England, May 21, 1950. *Geophysical Memoir*, **99.** London (HMSO for the Meteorological Office), 38pp

Lamb, H.H. 1972, British Isles weather types and a register of the daily sequence of circulation patterns 1861 to 1971. *Geophysical Memoir*, **116.** London (HMSO for the Meteorological Office). The classification is continued to date in *Climate Monitor*, Norwich (University of East Anglia, Climatic Research Unit)

Lamb, H.H. 1977, *Climate: present, past and future, vol. 2: Climatic history and the future.* London (Methuen), 835pp

Lamb, H.H. 1979, Climatic variation and changes in the wind and ocean circulation: the Little Ice Age in the northeast Atlantic. *Quaternary Research*, **11**, 1–20

Lamb, H.H. 1980, Coastal fluctuations in historical times and their connexion with transgressions of the sea, storm floods and other coastal changes. In *Transgressies en occupatiegeschiedenis in de kustgebieden van Nederland en Belgie.* (Ed. A. Verhulst and M.K.E. Gottschalk.) Ghent (Belgisch Centrum voor Landelijke Geschiedenis), pp. 251–90

Lamb, H.H. 1982, *Climate, history and the modern world.* London (Methuen), 387pp

Lamb, H.H. 1986, Ancient units used by the pioneers of meteorological measurements. *Weather*, **41**(7), 230–4

Lamb, H.H. 1988, *Weather, climate and human affairs.* London (Routledge), 364pp

Lamb, H.H. and Johnson, A.I. 1959, Climatic variation and observed changes in the general wind circulation. Parts I and II. *Geografiska Annaler*, **41**, 94–134. Stockholm

Lamb, H.H. and Johnson, A.I. 1961, *Ibid.* Part III. *Geografiska Annaler*, **43**, 363–400. Stockholm

Lamb, H.H. and Johnson, A.I. 1966, Secular variations of the atmospheric circulation since 1750. *Geographical Memoir*, **110.** London (HMSO for the Meteorological Office)

Lamb, H.H. and Weiss, I. 1979, On recent changes of the wind and wave regime of the North Sea and the outlook. *Fachliche Mitteilungen* Nr. 194 (Geophys. Dienst der Bundeswehr). Traben-Trarbach (Amt für Wehrgeophysik)

Landsberg, S.Y. 1955, The morphology and vegetation of the Sands of Forvie. University of Aberdeen (Unpublished Ph.D. thesis)

Landsberg, H.E. 1958, *Physical climatology*, 2nd edn. Dubois, Pennsylvania (Gray)

Lowe, E.J. 1870, *Natural phenomena and chronology of the seasons.* London

Ludlum, D.M. 1963, *Early American hurricanes 1492–1870.* Boston, Mass. (American Meteorological Society), 198pp

Manley, G. 1974, Central England temperatures: monthly means 1659 to 1973. *Quarterly Journal of the Royal Meteorological Society*, **100**, 389–405. London

*Meteorological Office* 1952, *Climatological atlas of the British Isles.* London (HMSO for the Meteorological Office, publication Met. O. 488)

*Meteorological Office* 1968, *Tables of surface wind speed and direction over the United Kingdom.* London (HMSO for the Meteorological Office, publication Met. O. 792). (The data are for 1950–9.)

Middleton, W.E.K. 1969, see Knowles Middleton

Miegham, J. van 1936, Prévision du temps par l'analyse des cartes météorologiques. *Publication belge de Recherches Radio-scientifiques.* Paris (Gautier-Villars)

Mörner, N.-A. 1980, The northwest European 'sea level laboratory' and regional Holocene eustasy. *Palaeogeography, Palaeoclimatology, Palaeoecology,* **29,** 281–300. Amsterdam (Elsevier)

Mossman, R.C. 1898, The meteorology of Edinburgh. Part III. *Transactions of the Royal Society of Edinburgh,* **40**(iii), No. 21, 476–8

Palmén, E. 1928, Zur Frage der Fortpflanzungsgeschwindigkeit der Zyklonen. *Meteorologische Zeitschrift,* **45,** 96–99, Braunschweig

Paterson, W. 1880, *Extracts from the records of the Convention of the Royal Burghs.* (Preface by J.D. Marwick.) Edinburgh

Petersen, M. and Rohde, H. 1977, *Sturmflut.* Neumünster (Karl Wachholtz), 148pp

Peterson, E.W. and Larsen, S.E. 1984, Analyse af vindobservationer fra Fanø i perioden 1872–1980. *Vejret,* **6**(1), 12–17. Copenhagen

Pettersen, Sv. 1936, Contribution to the theory of frontogenesis. *Geofysiske Publikasjoner,* **XI,** No. 6

Pfister, Chr. 1984, *Klimageschichte der Schweiz 1525–1860.* (2 vols.) Berne and Stuttgart (Verlag Paul Haupt for Academica Helvetica)

Pränge, W. 1971, Geologisch-historische Untersuchungen an Deichbrüchen der 15. bis. 17. Jahrhunderts in Nordfriesland. *Nordfriesisches Jahrbuch*

Refsdal, A. 1930, Zur Theorie der Zyklonen. *Meteorologische Zeitschrift,* **47,** 294–305. Braunschweig

Ritchie, W., Rose, N. and Smith, J.S. 1978, Beaches of Northeast Scotland: the Sands of Forvie. In a *Report commissioned in 1977 by the Countryside Commission for Scotland.* University of Aberdeen, Dept. of Geography, pp. 195–206, 269–74

Rohde, H. 1964, Nachrichten über Sturmfluten früherer Jahrhunderte nach Aufzeichnungen Tönniger Organisten. *Die Küste,* **12,** 113–132. Heide i. H. (Westholsteinsche Verlag Boyens & Co.)

Rohde, H. 1977, Sturmfluthöhen und säkularer Wasseranstieg an der deutschen Nordseeküste. *Die Küste,* **30,** 52–143. Heide, i.H

Rowe, M. 1988, The storm of 16th October 1987 and a brief comparison with three other historic gales in southern England (1362, 1662, 1703). *Journal of Meteorology (G.T. Meaden),* **13**(129), 148–55

Short, Thomas 1749, *A general Chronological History of the Air, Weather, Seasons, Meteors . . .* London (T. Longman), in 2 vols. 496 and 535pp

Sinclair, Sir John 1791–9, *Statistical Account of Scotland.* Edinburgh

Smout, T.C. 1963, *Scottish trade on the eve of Union.* Edinburgh and London (Oliver & Boyd), 320pp

Tooley, M.J. 1974, Sea level changes during the last 9000 years in northwest England. *Geographical Journal,* **140,** 18–42. London

Tooley, M.J. 1978, Interpretation of Holocene sea-level changes. *Geol. Föreningen Stockholms Förhandlingar,* **100,** 203–212. Stockholm

Tooley, M.J. 1985, Climate, sea-level and coastal changes. Pp. 206–234 in *The climatic scene,* ed. M.J. Tooley and G.M. Sheail. London (Allen and Unwin), pp. 206–34

Walton, K. 1956, Rattray: a study in coastal evolution. *Scottish Geographical Magazine,* **72,** 85–96

Weber, G.-R. 1990, Tropospheric temperature anomalies in the northern hemisphere 1977–1986. *International Journal of Climatology,* **10,** 3–19

Weiss, I. and Lamb, H.H. 1970, Die Zunahme der Wellenhöhen in jüngster Zeit in den Operationsgebieten der Bundesmarine, ihre vermutlichen Ursachen und ihre voraussichtliche weitere Entwicklung. *Fachliche Mitteilungen,* **160.** Porz-Wahn (Geophys. Beratungsdienst der Bundeswehr, Luftwaffenamt)

Wheeler, D.A. 1988, Sailing ships' logs as weather records: a test case. *Journal of Meteorology,* **13**(128), 122–5. Bradford-on-Avon (G.T. Meaden)

Woebcken, C. 1924, *Deiche und Sturmfluten an der deutschen Nordseeküste.* Bremen

# Authorship index

American Meteorological Society  64
Archives, local  xi
Arends, F.  3

Barker, Thomas  3
Breslau (today Wrocław), records in medical
 journals  73, 78
Bristol, city archives  xi, 64
Brooks, C.E.P.  18, 29, 60, 62, 64

Climatic Research Unit  xi, 5

Danish Meteorological Institute  x, 6, 53,
 64, 94
Defoe, Daniel  3–4, 48, 55–6, 59–63, 71
Derham, Rev. William  5, 60, 72
Deutsche Seewarte  6
Drake, Sir Francis  40
Dundee, city archives  xi, 60, 64

Evelyn, John  39

Fitzroy, Admiral R.  6n, 135, 196
Flohn, H.  34, 198
Folland, C.  34
Frydendahl, K.  xi, 5n, 53, 64, 72

Gottschalk, M.K.E.  3, 37–8, 44, 46, 56–7,
 195, 198
Gram-Jensen, I.  16, 17, 195

's Gravenhage (The Hague), Algemeen
 Rijksarchief  xi, 64
Grude, M.A.  19, 198

Hakluyt, R.  44
Hauerbach, P.  xi, 19, 23, 198
Holyoke, E.A.  5–6

Jelgersma, S.  18, 198
Jenkinson, A.F.  24, 198

Kirch family  197

Lamont, John  xi, 47–9, 195
Ludlum, D.M.  62, 195, 198

Manley, Professor Gordon  3, 29, 198
Mejer, Johannes  16
Merle, Rev. William  29
Meteorological Office  xi, 3, 7, 23–4, 135
Mossman, R.C.  50, 195, 199

Namias, J.  53n, 188–9
National Maritime Museum, Greenwich  xi,
 64
Norwegian Meteorological Institute  xi, 197

Palmén, E.  22, 41–2, 86, 158, 195
Paris, Observatory of  xi, 60, 73

Pfister, Chr.  xi, 24, 199
Pytheas  3

Rohde, H.  xi, 3, 17, 23, 46, 57, 85, 90, 93,
 101, 120, 127, 195
Royal National Lifeboat Institution  xi, 129,
 131
Royal Society, London  x–xi, 60, 64, 72

*Societas Meteorologica Palatina*,
 Mannheim  72, 94, 135, 197
Swedish Meteorological Service (*Statens
 Meteorologiska och Hydrografiska
 Anstalt*  xi, 24

*Theatrum Europaeum*  55–6
Tidal Institute, Birkenhead (Institute of
 Oceanographic Sciences)  xi, 4, 53
Tooley, M.J.  18, 199
Towneley, Richard  5, 60–1
Tycho Brahe  30, 42, 195, 198

Vassie, J.  xi, 4, 53

Wheeler, D.A.  xi, 116, 131, 196
White, Gilbert  3, 196
Wishman, E.  xi, 118–19, 197

# Geographical index

# Subject index